COMPUTER-AIDED CONTROL SYSTEMS DESIGN

Practical Applications Using MATLAB® and Simulink®

Cheng Siong Chin

COMPUTER-AIDED CONTROL SYSTEMS DESIGN

Practical Applications Using
MATLAB® and Simulink®

CRC Press
Taylor & Francis Group
Boca Raton   London   New York

CRC Press is an imprint of the
Taylor & Francis Group, an **informa** business

CRC Press
Taylor & Francis Group
6000 Broken Sound Parkway NW, Suite 300
Boca Raton, FL 33487-2742

First issued in paperback 2017

© 2013 by Taylor & Francis Group, LLC
CRC Press is an imprint of Taylor & Francis Group, an Informa business

No claim to original U.S. Government works

ISBN 13: 978-1-4665-6851-8 (hbk)
ISBN 13: 978-1-138-07349-4 (pbk)

Library of Congress Cataloging-in-Publication Data

Chin, Cheng Siong.
  Computer-aided control systems design : practical applications using MATLAB and Simulink / Cheng Siong Chin.
    p. cm.
  Includes bibliographical references and index.
  ISBN 978-1-4665-6851-8 (hardback)
  1. Adaptive control systems--Design and construction--Data processing. 2. Automatic control--Computer simulation. 3. Control theory--Mathematics. I. Title.

  TJ217.C478 2012
  629.8'9553--dc23   2012028124

Visit the Taylor & Francis Web site at
http://www.taylorandfrancis.com

and the CRC Press Web site at
http://www.crcpress.com

Contents

Foreword

Simulation, as a discipline, provides infrastructure for solving challenging problems for scores of application areas in cases where experimentation and experience are needed. Both features can be offered under all conceivable realistic as well as extreme conditions. For the first aspect, simulation is goal-directed experimentation with models of dynamic systems. With its ability to provide experience, simulation plays a vital role in training by providing possibilities to develop/enhance competence in one of the three types of skills, namely motor skills, decision making and communication skills, and operational skills. The last one is also very important in training operators of control systems.

Computer simulation and modeling are important aspects of modern applied control engineering as they are vital in all areas of simulation-based science and engineering. Technical journals and conference proceedings abound with examples of design of several types of engineered systems. Software for the engineering community has helped to shorten design times while accurately predicting the behavior of systems.

This book collects computerized modeling and simulation knowledge into a compilation geared to many users. The two important applications are elaborated in support of computer simulations on practical and very important industrial systems such as the ALSTOM gasifier system in power stations and underwater robotic vehicle (URV) in marine industry. The steps and codes required in basic control systems design and analysis are supported by widely used specialized software such as MATLAB™ and Simulink™ throughout the chapters. A related area is in the analysis of experimental data. In some engineered systems, the model can be developed and verified through computational-fluid dynamic software such as ANSYS-CFX™. This can be seen in the chapter on the modeling of remotely-operated vehicles (ROVs). It provides an interesting documentation on the steps involved in modeling and verification of the ROV model prior to control systems design and implementation. Hence, the chapters presented here provide a context for supporting all of these activities and help researchers as well as students find the needed relations without delving into many different papers, books, and handbooks.

Finding innovative solutions to challenging problems is very important. However, providing infrastructure to offer innovative solutions to some current as well as imminent and long term challenging problems is even more important.

This book, by providing examples in important applications for simulation-based experimentation and simulation-based experience in applied control engineering, is a valuable addition to the literature. It reflects where our knowledge is firm enough to provide mathematical description of systems as well as design, simulation, analysis, and implementation of control systems. Areas where the knowledge is yet shallow are targeted for future studies. This book provides a fertile delineation of areas needing attention. It is written for those who may be already familiar with or newly exposed to the concepts associated with applied control engineering. It will be most useful for researchers, engineers, faculty, and students.

Dr. Tuncer Ören
SCS, Hall of Fame – Lifetime achievement award, 2011
Professor Emeritus of Computer Science
School of Information Technology and Engineering
University of Ottawa, Ottawa, Ontario, Canada

Preface

The countless technological advances of the 20th century require that future engineering education emphasizes bridging the gap between theory and the real world. To help the reader achieve this goal, computer-aided applied control design (CAACD) is used throughout the text to encourage good habits of computer literacy while keeping the mathematical complexity to a minimum. Each CAACD uses fundamental concepts to ensure the viability of a computer solution. This edition emphasizes applying control fundamentals to practical industry systems such as the ALSTOM gasifier system in power stations and underwater robotic vehicles (URVs) in the marine industry. The software commonly adopted are MATLAB® and Simulink®. One of the fundamental aims in preparing the text has been to work from basic principles such as understanding the control engineering and recognizing that powerful software packages such as MATLAB and Simulink exist to aid the control systems design. At the time of this writing, MATLAB, its Toolboxes and Simulink have emerged, becoming the industry standard control system design package.

This book has been prepared with particular attention to the needs of those who seek a foundation in applied control as well as an ability to bridge the gap between control and its real-world applications. For practicality, the choice and emphasis of material are guided by the basic objective of making an engineer or student capable of dealing with practical control problems in industry. The book is also intended to be a reference course for practicing engineers, undergraduates, students undertaking master's degrees, and an introductory text for PhD research students who want to acquire knowledge in basic control systems design, analysis, and implementation using MATLAB and Simulink. The first part of the book can be used as a textbook for a first course in MATLAB, or a supplementary textbook for courses in which calculations are performed in MATLAB. For more information about MATLAB, I encourage you to use the help and demo facilities in the *User's Guide* supplied with the software.

OUTLINE OF THE BOOK

The book is organized into five chapters and a few appendices. The main chapters include sections on dynamic modeling, control structure design, controller design, final implementation, and testing. These chapters contain applications of various real-life systems such as the power plant and the URV. As it is not possible to cover every single subject from dynamic modeling to final implementation and testing, each of these chapters therefore emphasizes certain topics in the control systems design process. The software adopted throughout the chapters is MATLAB and Simulink.

In Chapter 1, essential ideas of applied control engineering are introduced. It also contains a brief history of control, from the ancient beginnings of process control to the contributions of modern and robust control applications. This brief history intends to introduce students to the origins of this field and the key figures who contributed to its development. Also, the types of closed-loop systems used in industry are shown. As applied control engineering is the main focus of the book, the control system design process for most systems is briefly explained.

Chapter 2 contains a brief presentation and a hands-on introduction to MATLAB and Simulink. A few methods to model a second-order system are shown through examples. It also covers closed-loop control of the second-order system using the Proportional-Integral-Derivative (PID) controller and its parameters tuning using the Ziegler–Nichols method.

Chapter 3 covers analysis, model order reduction, and controller design for a power plant such as the ALSTOM gasifier system. The chapter emphasizes the control structure design and controller design based on the given model. The inherent properties of this highly coupled and numerically ill-conditioned multivariable system are studied. Methods are introduced to achieve suitable input-output (I/O) pairings and numerical conditioning for the gasifer system. Model order reduction (MOR) methods such as modal truncation and Schur balanced truncation are applied. Few controllers are designed using the LQR, LQG/LTR, H-infinity, and H_2 techniques. The robustness of these controllers at different load conditions is assessed for its feasibility.

Chapter 4 shows modeling, simulation, and control of a remotely operated vehicle (ROV) for pipeline tracking. The chapter further emphasizes dynamic modeling, controller design, implementation, and testing of the ROV. It consists of dynamic modeling involving computational fluid dynamic software such as WAMIT®, ANSYS-CFX™, and experiments to verify the theoretical models obtained. The control systems design and simulations are performed using MATLAB and Simulink before actual testing. Implementations and testing in a swimming pool were performed to verify the PID control systems design.

Chapter 5 covers the nonlinear subsystem modeling and linearization of the ROV at vertical plane equilibrium points. A few controllers such as PID control, velocity state feedback linearization control, sliding mode control, cascaded control, and fuzzy logic control are designed for a nonlinear model of the ROV. This is followed by analyzing the open-loop characteristics of a linearized ROV's system and the use of a few linear control methodologies such as the LQG/LTR and H-infinity control methods on the linear model of the ROV.

MATLAB and Simulink are registered trademarks of The MathWorks, Inc. For product information, please contact

The MathWorks, Inc.
3 Apple Hill Drive
Natick, MA 01760-2098 USA
Tel: 508 647 7000
Fax: 508-647-7001
E-mail: info@mathworks.com
Web: www.mathworks.com

Acknowledgments

Finally, I wish to acknowledge my great debt to all those who have contributed to the development of computer-aided applied control engineering into the exciting field it is today, and specifically to the considerable help and education I have received from students and my colleagues. In particular, I would first like to thank my late master's thesis advisor, Professor Neil Munro, who made control engineering come alive by using computer-aided software. I benefited from his knowledge in robust control and computer-aided control engineering using Mathematica®. I would also like to thank Professor Emeritus Howard Rosenbrock for his inspiration on Direct Nyquist Array (DNA) in applied control engineering during his days at the University of Manchester. I would also like to express my sincere gratitude to my PhD supervisor, Associate Professor Micheal Lau and cosupervisor, Associate Professor Eicher Low at Nanyang Technological University (NTU) for their continuous support and assistance during tough times. There are many people who through their friendship and support have made this work an enjoyable and rewarding experience. I would also like to thank the editorial staff and reviewers for providing their comments to the text.

I would like to express my gratitude to my family. In particular, I would like to thank Irene (my wife), Giselle (late eldest daughter), Millicent (second daughter), Tessalyn (third daughter), and Alyssa (youngest daughter) for their support and understanding, without which this undertaking would not have been possible. I would like to share my joy and results with my strong and kind daughter, Giselle. Although, she passed away at the age of four after a few consecutive open-heart surgeries, without her presence and attitude toward life, many things would not have happened.

Dr. Cheng Siong Chin
PhD, MSc, BEng, Dip.Mech
CEng, EUR ING, MIET, SMIEEE, MIMarEST

1 An Overview of Applied Control Engineering

1.1 HISTORICAL REVIEW

In this section, we give a brief historical review of the major developments in the area of modern control systems. We shall attempt to discuss the development of classical control and its evolution into modern and robust control. The emphasis of our discussion is directed toward applying the control systems design method within which the framework of this book is based.

Applied control engineering is the engineering discipline that applies control theory to design systems with predictable behaviors. The practice uses sensors to measure the output performance of the device being controlled and those measurements can be used to give feedback to the input actuators that can make corrections to achieve the desired performance. When a device is designed to perform without the need of human inputs for correction, it is called *automatic control.*

Going backward in time, the Romans did use some elements of control theory in their aqueducts. Indeed, ingenious systems for regulating valves were used in these constructions in order to keep the water level constant. This certainly was a successful device as the water clocks were still being made in Baghdad when the Mongols captured the city in AD 1258. A variety of automatic devices have been used over the centuries to accomplish useful tasks. The latter includes the automata, popular in Europe in the 17th and 18th centuries. In 1788, J. Watt adapted these ideas when he invented the steam engine and this constituted a magnificent step in the industrial revolution. The British astronomer G. Airy was the first scientist to analyze mathematically the regulating system invented by Watt. But the first definitive mathematical description was given only in the works by J. C. Maxwell (who discovered the Maxwell electromagnetic field equations) in 1868, where some of the erratic behaviors encountered in the steam engine were described and some control mechanisms were proposed. He was able to explain instabilities exhibited by the flyball governor using differential equations. This demonstrated the importance and usefulness of mathematical models, and methods in understanding complex phenomena, and it signaled the beginning of mathematical control and systems theory. Elements of control theory had appeared earlier but not as dramatically and convincingly as in Maxwell's analysis.

The issue of finding stability criteria [1–2] for certain linear systems was discussed first. This work was then extended to determine the stability of nonlinear systems by Lyapunov [3]. In the 1930s, work on feedback amplifier design at the Bell Telephone Laboratories was based on the concept of frequency response presented in the paper on "Regeneration Theory" [4], which basically described how to determine system stability using the frequency domain approach. This was later extended [5] during the next decade to give rise to one of the most widely applied control system design methodologies [6]. Based on the Root Locus method [2,7], a few designed techniques that allowed the roots of the characteristics equation to be displayed in a graphical form were subsequently proposed.

The invention of computers in the 1950s gave rise to the application of state-space equations that use vector matrix notation for machine computation. The concept of optimum [8] design was first proposed. The method of performing dynamic programming [9] was then developed at the same time as the maximum principle [10]. At the initial conference of the International Federation of Automatic Control, the concept of observability and controllability [11] was introduced. Around the same period, Kalman demonstrated that when the system dynamic equations are linear, performance criterion is quadratic and can be controlled using the LQ method. With the concept of the Kalman filter [12], which combined with an optimal controller, a linear-quadratic-Gaussian (LQG) control was introduced.

The 1980s showed great advances in control theory for the robust design of systems with uncertainties in their dynamic equations. The works on H-infinity norm and μ-synthesis theory [13–15] demonstrated how uncertainty can be modeled in the system equations. A decade later, the concept of intelligent control systems was developed. An intelligent machine [16] that is able to give a better behavior under uncertainty condition was introduced. Intelligent control theory has the ideas laid down in the area of Artificial Intelligence (AI). Artificial Neural Networks [17–19] uses many simple elements operating in parallel to emulate how their biological counterparts are used. Subsequently, the idea of fuzzy logic [20] was developed to allow computers to model human vagueness. The fuzzy logic [21–24] controllers offer some form of robust control without the requirement to model the system dynamic behavior.

Thus, control theories have made significant strides in the past 100 years in history. New mathematical techniques made it possible to more accurately control significantly more complex dynamical systems than the original flyball governor. These techniques include developments in optimal control in the 1950s and 1960s, followed by progress in stochastic, robust, adaptive, and optimal control methods in the 1970s and 1980s, and intelligent control in 1990s. Applications of control methodology have helped to make efficient power generation, space travel, communication satellites, aircraft, and underwater exploration possible.

1.2   COMPUTER-AIDED CONTROL SYSTEM DESIGN

One of the aspects of control systems that has received considerable attention is that of developing efficient and stable computational algorithms. Parallel with the advances in modern control, computer technology has made its own progress and has played a vital role in implementing the control algorithms. As a result, the field of control has been influenced by the revolution in computer technology. Now, most control engineers have easy access to a powerful computer package for system analysis and design. In fact, computers have become an integral part of control systems. With the progress in computing ability, the classes of problems which can be modeled, analyzed, and controlled are considerably larger than those previously treated.

One of the main goals of this book is to establish a relation between applied control engineering and computer software engineering. Applications of computers in control system design and/or implementation is commonly called *computer-aided control system design* (CACSD) or generally known as *computer-aided applied control engineering* (CAACE). Using a computer-aided design approach, all principles and techniques of control can be demonstrated in fairly simple fashion. The software in creating and solving problems is not limited to a particular one, rather a collection of powerful available packages. Even if the reader does not have access to these packages, it is still worthwhile to study the numerical results that have been presented. In this way certain trade-offs, trends, comparisons with experiment results, and other results will become apparent.

A major breakthrough in computer-aided control system design was the creation of a "matrix laboratory" for linear algebra. This software called *MATLAB* although not initially intended for control system design, was turned into a stepping-stone for many powerful CAACE programs in a relatively short period of time. In parallel to this effort, several other CAACE programs such as Mathematica®, KEDDC™ and TIMDOM™ have been developed for a wide range of problems and classes of systems.

An important part of CAACE is simulating a dynamic system model to obtain theoretical behavior of the system before actual implementation on an application. However, a problem such as incompleteness and uncertainty in the dynamic model could happen. The system dynamics may change with time and thus a fixed control method does not work. For example, the mass of an airplane is different before and after the flight journey. This affects the dynamic model as the mass changes with time. Measurements may be contaminated with noise and external disturbance effects. Sensors, which provide accurate data, because of these difficulties in measuring the output data produce highly random and irrelevant information. Sometimes, the finest controller on a miserably designed system may not deliver the desired performance. However, advanced controllers are able to eke out better results for a badly designed system. But on this system, there is a definite end point, which can be approached by instrumentation.

With this in mind, there should be a unified approach to the design of the control systems. The first step is to perform a simulation of the theoretical dynamic system to obtain an insight to its behaviors and try to remodel it closer to the actual response before a controller is designed for the model. If modification of the model is difficult or sometimes unfeasible, the controller has to withstand the model inaccuracy without compromising the desired performance. The control system design is actually a combination of experience and techniques. Experience, however, only comes with time as one is exposed to more control applications. Usually, the techniques of designing a control system involve mostly a team of engineers. The process often emerges in a step-by-step design procedure as follows:

1. Study the system to be controlled.
2. Obtain information about the control objectives.
3. Simulate the system; simplify the model if necessary.
4. Analyze the resulting model.

5. Decide variables to be controlled.
6. Decide on measurements: what sensors and actuators will be used and where will they be placed?
7. Select the control configuration.
8. Decide on the type of controller to be used.
9. Decide on performance specifications, based on the overall control objectives.
10. Design a controller.
11. Analyze the resulting controlled system to see if the specifications are satisfied; and if they are not satisfied, modify the specifications or the type of controller.
12. Simulate the resulting controlled system on a computer or plant using hardware-in-the-loop.
13. Repeat step 3 or all previous steps, if necessary.
14. Choose hardware and software, and implement the controller.
15. Test and validate the control system, and fine-tune the controller if necessary.

Control engineering courses and textbooks usually focus on steps 10 and 11 in the above procedure; that is, on methods for controller design and control system analysis. Interestingly, some applications are designed without performing the simulation with the hardware or the controlled system. How the control algorithm can be coded in hardware after it completes the simulation is normally omitted in the control engineering course. A special feature of this book is the provision of computer-aided design steps for simulation and its subsequent implementation on the real-life applications such as ALSTOM gasifier and URV. This book also explains some tried and test applications using the available control system methods on these applications.

1.3   CONTROL SYSTEM FUNDAMENTALS

Before discussing the computer-aided applied control engineering applications, it is vital to define the term called *system*. Examples of a system can be physical systems such as underwater robotic vehicles, ships, submarines, planes, and robots. But, most systems have things in common. They need outputs and inputs to be specified. In the example of the underwater robotic vehicles, the inputs are the voltage and current to the thrusters, and the outputs are the position, velocity, and acceleration of the vehicles. The system to be controlled is known as the *plant* and is represented by a block diagram. The control engineers may have direct control on the inputs, and they are known as *controllable inputs*. But, there are some inputs over which the control engineers have no control and they are called *disturbance inputs*. A typical closed-loop system is shown in Figure 1.1; it consists of the forward path controlled by a controller and the feedback path.

As seen in Figure 1.1, to attenuate the disturbance, a controller (or embedded controller) is used to provide control signal to the plant. The most popular controller used in industry is the Proportional-Integral-Derivative (PID) controller. It generates P-action, I-action, and D-action and combines them to produce the control signal. The derivative mode is added to the PI controller to make the response

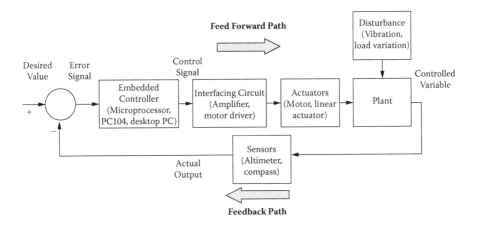

FIGURE 1.1 Closed-loop control system block diagram.

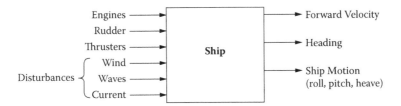

FIGURE 1.2 A ship as a control system.

speed faster and less oscillatory. This type of control provides zero offset and faster response but can be quite poor in attenuating the disturbance. The control signal will actuate the plant via an interfacing circuit (e.g., amplifier or digital-to-analog converter) and actuators such as motors. To obtain the actual outputs, sensors are used. For example, an altimeter measures the depth of the water and a magnetic compass measures the heading angle of the vehicle. Difference between the actual outputs and desired value (or set point) is called *error signal*. The error signal is used for an input to the controller to generate control signal.

In control system design, the responses of system outputs to the system inputs are important to the engineers. The control engineers often try to determine the system response by using a mathematical model that represents the system behavior. With the system model obtained, the system outputs are computed. Comparisons can be made with the theoretical results to see whether it matches the mathematical model. In order to obtain the system inputs, instrumentation and measurement devices such as communication cables, data acquisition cards, measurement devices, and a microprocessor are needed.

In the example of a ship seen in Figure 1.2, the fin or rudder and diesel engines are the control inputs, which can be changed to control the outputs such as the ship's heading and surge velocity. The wave, wind, and ocean current are disturbance inputs that create errors in the controlled variables such as the roll, pitch, heave, forward, sway, and heading or turning velocity.

1.3.1 OPEN-LOOP SYSTEMS

The control systems are broadly classified as either open-loop or closed-loop systems. An open-loop control system is controlled directly, and only, by an input signal. It does not compare the actual output with the desired value to determine the control action. Instead, a calibrated setting—previously determined by some sort of calibration procedure or calculation is used to obtain the desired result. One common example is an amplifier and a motor. The relationship between speed and voltage input is determined through the calibration. The amplifier receives a low-level input voltage signal and amplifies to drive the motor to the desired output speed. The open-loop control system is shown in basic block diagram form in Figure 1.3. The output of the amplifier is proportional to the amplitude of the input signal. The phase (ac system) and polarity (dc system) of the input signal determines the direction that the motor shaft will turn. After amplification, the input signal is fed to the motor, which moves the output shaft (load) in the direction that corresponds with the input signal. The motor will not stop driving the output shaft until the input signal is reduced to zero or removed. This system usually requires an operator who controls speed and direction of movement of the output by varying the input. The operator could be controlling the input by either a mechanical or an electrical linkage (i.e., potentiometer). The open-loop system could be very vulnerable to any disturbances as there is no feedback control system.

An example of an open-loop system is a cellular phone. The loop begins when an incoming call causes the phone to ring or when a person dials out using the phone. After the phone has been turned on, it will make a connection with a satellite. The phone will then continue to transmit to that satellite until the connection is broken by a human operator. Because the phone is unable to turn itself off after the necessary functions (i.e., the desired value) have been performed, it is an open loop.

The firing of a rifle bullet is another example of an open-loop control system. The desired result is to direct the bullet to the bull's eye. The actual result is the direction of the bullet after the gun has been fired. The open-loop control occurs when the rifle is aimed at the bull's-eye and the trigger is pulled. Once the bullet leaves the barrel, it is on its own: If a gust of wind comes up, the direction will change and no correction will be possible.

Besides, the open-loop system is vulnerable to disturbance. The advantage of using the open-loop control is that it is less expensive than closed-loop control: It is not necessary to measure the actual result. In addition, the controller is much simpler

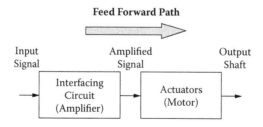

FIGURE 1.3 Open-loop system block diagram.

because corrective action based on the error is not required. Often, the human operator needs to correct the slowly changing disturbances by manual adjustment. In this case, the operator is actually closing the loop by providing the feedback signal.

1.3.2 CLOSED-LOOP SYSTEMS

Closed-loop control or feedback is the action of measuring the difference between the actual output and desired value, and using that difference to drive the actual output toward desired value. The term *feedback* comes from the direction in which the measured output travels in the block diagram. The signal begins at the output of the controlled system and ends at the input of the controller. Closed-loop control systems are the type most commonly used in industry because they control with greater accuracy than open-loop systems in Figure 1.3.

The block diagram of the closed-loop system to monitor a room temperature is illustrated in Figure 1.4. The actual output from a thermocouple is compared with the desired temperature set by the user. Any possible error causes the controller to generate a control signal to the gas solenoid valve for adjusting the gas flow to the burner. The desired value is determined from manual adjustment of a potentiometer as seen in Figure 1.4. The physical values of the signals can be seen within the brackets in Figure 1.4. Steady conditions exist if the desired and actual temperatures are similar. The system can work in two basic modes:

- *Bang-bang type of control:* The gas valve is either fully closed or open. This form of control produces an oscillation of the actual temperature about the desired or set point temperature. However, it is usually for low-cost applications (i.e., domestic heating systems).
- *Proportional type of control:* The linear movement of the valve is linearly proportional to the error signal. This gives a continuous modulation of the heat input to give a precise temperature control. This method is used for temperature control where accuracy is a concern over the cost.

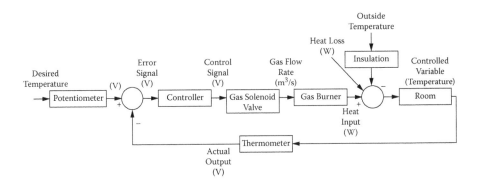

FIGURE 1.4 Closed-loop system block diagram for room temperature control.

1.4   EXAMPLES OF CONTROL SYSTEMS

1.4.1   Ship Control System

The ship control system is designed to control a vessel on a desired heading (or heading control) while subjected to disturbances such as waves and wind. The main components of the ship control system can be seen in Figure 1.5. The actual heading is measured by a magnetic compass and compared with the desired heading. The ontroller determines the commanded rudder angle and sends a control signal to the steering gear to steer the ship. The actual rudder angle is measured by an angle sensor and compared with the desired rudder angle, to form a control loop. The rudder gives a moment on the hull to change the actual heading toward the desired heading under the wind and wave disturbances that may hinder the control efforts.

1.4.2   Underwater Robotic Vehicle Control System

Another closed-loop system of interest in underwater is the URV. URVs have experienced tremendous growth in underwater pipeline inspection, ocean surveying, and underwater rescuing. They are essential at depths where human diving is impractical. The URV has evolved into different classes based on the vehicle's application and autonomy. The URV is classified into remotely operated vehicle (ROV) and autonomous underwater vehicle (AUV). The guidance and control of the ROV requires the operator to have a high degree of expertise, and performances are highly affected by the operator's lack of concentration and weariness. In order to simplify the task of the human operator (in the case of ROVs) or of the mission control (in the case of AUVs), effort has focused on the realization of systems that could guarantee the automation of elementary behaviors, such as the regulation of state variables for assigned set points.

They are many parameters to control in the ROV. As seen in Figure 1.6, the problem of depth regulation for ROVs is used for illustration. The controller can guarantee the suppression or at least the limitation of the overshoot in the system response. In fact, overshoot in the system response to depth set point are particularly dangerous when the vehicle operates in a cluttered environment, like that found in the vicinity of offshore structures and submersed industrial installations or during

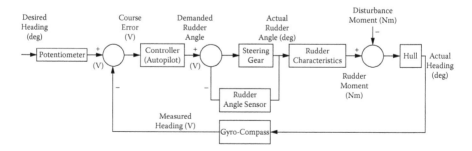

FIGURE 1.5   Closed-loop system block diagram for ship control system.

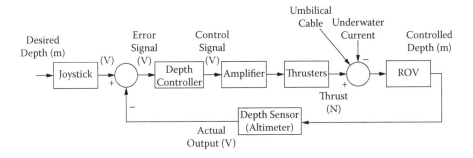

FIGURE 1.6   Closed-loop system block diagram for ROV control system.

archeological activities. Clearly, an overshoot in the vehicles' vertical trajectory, in these circumstances, may cause damage to both the vehicle and the inspected structure. In this case, a PID controller is used.

The important components of the control system for the ROV are seen in Figure 1.6. The operating depth is measured by an altimeter or sonar depth sensor and compared with the desired depth. The controller determines the commanded depth and sends a signal to the thrusters through the onboard amplifier. The thrusters are configured such that for each degree of freedom, the ROV requires certain thrusters. The real depth is measured by the altimeter or sonar depth sensor, and the readings are compared with the desired depth value. This forms a control loop control system. The thruster provides motion to the ROV to reach the desired depth while the ocean current and the umbilical cable attached to the ROV create moments that may hinder this control action. Electronic noise can affect the output of the sensor. For real-time depth monitoring, the monitor screen with the graphic user interface is used to display the real-time behavior of the ROV in water.

1.4.3   Unmanned Aerial Vehicle Control System

Another closed-loop control system that operates in the air is unmanned aerial vehicles (UAVs). They are capable of flight without an onboard pilot. It can be controlled remotely by an operator, or can be controlled autonomously via preprogrammed flight paths. Such aircraft have already been implemented by the military for recognizance flights. Further use for UAVs by the military, specifically as tools for search and rescue operations, warrants continued development of UAV technology. For example, a quad-rotor helicopter is one type of UAV. It is an aircraft whose lift is generated by four rotors. Control of such a craft is accomplished by varying the speeds of the four motors relative to each other. Quad-rotor crafts naturally demand a sophisticated control system in order to allow for balanced flight.

The main elements of the quad-rotor helicopter control system are shown in Figure 1.7. The control system has a cascaded loop that consists of an inner loop for speed control and an outer loop for altitude control. The actual propellers' speeds are measured by tachometers and compared with the desired speed. The controller computes the demanded speed based on the altimeter feedback and sends a control signal

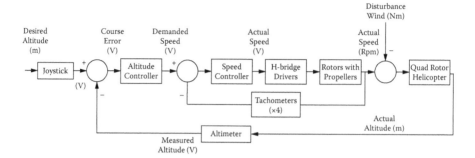

FIGURE 1.7   Closed-loop system block diagram for quad-rotor helicopter control system.

to the inner loop for speed control. Hence, the actual speed and altitude are moni-tored by the tachometers and altimeter, respectively. The value is compared with the desired value, to form a single control loop. The rotors provide a lift force by varying the speed of the propellers on the quad-rotor helicopter to reach the desired altitude despite the external wind effect that can hinder this action. Practically, it is quite difficult to obtain good feedback due to the sensitivity of the sensors used. Electronic noise could affect the output of each sensor.

1.5   CONTROL SYSTEM DESIGN

The availability of powerful low-cost microprocessors has spurred great advances in the theory and applications of control systems. In terms of theory, major strides have been made in the areas of feedback linearization, sliding control, nonlinear adapta-tion and other techniques. In terms of applications, many practical control systems have been developed, ranging from digital "fly-by-wire" flight control systems for aircraft, to "drive-by-wire" underwater robotic vehicles, to advanced medical robotic and space systems. As control systems have become more complicated both in design and requirements, the modeling and simulation become important to aid in control systems design. As a result, the subject of computer-aided control system design is occupying an increasingly important place in automatic control engineering, and has become a necessary part of the fundamental background of control engineers prior to control applications.

   With the information and knowledge available to the control engineers, the next step is to design the control system. Some problems such as the system uncertainties and the poor measurement caused by external disturbances and noise effects could exist. However, there is a common approach that can be used to design typical control systems. The steps in the approach are shown in Figure 1.8.

   As observed in Figure 1.8, after the control objectives are defined, the variables to be controlled are decided. To provide a numerical solution to the dynamic model of the plant, the plant's model is modeled and simulated in MATLAB and Simulink. Prior to controller design, the control structure design has to be performed. It includes selecting the input and output pairs to effectively reduce the coupled effects, reducing the order or states of the system without affecting the time and

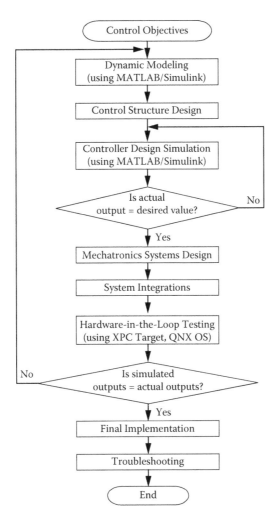

FIGURE 1.8 Steps in designing a control system using computer-aided design software.

the frequency response of the original system, improving the numerical condition of the system model in order to provide numerically stable results and lastly, performing a sanity check on the open-loop system stability and response.

Very often, other software such as computational fluid dynamic (CFD) together with other software such as AutoCAD™, SolidWorks™, and Pro/ENGINEER®* are used to obtain the solid model representation of the plant and other parameters such as: mass, moment of inertia, added mass coefficients, and damping force coefficients. The controller is then designed to control the plant such that it meets the control objectives. After the control system is designed for the plant, the selected hardware and control algorithm need to be tested. To accomplish this, hardware-in-the-loop

---

* http://www.ptc.com/support/legalagreements/proe-agreements.htm

testing helps to simulate the system in real time. Sometimes, actual implementation using software such as xPC Target, QNX, DOS, Linux operating systems and hardware (i.e., sensors and actuators) are used. The test objectives are to check whether the control system design is satisfactory. This is followed by final integration with mechanical components (i.e., enclosure design, onboard power supply, user input devices, and graphic-user-interface). Troubleshooting on the control system has to be done if issues occur after implementation.

2 Introduction to MATLAB and Simulink

2.1 WHAT IS MATLAB AND SIMULINK?

MATLAB is a program for numerical computation. It is widely used for control systems analysis and design. There are many toolboxes available in MATLAB. But in in these sections, we will make extensive use of the Control Systems Toolbox. MATLAB is supported on Unix, Macintosh, and Windows environments. In this section, a brief introduction to MATLAB is given. Most functions and conventions are logical and easy to use for numerical computation.

Simulink is a program with graphical programming facilities for simulating dynamic systems. As an extension to MATLAB, Simulink adds many features specific to dynamic systems while retaining MATLAB general purpose functionality. For example, complex systems also containing nonlinearities can be built and analyzed easily in graphical form. Simulink has two phases of use, model definition and model analysis. First, a model has to be defined or a previously defined model is recalled. Then, that model is analyzed. The progress of a simulation can be viewed while the simulation is running, and the final results can be made available in MATLAB workspace when a simulation is completed.

In this chapter, we will learn how to model and analyze a physical system such as a simple mass-spring-damper system or a second-order differential equation in MATLAB and Simulink. This tutorial also provides canonical steps, and syntax toward solving the problems. It also covers closed-loop control of the second-order systems using the Proportional-Integral-Derivative (PID) controller and its parameters tuning using the Ziegler–Nichols method [25].

2.2 MATLAB BASIC

2.2.1 VECTOR

We begin by doing something simple, like a row vector. Enter each element of the vector by giving a space between each value. Then, set it equal to a variable 'a.' For example, enter the following command into the MATLAB command prompt:

```
>> a = [0.1 0.2 0.3 0.4 0.5 0.6]
```

MATLAB will return:

```
a =
0.1 0.2 0.3 0.4 0.5 0.6
```

Next, to create a column vector, simply type the following:

```
b = [0.1;0.2;0.3;0.4;0.5;0.6;0.7]
b =
    0.1
    0.2
    0.3
    0.4
    0.5
    0.6
    0.7
b = a'
b =
    0.1
    0.2
    0.3
    0.4
    0.5
    0.6
    0.7
```

If you want to create a row vector with elements starting from 0 and 10 with an increment of 0.1, the following command is used:

```
>> t = 0:.1:10
t =
0 0.1 0.2 0.3 0.4 0.5 0.6 0.7 0.8 0.9 0.10
```

It is vital to note that in MATLAB, the vectors are indexed from 1. This means that location of the first element is 1. The following command is used:

```
>> t(1)
ans =
0
```

If you would like to add 0.2 to each of the elements in the vector 'a', the following command is used:

```
>> b = a + 0.2
b =
0.3 0.4 0.5 0.6 0.7 0.8 0.11 0.10 0.9
```

Now, if you want to add (or subtract) two row vectors, if the two vectors are the same length, the following command is used:

```
>> d = a + b
d =
0.4 0.6 0.8 0.10 0.12 0.14 0.20 0.18 0.16
```

Multiplication of vectors must be performed using dot star (.*) so the corresponding elements of each vector are multiplied. The following command is used:

```
>> d = a.*b
d =
0.03 0.08 0.15 0.24 0.35 0.48
```

To create a vector of powers of 2, you would need to use a dot caret (.^). Note that the use of a semicolon makes the output invisible. The following command is used:

```
>>power = 1:5;
>>power_of_2 = 2.^powers
power_of_2 =
2 4 8 16 32
```

2.2.2 MATRICES

Creating a matrix into MATLAB is the same as entering row and column vectors except each row of elements is separated by a semicolon (;). We can multiply the two matrices D1 and C1 together. The following command is used:

```
>> C1 =[ 1 2 3 ; 4 5 6; 9 9 9]
C1 =
     1 2 3
     4 5 6
     9 9 9
>> D1 =[ 8 8 8 ; 4 5 6; 9 9 9]
D1 =
     8 8 8
     4 5 6
     9 9 9
>> E1 = C1 * D1
E1 =
     43 45 47
     106 111 116
     189 198 207
>> E1 = D1 * C1
E1 =
     112 128 144
     78 87 96
     126 144 162
```

Another option for matrix manipulation is to multiply the corresponding elements of two matrices using the (.*) operator. The following command is used:

```
>> F1 = [1 2; 3 4]
F1 =
1 2
3 4
```

```
>> G1 = [2 3;4 5]
G1 =
2 3
4 5
>> H1 = F1 .* G1
H1 =
2 6
12 20
```

You can determine the cube of the matrix using the following command. The following command is used:

```
>> F3 = F1^3
F3 =
37 54
81 118
```

If you want to cube each element in the matrix, just use the element-by-element method. The following command is used:

```
>> F3 = F1.^3
F3 =
1 8
27 64
```

You can determine the inverse of a matrix:

```
>> invX = inv(F1)
invX =
-2.0000 1.0000
1.5000 -0.5000
```

You can determine the eigenvalue of a matrix:

```
>> eigenvalueF = eig(F1)
eigenvalueF =
-0.3723
5.3723
```

To determine the coefficients of the characteristic polynomial of a matrix, the poly function is used. It forms a vector with the elements that are the coefficients of the characteristic equation.

```
>> p1 = poly(F1)
p =
1.0000 -5.0000 -2.0000
```

Recall the eigenvalues of a matrix are similar to the roots of its characteristic polynomial. The result of using the `roots` command is similar to the result of using the `eig` command on matrix F1.

```
>> roots(p1)
ans =
5.3723
-0.3723
```

2.2.3 PLOT GRAPH

It is simple to do plotting in MATLAB. If you wanted to plot a sine and a cosine wave as a function of time, create a time vector and then compute the sine and cosine value at each time.

```
>> t = 0:0.22:7;
>> y1 = sin(t);
>> y2 = cos(t);
>> plot(t,y1,'-',t,y2,'--')
>> xlabel('time(s)')
>> ylabel('y')
>> title('Sine vs Cosine Function')
>> legend('sine','cosine')
```

The '–' and '––' indicate the different style to differentiate between the sine and cosine function. The plot contains one period of a sine and cosine wave. For clarity, we can include the following labels and legend in Figure 2.1.

2.2.4 POLYNOMIALS

A polynomial is represented by a vector. To create a polynomial in MATLAB, enter each coefficient of the polynomial into the vector in descending order. If the polynomial has missing coefficients, you must enter zeros in the appropriate place within the vector.

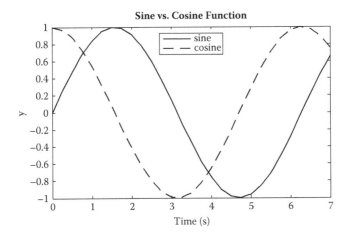

FIGURE 2.1   Sine and cosine curves.

For example, $s^4 + 1$ would be interpreted in MATLAB as:

```
>> y1 = [1 0 0 0 1]
```

Similarly, you can use the inline function in MATLAB.

```
>> y1 = inline('s^4+1')
```

You can determine the value of a polynomial using the polyval function. To determine the value of the above polynomial at s = 2, type the following command.

```
>> polyval([1 0 0 0 1],2)
ans =
17
```

For example, to plot the polynomial $(s^4 + 3s^3 - 15s^2 - 2s + 9)$ ranging from −6 to 4, the commands used are as follows:

```
>> t = -6:.2:4;
>> x = [1 3 -15 -2 9];
>> plot(t, polyval(x,t))
>> grid on
```

As observed in Figure 2.2, the function crosses zero four times. It has four roots to the equation. To find the exact locations of the roots, use the roots command.

```
>> roots([1 3 -15 -2 9])
ans =
-5.5745
2.5836
-0.7951
0.7860
```

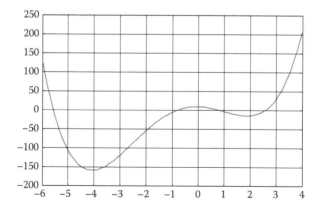

FIGURE 2.2   Polynomial curve.

2.2.5   M-FILES AND FUNCTION

M-files enable the user to write functions and later execute them in the MATLAB command prompt. This is especially useful because control loop statements such as IF-ELSE, FOR, and WHILE can be incorporated to operate like a C program. Take note that the M-file syntax is the same as the syntax used in the command window of MATLAB.

To create an M-file, go to FILE menu and choose NEW M-file. Save the file with extension '.m'. To run the file, type in the name of the file in the MATLAB command prompt. Note that the file must be in the current directory to be called from the command window. The current directory is shown at the top of the MATLAB window screen.

The new M-file will be used to plot sine and cosine that were done previously. Instead of issuing few commands at a time, we can group them under a function. The same command `plotgraph(t)` can be issued whenever the graph needs to be plotted. The same plot is shown. Now, you need to type one single line of command with the input parameter, t.

The M-file should look like this. Save the file as plotgraph.m.

```
function plotgraph(t)
% use function to plot sine and cosine curves
y1 = sin(t);
y2 = cos(t);
plot(t,y1,'-', t,y2,'--')
xlabel('time(s)')
ylabel('y')
title('Sine vs Cosine Function')
legend('sine','cosine')
```

Then, type the following command at MATLAB to call the function.

```
>>t = 0:0.25:7;
>>plotgraph(t);
```

In MATLAB, the percent sign (%) is to comment your M-file code. This is useful during troubleshooting when trying to locate a problem.

A help database for each function is available in MATLAB if you are not able to recall the exact command. In addition, there are a few methods to get assistance from MATLAB such as entering demo in the command prompt. For example, you can search the function names and document files by using the following command at MATLAB:

```
>> help plot
```

or

```
>> doc plot
```

2.3   SOLVING A DIFFERENTIAL EQUATION

In this section, we will use the knowledge described in the previous sections to solve the second-order dynamic response of a mass-spring-damper system. This will be done using an M-file. The differential equation used is as follows.

$$M1\ddot{z} + B1\dot{z} + K1z = F1 \tag{2.1}$$

where
$M1 = 550$, $B1 = 7000$, $K1 = 90000$, $F1 = 21610$, $z(0) = 0.3$ and $\dot{z}(0) = -4$.
It is good practice to start the M-file with a heading the same as the function name.

```
% diffeq.m
M1 = 550;
B1 = 7000;
K1 = 90000;
F1 = 21610;
```

The homogeneous solution plus the particular solution can be written as follows:

$$z(t) = z_k(t) + z_p(t) \tag{2.2}$$

The particular solution is assumed to be constant. By dividing the whole equation by K and solving for zp, it becomes:

$$z_p = F1/K1 \tag{2.3}$$

```
>> zp = F1/K1
zp =
0.2401
```

The homogeneous solution is the solution for which the right side of the equation in (2.1) is zero. Assume the homogeneous solution of a second-order equation is a constant time exponential to the power of time. Equation (2.1) becomes:

$$M1 \cdot \ddot{z}_k + B1 \cdot \dot{z}_k + K1 \cdot z_k = 0 \tag{2.4}$$

where

$$z_k = C \cdot e^{\lambda t}, \quad \dot{z}_k = \lambda C \cdot e^{\lambda t} \quad \text{and} \quad \ddot{z}_k = \lambda^2 \cdot C \cdot e^{\lambda t}$$

Dividing (2.4) throughout by $C \cdot e^{\lambda t}$ gives a quadratic equation. Using the `roots` command, the values for λ can be determined.

$$M1\lambda^2 + B1\lambda + K1 = 0 \tag{2.5}$$

```
>> lambda = roots([M1 B1 K1])
lambda =
-6.3636 + 11.0969i
-6.3636 - 11.0969i
```

We can observe that λ are complex and hence we have an oscillatory response. To find the values of the constants in the homogeneous equation, the boundary conditions are needed. For example:

$$z(0) = C_1 + C_2 + z_p = 0.3$$
$$C_1 + C_2 = 0.3 - z_p \tag{2.6}$$
$$\dot{z}(0) = \lambda_1 C_1 + \lambda_2 C_2 = -4$$

The result is a set of linear equations. This set of linear equations can be written as a vector-matrix equation. The inverse function is used to solve for the C, which is a 2×1 matrix containing the constants C_1 and C_2. The MATLAB commands to solve for C are as follows:

```
>>A = [1 1; lambda(1) lambda(2)];
>>B = [(.3-zp);-4];
>>C = inv(A)*B
C =
0.0299 + 0.1631i
0.0299 - 0.1631i
```

With value of C obtained, we are ready to plot the graph for the total solution,

$$z = z_p + C_1 e^{\lambda_1 t} + C_2 e^{\lambda_2 t} \tag{2.7}$$

We used the following function file to execute the above-mentioned steps. The inputs to the function are $M1$, $B1$, $K1$, and $F1$. The time response of the mass-spring-damper system can be seen in Figure 2.3.

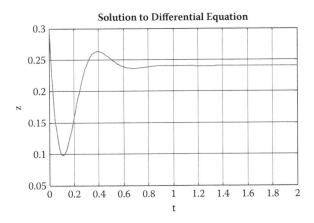

FIGURE 2.3 Time response to the differential equation using MATLAB equations.

```
function [z, t]=diffeq(M1,B1,K1,F1)

% MATLAB Basics Tutorial
% Put in the values for the constants that will be used.
% Type >> M1 = 550;B1 = 7000; K1 = 9000; F1 = 21610;
% Type >> [z, t]= diffeq(M1,B1,K1,F1);

lambda = roots([M1 B1 K1])

A = [1 1; lambda(1) lambda(2)]
zp = F1/K1
B = [(.3-zp);-4]

C = inv(A)*B

t = 0:.01:2;
z = zp + C(1)*exp(lambda(1)*t) + C(2)*exp(lambda(2)*t);
        % plotting and setting up window
plot(t,z)
grid on

title('Solution to Differential Equation')
xlabel('t')
ylabel('z')
```

2.3.1　MATLAB OPEN-LOOP TRANSFER FUNCTION MODELING

Another method to solve the differential equation is to use the transfer function approach. The modeling equation gathered from the free body diagram is in the time domain. Some analyses are easier to perform in the frequency domain. In order to convert to the frequency domain, apply the Laplace transform to determine the transfer function of the system.

The Laplace transform converts linear differential equations into algebraic expressions, which are easier to manipulate. The Laplace transform converts functions with a real dependent variable (such as time) into functions with a complex dependent variable (such as frequency, often represented by s).

The transfer function is the ratio of the output Laplace transform to the input Laplace transform assuming zero initial conditions. Many important characteristics of dynamic or control systems can be determined from the transfer function.

The general procedure to find the transfer function of a linear differential equation from input to output is to take the Laplace transforms of both sides assuming zero conditions, and to solve for the ratio of the output Laplace over the input Laplace. Note that the initial condition of the step response is different.

The Laplace transform allows a linear equation to be converted into a polynomial. The most useful property of the Laplace transform for finding the transfer function is the differentiation theorem. In order to convert the time dependent governing equation to the frequency domain, perform the Laplace transform to the input and output functions and their derivatives. These transformed functions must then be substituted back into the governing equation assuming zero initial conditions. Because the transfer function is defined as the output Laplace function over the input Laplace function, rearrange the equation to fit this form.

As shown, the transfer function of the second-order example can be obtained through the steps below.

The notation of the Laplace transform operation is $L\{\}$.

$$\begin{aligned}
L\{z(t)\} &= Z(s) \\
L\{\dot{z}(t)\} &= sZ(s) - z(0) \\
L\{\ddot{z}(t)\} &= s^2 Z(s) - sz(0) - \dot{z}(0) \\
L\{f(t)\} &= F(s)
\end{aligned} \tag{2.8}$$

When finding the transfer function, 'zero' initial conditions must be assumed, so $z(0) = \dot{z}(0) = 0$.

Taking the Laplace transform of the governing equation results in:

$$F(s) - K[Z(s)] - B[sZ(s)] - M[s^2 Z(s)] = 0 \tag{2.9}$$

Collecting all the terms involving $Z(s)$ and factoring leads to:

$$Ms^2 Z(s) + BsZ(s) + KZ(s) = F(s) \tag{2.10}$$

The transfer function is defined as the output Laplace transform over the input Laplace transform, and so the transfer function of the second-order system is:

$$\frac{Z(s)}{F(s)} = \frac{1}{Ms^2 + Bs + K} \tag{2.11}$$

In order to enter a transfer function into MATLAB, the variables must be given in numerical value because MATLAB cannot manipulate symbolic variables without the symbolic toolbox. Enter the numerator and denominator polynomial coefficients separately as vectors of coefficients of the individual polynomials in descending order. The syntax for defining a transfer function (for the mass-spring-damper system) in MATLAB is as follows.

```
>>TF = tf(1, [550 7000 90000])
```

For step with magnitude other than one, the step response in Figure 2.4 can be obtained as follows.

```
>> step(21610*TF);
```

MATLAB can find the state-space representation (denoted by system matrix A, input vector B, output vector C, and feed forward vector D) directly from the transfer function in a few ways. The state-space model represents a physical system as first-order coupled differential equations. This form is better suited for computer simulation than an nth order input-output differential equation:

$$\begin{aligned}
\dot{x} &= Ax + Bu \\
y &= Cx + Du
\end{aligned} \tag{2.12}$$

where y is the output equation, and x is the state vector.

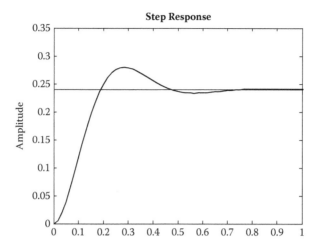

FIGURE 2.4 Time response to the differential equation using Step method.

To find the state-space representation of the system from the numerator and denominator of the transfer function in the form, use MATLAB `tf2ss` command:

```
>> [A, B, C, D] = tf2ss(1,[550 7000 90000])
A =
 -12.7273 -163.6364
  1.0000 0
B =
  1
  0
C =
  0 0.0018
D =
  0
>>
```

In order to find the entire state-space system in addition to the separate matrices from the transfer function, use the following command:

```
>> sysss = ss(A,B,C,D)
a =
 x1 x2
 x1 -12.73 -163.6
 x2 1 0
b =
 u1
 x1 1
 x2 0
c =
 x1 x2
 y1 0 0.001818
```

```
d =
 u1
 y1 0
Continuous-time model.
```

2.3.2   SIMULINK OPEN-LOOP TRANSFER FUNCTION MODELING

Another method to solve the differential equation is to use Simulink's transfer function block diagram. Simulink is a graphical extension to MATLAB for
modeling and simulation of systems. In Simulink, systems are drawn on screen
as block diagrams. Many elements of block diagrams are available, such as transfer functions, summing junctions, and so forth, as well as virtual input and output
devices such as function generators and oscilloscopes. Simulink is integrated with
MATLAB and data can be easily transferred between the programs. In these sections, we apply Simulink to the mass-spring-damper system to model and simulate
the system.

The idea behind these sections is that you can view them in one window while
running Simulink in another window. Some images in these sections are not live—
they simply display what you should see in the Simulink windows.

Simulink is started from the MATLAB command prompt by entering the
following command:

```
>> simulink
```

Alternatively, you can hit the Simulink button at the top of the MATLAB window
(Figure 2.5). When it starts, Simulink brings up the Simulink Library Browser
(Figure 2.6).

Open the modeling window by selecting New and followed by Model from the
File menu on the Simulink Library Browser as shown in Figure 2.6 above.

This will bring up a new untitled modeling window shown in Figure 2.7.

In Simulink, a model is a collection of blocks, which, in general, represents a
system. In addition, to drawing a model into a blank model window, previously saved
model files can be loaded either from the File menu or from the MATLAB command
prompt.

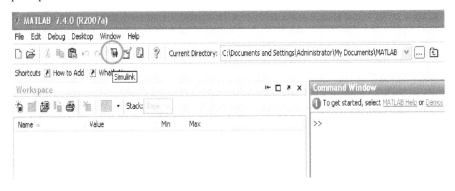

FIGURE 2.5   MATLAB screen.

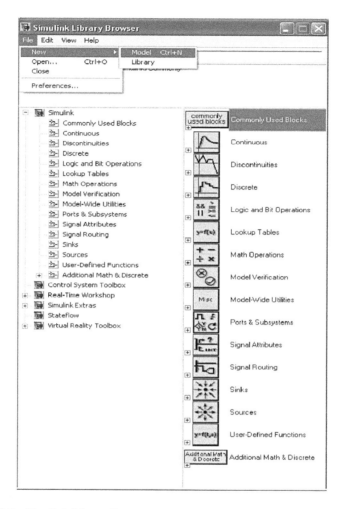

FIGURE 2.6   Simulink Library Browser screen.

A new model can be created by selecting New from the File menu in any Simulink window (or by hitting Ctrl+N). There are several general classes of blocks (just to name a few):

- Continuous
- Discontinuous
- Discrete
- Look-Up Tables
- Math Operations
- Model Verification
- Model-Wide Utilities
- Ports & Subsystems
- Signal Attributes

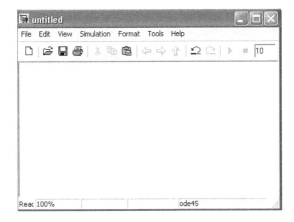

FIGURE 2.7   New untitled modeling window.

- Signal Routing
- Sinks: Used to output or display signals
- Sources: Used to generate various signals

Lines transmit signals in the direction indicated by the arrow. Lines must always transmit signals from the output terminal of one block to the input terminal of another block. One exception to this is splitting the signal to two destination blocks, as shown in Figure 2.8.

Lines can never inject a signal into another line; lines must be combined through the use of a block such as a summing junction.

A signal can be either a scalar signal or a vector signal. For Single-Input, Single-Output (SISO) systems, scalar signals are generally used. For Multi-Input, Multi-Output (MIMO) systems, vector signals are often used, consisting of two or more scalar signals. The lines used to transmit scalar and vector signals are identical. The type of signal carried by a line is determined by the blocks on either end of the line.

The simple model (as shown in Figure 2.8) consists of three blocks: Step, Transfer Fcn, and Scope using the building blocks in Simulink's Block Libraries. The Step is a source block from which a step input signal originates. This signal is transferred through the line in the direction indicated by the arrow to the Transfer Function linear block. The Transfer Function modifies its input signal and outputs a new signal on a line to the Scope. The Scope is a sink block used to display a signal. It works just like an oscilloscope. The simulation time can be changed from 10 to 2 seconds.

A block can be modified by double-clicking on it. For example, if you double-click on the Transfer Fcn block in the simple model, you will see the following dialog box in Figure 2.9.

This dialog box contains fields for the numerator and the denominator of the block's transfer function. By entering a vector containing the coefficients of the desired

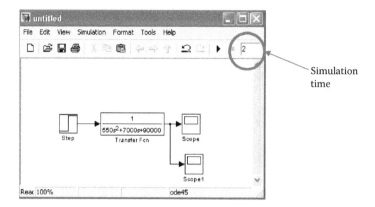

FIGURE 2.8   Transfer function block diagram with split signals.

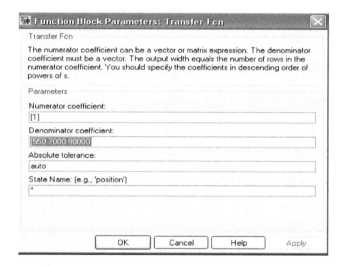

FIGURE 2.9   Transfer function block parameters.

numerator and denominator polynomial, the desired transfer function can be created. For example, the denominator is [550 7000 90000] and numerator is [1] or 1.

The Step block can also be double-clicked, bringing up the following dialog box as shown in Figure 2.10.

The default parameters in this dialog box generate a step function occurring at time = 0 sec, from an initial level of zero to a final value of 21610 (in other words, a 21610 step at t = 0). Each of these parameters can be changed. In this case, change the step time to 0 and the final value to 21610. Close this dialog before continuing.

Then, to start the simulation, either select Start from the Simulation menu (as shown in Figure 2.11) or hit Ctrl+T in the model window.

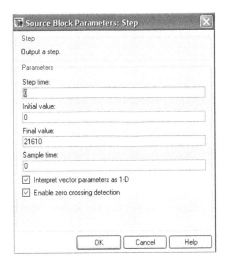

FIGURE 2.10   Source block parameters for Step input.

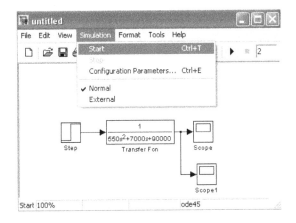

FIGURE 2.11   Simulation menu to start simulation.

After a simulation is performed, the signal, which feeds into the scope, will be displayed in this window. If it does not, just double-click on the block labeled Scope. The only function we will use is the Autoscale button, which appears as a pair of binoculars in the upper portion of the window. Click the Autoscale button to automatically scale the response in Figure 2.12.

2.3.3   SIMULINK OPEN-LOOP SYSTEM MODELING

The last method to solve the differential equation is to use a few of Simulink's block diagrams instead of a single transfer function block diagram as shown earlier.

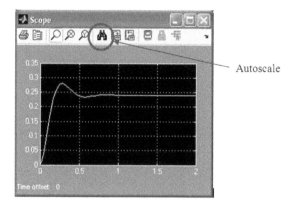

FIGURE 2.12   Scope display after simulation.

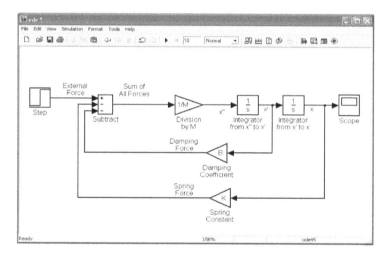

FIGURE 2.13   Final Simulink block diagram to represent mass-spring-damper system.

In this section, you will learn how to build systems in Simulink using the building blocks in Simulink's Block Libraries. You will build the final Simulink model as shown in Figure 2.13.

First, you will gather all the necessary blocks from the block libraries. Then you will modify the blocks so they correspond to the blocks in the desired model. Finally, you will connect the blocks with lines to form the complete system. After this, you will simulate the complete system to verify that it works.

For a system represented by an nth order input/output ordinary differential equation it is necessary to integrate the highest derivative n times to obtain the output. For this reason, the preferred form of the differential equation is to solve for the highest derivative as a function of all of the other terms. For this reason, it is useful to rewrite the equation of motion in the following form:

$$\ddot{z} = 1 / M[B\dot{z} + Kz + F(t)] \qquad (2.13)$$

Follow the steps below to collect the necessary blocks:

To start a new model in Simulink, click File on the Simulink Library Browser window's toolbar, click on New and then click on Model. You will get a blank model window. Click on File and Save As: ode.mdl.

Click on the Library Browser button ⊞ to open the Simulink Library Browser. Click on the Sources option under the expanded Simulink title to reveal possible sources for the model in Figure 2.14.

From the Simulink Library Browser, drag the two integrators from the Continuous option found under the Simulink in Figure 2.15.

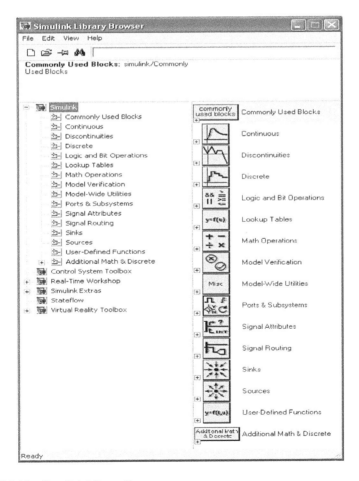

FIGURE 2.14   Simulink Library Browser.

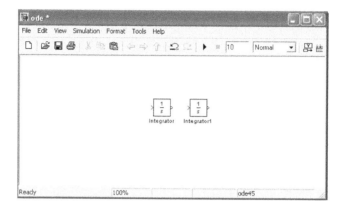

FIGURE 2.15   Simulink block diagram for two integrators.

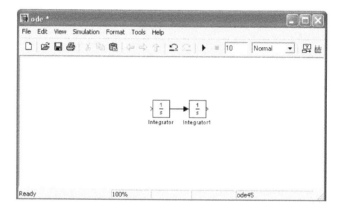

FIGURE 2.16   Simulink block diagram for two connected integrators.

Next, connect the two integrators by dragging a line between them. Click on the output arrow of the first integrator and drag it to the input arrow of the second integrator as shown in Figure 2.16.

Label this line x' by double-clicking on the line and typing in the textbox as shown in Figure 2.17.

Next, add a line going into the first integrator and a line coming out of the second integrator. To add the line to the input of the first integrator, click the input of the first integrator and drag it to the left. And for the output of the second integrator, click on the output and drag it to the right. Label the line attached to the input of the first integrator as x″, and label the line attached to the output of the second integrator as x. This can be seen in Figure 2.18.

It is useful to label the integrators themselves. To rename any block, click on the block's label and type in the text box as shown in Figure 2.19.

Now let's construct the signal representing the damping force. The damping force is the x' multiplied by the damping constant. We can pull x' from the output

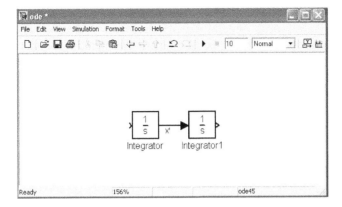

FIGURE 2.17   Label for Simulink block diagram.

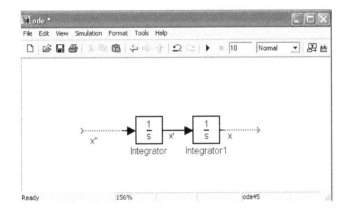

FIGURE 2.18   Label for block diagram outputs.

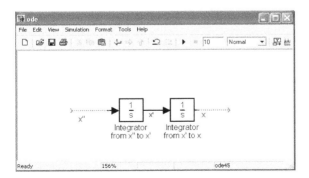

FIGURE 2.19   Label for Simulink block diagram.

of the integration from x″ to x′ block. The Simulink equivalent to multiplication is the Gain block. From the Simulink Library Browser window add a Gain block by expanding Simulink and then clicking on Math Operations and then dragging the Gain block into the model as shown in Figure 2.20.

The number inside the triangle indicates the gain of the Gain block. To change the gain of this Gain block from 1 to B, right click on the Gain block and click on Gain parameters. A new window will pop up labeled Block Parameters: Gain. Change the value in the field for Gain from 1 to B as shown in Figure 2.21.

The input of the Gain block is currently on the left side. To make the line connection simpler, we would like the input of the Gain block to be on the right. To flip a Gain block, right click on the block and select Format and select Flip block as shown in Figures 2.22 and 2.23.

Also, it may be useful to relabel this Gain block as Damping Coefficient; do so by clicking on the title and typing in the text box that opens. The damping force is now

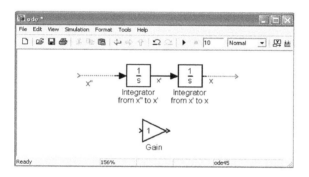

FIGURE 2.20   Gain block for damping coefficient.

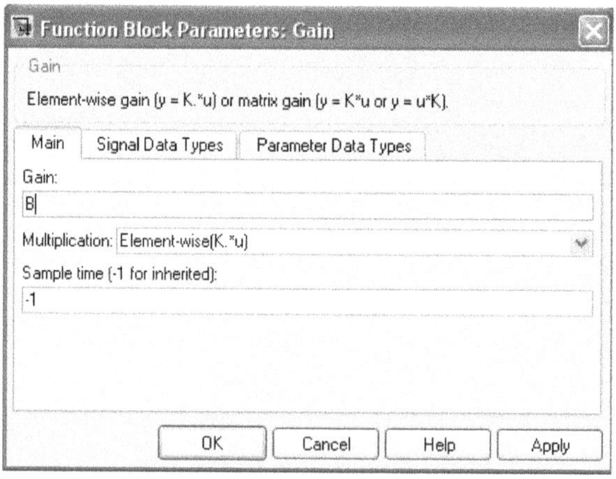

FIGURE 2.21   Function block parameters for damper block.

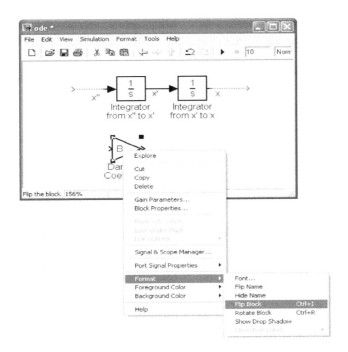

FIGURE 2.22   Flip damper coefficient Gain block.

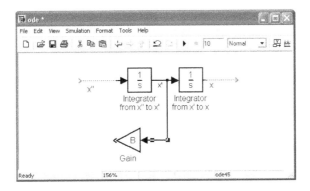

FIGURE 2.23   Connect Gain block for damper coefficient.

the output of the Damping Coefficient block. Drag a line out of the output and label it Damping Force as shown in Figure 2.24.

The spring force is determined by multiplying x with the spring constant, K. This can be done using the same methods prescribed above for creating the Damping Force resulting in the following addition to the model as shown in Figure 2.25.

In the governing equation, there are three forces that are summed together to form the total force on the mass, one external force and two internal forces,

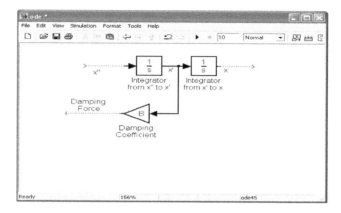

FIGURE 2.24   Label Gain block for damper coefficient.

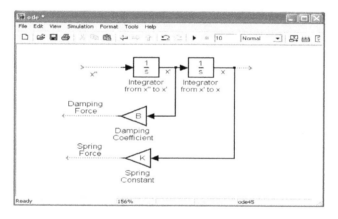

FIGURE 2.25   Connect Gain block for spring constant.

namely, the Damping Force and the Spring Force created above. Summation in Simulink is accomplished by using the Sum block. From the Simulink Library Browser window, expand Simulink and click on Math Operations and then scroll down in the right column to find Sum. Drag the Sum block into the model as shown in Figure 2.26.

The default Sum block has two positive inputs. The system we are modeling has one positive input and two negative inputs. To change the number and polarity of inputs, right click on the Sum block and select Sum parameters. A new window will open up titled Block Parameters: Sum. In the field marked List of Signs change the value from +− to +−− as shown in Figure 2.27. Selecting OK will return you to the model.

The default size for the Sum block is too small to comfortably hold the three inputs. To change the size, click on the Sum block, drag on one of the corner blocks to enlarge it.

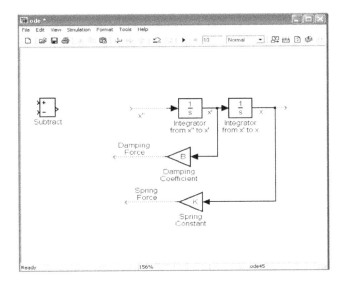

FIGURE 2.26   Label Subtract block.

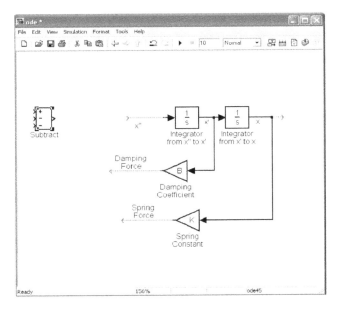

FIGURE 2.27   Modify Subtract block for three inputs.

Now, to connect the Damping Force and Spring Force to the two negative input terminals of the Sum block, drag the end of the arrow labeled Damping Force to a negative input of the Sum block. Similarly, do it for the Spring Force in Figure 2.28.

The third input of our Sum block is reserved for the external force. Simply draw a line out of the positive input of the Sum block and label it as External

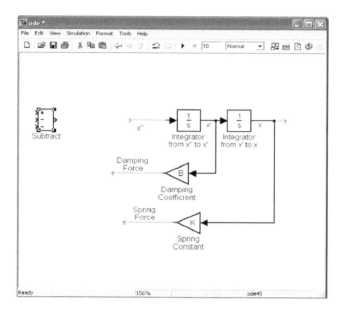

FIGURE 2.28 Connect inputs to Subtract block.

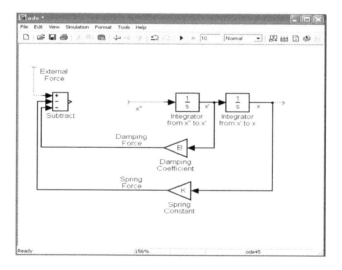

FIGURE 2.29 Connect external input force to Subtract block.

Forces. As seen later, the output of a source will attach to this arrow as shown in Figure 2.29.

The output of the Sum block is the Sum of all of the forces acting on the system. Draw a line out of the input and label it as Sum of All Forces as shown in Figure 2.30.

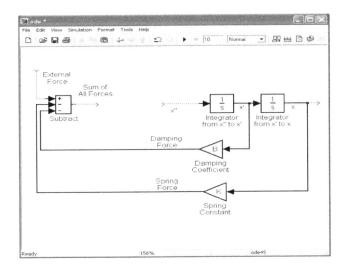

FIGURE 2.30 Connect Subtract block output.

From Newton's law, the sum of all forces is equal to Mx″. So, dividing the sum of all forces by M results in x″. To multiply the Sum of All Forces by 1/M, add a Gain block with value 1/M. From the Simulink Library Browser window, expand Simulink and select Math Operations and drag the Gain block into the model from the right column.

Change the value of the Gain block from 1 to 1/M by right clicking on the Gain block, selecting Gain parameters and changing the value for Gain in the Block Parameters.

The block is too small to display the value of the gain. Enlarge the block slightly by clicking on it and dragging the corner triangle. Also, it may be beneficial to rename this Gain block as Division by M as shown in Figure 2.31.

Once the model has been created in MATLAB, it is easy to simulate the response to a step input. In order to simulate the step response, you need to add a source to provide the external force, and you need a sink to view the response of the system. In the Simulink Library Browser window, expand Simulink and click on Sources and then drag the Step source ⊓ from the right column into the model. Connect the tail of the arrow labeled External Forces to the Step source output as shown in Figure 2.32.

The default parameters for the Step source are a Step time of 1, an Initial value of 0, a Final value of 1, and a Sample time of 0. To change the parameters of the Step source, right click on the Step source and select Step parameters. In the Block Parameters: Step window; change the parameters to whatever is desired.

To find the response to a step input of magnitude 21610 with the step time t = 0, change the parameters to the following: a Step time of 0, an Initial value of 0, a Final value of 21610, and a Sample time of 0.

To monitor the value of x, add a Scope sink. Go to the Simulink Library Browser window, expand Simulink, click on Sinks, and then drag the Scope sink ⊟ from

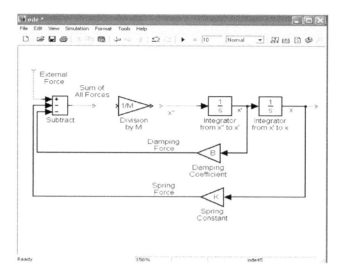

FIGURE 2.31   Gain block for 1/M.

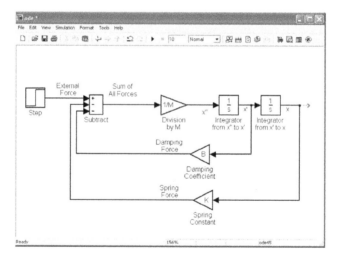

FIGURE 2.32   Connect Gain block for 1/M.

the right column into the model. Connect the head of the arrow labeled x to the Scope sink as shown in Figure 2.33.

Now, double-click on the Integrator to change the initial condition: $z(0) = 0.3$ where z is the position as shown in Figure 2.34.

Now, double-click on the Integrator to change the initial condition: $\dot{z}(0) = -4$ where $\dot{z}$ is the velocity as shown in Figure 2.35.

To run the simulation, click Simulation in the toolbar and select Start, or equivalently hit Ctrl+T on the keyboard, or click the ▸ button on the toolbar. To view

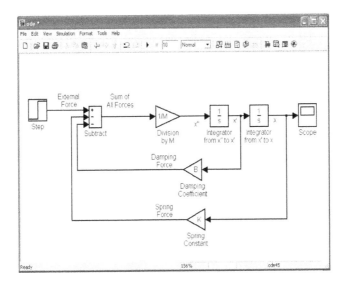

FIGURE 2.33   Scope for mass-spring-damper system.

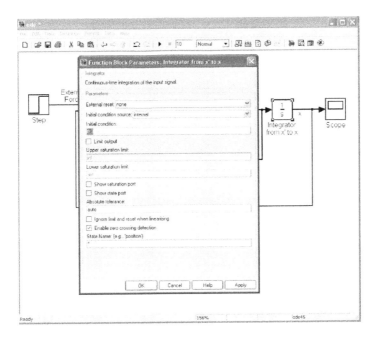

FIGURE 2.34   Change initial condition for position.

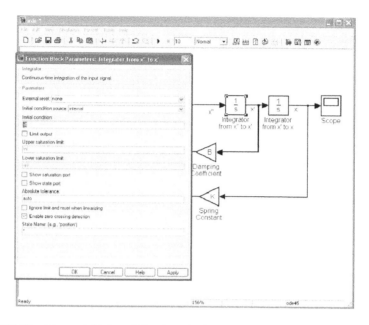

FIGURE 2.35   Change initial condition for velocity.

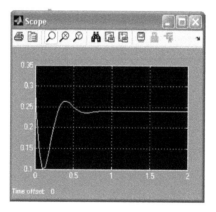

FIGURE 2.36   Time response using Simulink block diagrams.

the output of the Scope sink, which is monitoring the value of x, double-click on the Scope in the model. The following plot of the step response will appear in a new window as shown in Figure 2.36.

In comparison with the previous method, to solve the differential equation for the spring-mass-damper system, the response is output to the workspace. In the Simulink Library Browser window, expand Simulink and click on Sinks and then drag the To Workspace from the right column into the model as shown in Figure 2.37.

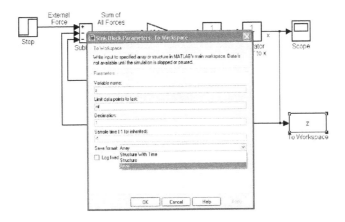

FIGURE 2.37 Output to workspace.

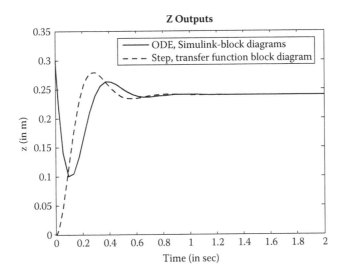

FIGURE 2.38 Time response comparison between different methods.

Double-click on the To Workspace block diagram to change the Variable name to z and Save format to Array as shown.

Now run the simulation again and plot the graphs using the `plot` command in MATLAB. For comparison, graphs resulting from other methods can be plotted on the same graph as shown in Figure 2.38. Note that the figure () will automatically assign a new figure number for the plot.

```
>>figure()
>>plot(tout,z,'-')
```

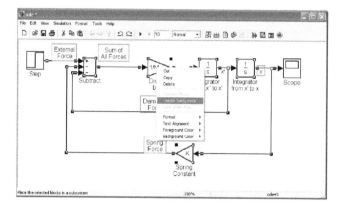

FIGURE 2.39   Create subsystem for mass-spring-damper system model.

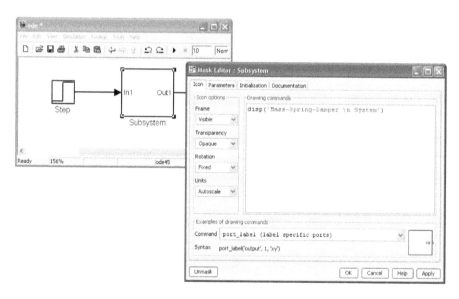

FIGURE 2.40   Mask subsystem for mass-spring-damper system model.

It is possible to create a subsystem that contains certain block diagrams. The subsystem can be created by selecting the block diagrams (except the Step and Scope block sets) and right-click on Create Subsystem. It makes the model neater and easy to troubleshoot as shown in Figure 2.39.

The subsystem can be further masked by right-clicking and selecting Masked subsystem as seen in Figure 2.40. Type the following command into the space provided in Drawing Commands. It results in the subsystem as seen in Figure 2.41.

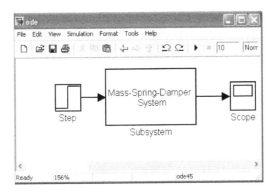

FIGURE 2.41 Masked subsystem.

2.4   SIMULINK CLOSED-LOOP CONTROL SYSTEM DESIGN

After completing the open-loop simulation using MATLAB and Simulink, a closed-loop control system can be designed. In this section, PID parameters tuning using the Ziegler–Nichols method is used. A closed-loop control system is one in which the output signal has a direct effect upon the control action, that is, closed-loop control systems are feedback control systems. The system error signal, which is the difference between the input signal and the feedback signal, is fed back to the controller so as to reduce the output error and thus bring the output of the system to the desired value. In other words, the term *closed-loop* implies the use of feedback action in order to reduce the system error. Figure 2.42 shows the block diagram of the closed-loop control system.

A Proportional-Integral-Derivative (PID) controller is one in which the outputs of the P-action, I-action and D-action are added together to produce the controller output. This type of control provides zero offset and faster response. Derivative mode is added to the PI controller to make the response speed faster and less oscillatory.

In order to get the best performance from the three terms in the controller, the amount of each action (P-action, I-action, D-action) has to be selected carefully. If a perfect model of the plant was available, then the selection process could be done through simulation or other analytical techniques. However, if the mathematical model of the plant is unknown, then an analytical approach to the design of a PID controller is not possible. Then, we must resort to an experimental approach to the design of PID controllers. There are many methods to tune the PID controller. One of the methods is to use the Ziegler–Nichols method.

2.4.1   PID TUNING USING SIMULINK

The process of selecting the controller parameters to meet given performance specifications is known as *controller tuning*. Ziegler and Nichols suggested a rule [25] for tuning PID controllers (meaning to set values of K_p, T_i and T_d), which is

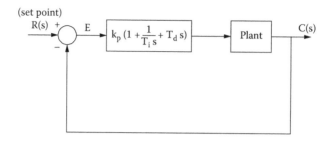

FIGURE 2.42   Closed-loop control system using PID controller.

TABLE 2.1
Parameters Used in the Ziegler–Nichols Method

Type of Controller	K_p	T_i	T_d
P	$0.5\,K_C$	∞	0
PI	$0.45\,K_C$	$0.85\,T_C$	0
PID	$0.6\,K_C$	$0.5\,T_C$	$0.125\,T_C$

based on the experimental value of K_p that results in marginal stability with the P-action acting alone.

Initially the line is operated using only a P controller. Its gain is increased until a continuous oscillation of the controlled variable is produced. This value of gain is the critical gain K_c. Also, the periodic time of the oscillation T_c (critical period) must be established. Based on these values, the controller can then be tuned based on the Ziegler–Nichols recommended settings for process controllers shown in Table 2.1:

Note that the Ziegler–Nichols method only provides the initial setting for the PID controller. In order to have the optimal result, fine-tuning is required.

For example, the transfer function of the plant is given by

$$G_p(s) = \frac{1}{s^3 + 2s^2 + 2s + 1} \tag{2.14}$$

First, construct the system model in Simulink as shown in Figure 2.43. Change the following parameters:

For **Step Input** block	Step Time:	0	
For **Transfer Fcn** block	Numerator:	[1]	
	Denominator:	[1 2 2 1]	
Set the **PID Controller** block initially	Proportional:	1	(proportional gain, K_p)
	Integral:	0	(integral gain, K_i)
	Derivative:	0	(derivative gain, K_d)

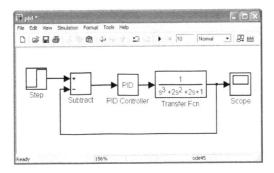

FIGURE 2.43 PID block diagram using Simulink.

Next, to find out the critical gain K_c and critical period T_c, follow the procedure below.

- Select the simulation.
- Observe the transient response in the Scope window.
- Increase the parameter: Proportional, in the PID Controller block, (suggested incremental value of Proportional is 0.5).
- Restart your simulation. Check the response curve. If it is not a continuous oscillation (sine waveform), repeat step with new K_p values until you get continuous oscillation in the transient response. When the sustained oscillation occurs, this K_p (proportional gain) value is called *critical proportional gain*, K_c.
- The value of critical gain $K_c=3$.
- The value of critical period $T_c=3s$.

After obtaining the K_c and T_c, we can use Table 2.1 to obtain the following parameters. Follow the steps to obtain these PID parameters.

- Using Table 2.1, the Ziegler–Nichols Criterion, determine the PID controller settings: $K_p=1.8$, $T_i=2.5s$ and $T_d=0.625s$.
- Set the PID Controller block according to the values found.
  - Proportional: 1.8 (Proportional gain, K_p)
  - Integral: 0.9 (Integral gain, $K_i=K_p/T_i$)
  - Derivative: 0.9 (derivative gain, $K_d=K_pT_d$)
- Start the simulation and the final value of the controlled variable in this case is around 0.98 V. Hence, it has a steady-state error of 0.02 V. The time response of the PID control system can be seen in Figure 2.44.

2.4.2 PID Tuning Using the SISO Tool

Instead of tuning the PID controller gains using Simulink, a single-input, single-output (SISO) tool in MATLAB can be used. To use the SISO tool in MATLAB, enter the plant's transfer function to be used. In this case, the transfer function used for the plant is shown in (2.14).

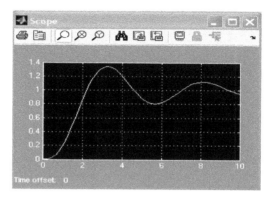

FIGURE 2.44   Time response from Scope in Simulink.

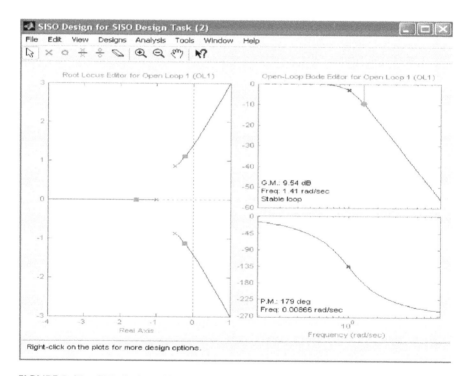

FIGURE 2.45   GUI display of SISO tool in MATLAB.

The commands are as follows:

```
>> TF = tf([1],[1 2 2 1]);
>> sisotool(TF)
```

The graphical user interface (GUI) for the SISO tool can be seen in Figure 2.45. It contains Bode plot and Root Locus plot for open-loop system analysis and control.

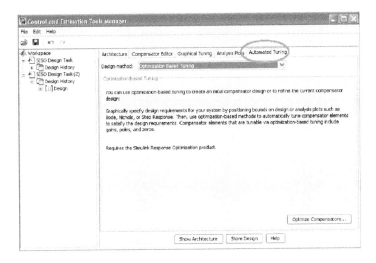

FIGURE 2.46   Select Automated Tuning method in SISO tool.

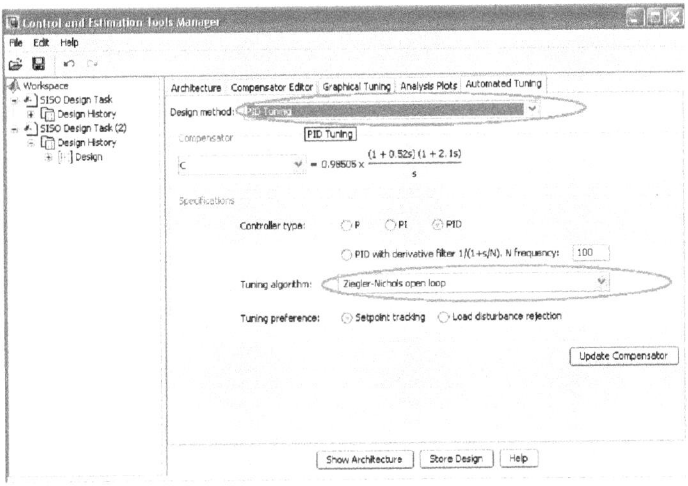

FIGURE 2.47   Display or Update controller in SISO tool.

For automated tuning, we use the Ziegler–Nichols method. Select Automated Tuning method as shown in Figure 2.46. Then, select PID Tuning and choose the Ziegler–Nichols method.

To begin with the controller design, click Update Compensator. The controller can be obtained as shown in Figure 2.47. As observed in the procedures, the tuning method is quite automatic without any involvements in defining the critical gain, critical period, and other parameters.

3 Analysis and Control of the ALSTOM Gasifier Problem

3.1 GASIFIER SYSTEM DESCRIPTION AND NOTATION

With an understanding of MATLAB and Simulink, the next step is to apply the knowledge onto real systems. There are many control systems in industry. The control systems design encompasses not only the PID controller (which we have studied previously), it also includes precontrollers design activities such as scaling the ill-conditioned system, selecting the optimal input and output pairs to reduce the level of coupling between them, and model-order reduction to reduce the size of the system to be controlled.

To demonstrate this, one of the selected applications is on the ALSTOM gasifier used in power generation. ALSTOM developed a benchmark problem based on actual experiment results for control system design. This chapter concentrates on the analysis and the control structure design on the ALSTOM gasifier system. So far, several papers [26–29] have been published on the control of this gasifier system. The ill-conditioned nature of the linear models of this gasifier system is still quite challenging. The difficulties in the ALSTOM gasifier problem are the high-order system, highly coupled in the inputs and outputs, and the numerically poor linear models that affects the control system design process. Throughout the chapter, MATLAB and Simulink are used for analysis and simulation of the control systems. Some of Simulink block diagrams are not live images. They are simplified to provide a clear picture of the block diagrams used during simulation.

First, the ALSTOM gasifier system and its notation used in the chapter are described. ALSTOM gasifier is a nonlinear multivariable system, having four outputs to be controlled with a high degree of cross coupling between them. A schematic gasifier can be seen in Figure 3.1.

The control inputs are ordered as the char extraction flow in kg/s (WCHR), air mass flow in kg/s (WAIR), coal flow rate in kg/s (WCOL), steam mass flow in kg/s (WSTM), and also the disturbance input in N/m² (PSINK). The outputs to be controlled are ordered as fuel gas calorific value in J/kg (CVGAS), bed mass in kg (MASS), fuel gas pressure in N/m² (PGAS), and fuel gas temperature in K (TGAS). By initially neglecting the effects of the input disturbances, PSINK, and noting that limestone mass flow in kg/s (WLS) absorbs sulphur in the WCOL with a fixed ratio of 1:10, this leaves effectively four degrees of freedom for control design. Hence, the gasifier becomes a square system, that is a four inputs (m = 4) and four outputs (l = 4) system.

The gasifier is a nonlinear system and is described by three state-space models obtained by linearization about three operating points: namely, the 100%, 50%,

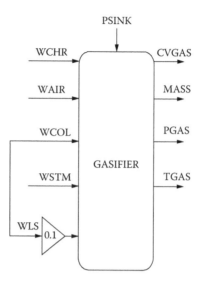

FIGURE 3.1 Schematic diagram of the gasifier.

and 0% load condition. In the following, $G_{100\%}$ will denote the plant model at the 100% load condition. For the three cases to be considered, the gasifier model is assumed to be of the following continuous linear time invariant (LTI) state-space form:

$$\dot{\mathbf{x}}(t) = \mathbf{A}\mathbf{x}(t) + \mathbf{B}\mathbf{u}(t)$$

$$\mathbf{y}(t) = \mathbf{C}\mathbf{x}(t) + \mathbf{D}\mathbf{u}(t)$$

(3.1)

where the states $\mathbf{x} \in \mathfrak{R}^n$, the inputs $\mathbf{u} \in \mathfrak{R}^m$, the outputs $\mathbf{y} \in \mathfrak{R}^l$, and the system matrix $\mathbf{A} \in \mathfrak{R}^{n \times n}$ and input matrix $\mathbf{B} \in \mathfrak{R}^{n \times m}$, the output matrix $\mathbf{C} \in \mathfrak{R}^{l \times n}$, and feed forward matrix $\mathbf{D} \in \mathfrak{R}^{n \times m}$. The order of the gasifier system to be studied is n = 25.

The aim of this benchmark challenge is to design a controller to satisfy a set of desired specifications in Table 3.1 on the system outputs, inputs and input rates, when a step or a sine wave pressure disturbance is applied to the linearized model, $G_{100\%}$. The controller to be designed should also regulate the outputs within the output constraints. In addition, the robustness of this controller is to be evaluated at the 50% and 0% load conditions.

A detailed description of the process (ALSTOM gasifier control) and desired performance specifications are available [26].

3.2 INHERENT PROPERTIES ANALYSIS

Before starting on the development of a controller, several tests are performed on the state-space model $G_{100\%}$. This enables the inherent properties of the gasifier, such as the open-loop stability, the ill-conditioning level, and the uncontrollable and unobservable modes of the linear system model of the system, to be determined. The following are the tests performed on the gasifier system at the 100% load condition.

TABLE 3.1

Control System Specifications

Inputs	Range (kg/s)	Rate (kg/s²)
WCHR	[0, 3.5]	0.2
WAIR	[0, 20]	1.0
WCOL	[0, 10]	0.2
WSTM	[0, 6]	1.0
Outputs	**Range**	
CVGAS (kJ/kg)	±10	
MASS (kg)	±500	
PGAS (kN/m²)	±10	
TGAS (K)	±273	

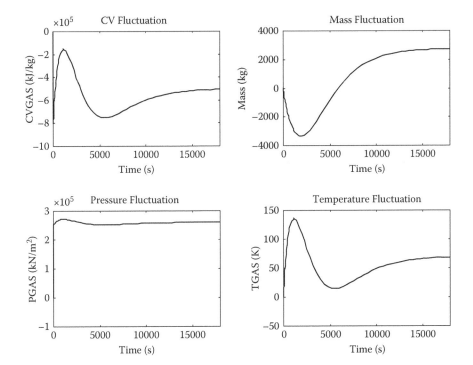

FIGURE 3.2 Open-loop steady-state response of gasifier system (outputs).

a. **Open-Loop Stability Test**. First, by examining the eigenvalues of the linear system model, it is found that the system is completely stable, since there are no eigenvalues with a positive real part. This phenomenon can also be observed in the open-loop time response in Figure 3.2, since all outputs settled to their steady-state values as time increased.

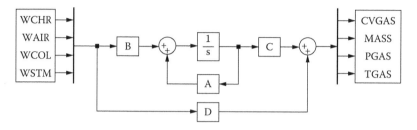

FIGURE 3.3   State-space system.

The time responses of the linear system at steady state are generated using the state-space configuration shown below (Figure 3.3).

```
% Load Gasifier *.mat file
run load_file
        % Form state-space matrices
sys=ss(A,B,C,D);
        % Form transfer function matrix, TFM G(s)
[Ass,Bss,Css,Dss]=ssdata(sys);
[G, num1,num2,num3,num4,den]=tfm(sys);
        % Input
WCHR_T=0.9;
WAIR_T=17.42;
WCOL_T=8.55;
WSTM_T=2.7;
PSINK_T=-0.2e5;
        % simulation time
sim_time=17200;
t_out=0:1:sim_time;
        % Input sequence for WCHR, WAIR, WCL, WSTM
WCHR=0.9;y_chr=step([0 1],[0 1],t_out)*WCHR;
WAIR=17.42; y_air=step([0 1],[0 1],t_out)*WAIR;
WCOL=8.55; y_col=step([0 1],[0 1],t_out)*WCOL;
WSTM=2.7; y_stm=step([0 1],[0 1],t_out)*WSTM;
        % Plot open-loop response
four_inp=[y_chr y_air y_col y_stm];
yGOL=lsim(G,four_inp, t_out);
figure();
subplot(2,2,1);
plot(t_out,yGOL(:,1),'r-');
ylabel('MASS (kg)');
xlabel('TIME(s)');
        % Repeats for other subplots
```

First, by examining the eigenvalues (i.e., using eig(**A**)) of the linear open-loop system model that is shown in Table 3.2, it is found that the system is completely stable, since there are no eigenvalues with a positive real part. This stable phenomenon can also be observed in the open-loop time

TABLE 3.2

Eigenvalues of the Linear System at 100% Load

Open-Loop Eigenvalues	
$-7.117 \times 10^{-4} \pm 4.365 \times 10^{-4}i$	-0.0568
$-3.753 \times 10^{-4} \pm 1.509 \times 10^{-4}i$	-0.0568
-3.252×10^{-4}	-0.0568
-7.172×10^{-4}	-0.0568
-6.977×10^{-3}	-0.0568
-0.0155	-0.0568
-0.0301	-0.108
-0.0568	-0.132
-0.0568	-0.396
-0.0568	-1.006
-0.0568	-33.125
-0.0568	

response in Figure 3.2, since all outputs settled to their steady state as time increased.

b. **RHP Zeros Test**. Since RHP zeros, in particular, are undesirable due to their effect on the system behavior such as nonminimum phase phenomena and high gain instability in the system time responses, a knowledge of their presence is essential. By examining the open-loop gasifier system model, two RHP zeros were detected. However, these RHP zeros are situated at high frequencies (659 and 1.11×10^{11} rad/s) for the $G_{100\%}$ model. Therefore, it can be expected that they will not have any significant effect on the system performance, which requires a much lower bandwidth.

```
G=pck(A,B,C,D);
z1 = szeros(G);
sort_z1=sort(z1)
```

c. **Condition Number Test**. The condition number plots of the open-loop system across the frequency range of interest are shown in Figure 3.4 and indicate that the condition number is very high in certain regions, in particular at 0.01 rad/s to 100 rad/s. This implies a wide spread of the singular values of the system at these frequencies. Thus, the gasifier system is indeed highly numerically ill conditioned.

```
w=logspace(-8,2,200);
figure()
svGp=sigma(sys,w);
semilogx(w,max(svGp)./min(svGp),w,max(svGp)./min(svGp));
ylabel('Condition number');
```

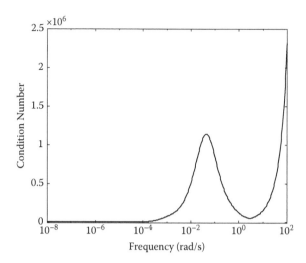

FIGURE 3.4   Condition number of the gasifier system.

d. **Reduction of the State-Space Equations**. Using the routines available
in MATLAB to test for controllability and observability, based on the
Kalman tests, gave extraneous results due to the ill-conditioned nature of
the matrix A. As well as using the Kalman tests, the following simple test,
which utilizes solely row and column operations, was adopted. It was found
that matrix A is reducible to a block upper triangular form, as shown below,
by using a suitable permutation matrix P,

$$\mathbf{PAP}^{\mathrm{T}} = \begin{bmatrix} \mathbf{A}_{11} & \mathbf{0} \\ \mathbf{A}_{21} & \mathbf{A}_{22} \end{bmatrix} \qquad (3.2)$$

where $\mathbf{A}_{11}$ and $\mathbf{A}_{22}$ are square matrices of dimensions 7×7 and 18×18, with
$\mathbf{A}_{21}$ being a 18×7 matrix, with $\mathbf{A}_{11}$ being diagonal. Since the corresponding
block in the B matrix is also zero, these modes are uncontrollable and we
can simply delete rows and columns 1–7 of the A matrix, row 1–7 of the
B matrix, and columns 1–7 of the resulting C matrix. The state-space model
matrices then become

$$\mathbf{A}_r = \begin{bmatrix} \mathbf{A}_{11} & \mathbf{0} \\ \mathbf{A}_{21} & \mathbf{A}_{22} \end{bmatrix} \quad \mathbf{B}_r = \begin{bmatrix} \mathbf{0} \\ \mathbf{B}_{21} \end{bmatrix} \quad \mathbf{C}_r = \begin{bmatrix} \mathbf{C}_{11} & \mathbf{C}_{12} \end{bmatrix} \quad \mathbf{D}_r = \mathbf{D} \quad (3.3)$$

The diagonal elements of $\mathbf{A}_{11}$ contain the uncontrollable modes of the
system, which are namely {−0.0567, −0.0567, −0.0567, −0.0567, −0.0567,
−0.0567, −0.000242}. Thus, deleting these rows and columns yields the
minimal realization of the gasifier system. This method of obtaining

the minimal realization is more reliable, since it does not involve any numerical calculations and simply involves elementary row and column operations. Hence, it is beneficial to use this approach in this gasifier system, which is numerically very ill conditioned.

Similarly, using the Dulmage–Mendelsohn permutation also enables the system matrix A to be reduced to a lower block triangular form. The following MATLAB commands are used to carry out these operations:

```
[p, q, r] = dmperm(A);
Ared = A(p, q);
[p, q, r] = dmperm(B);
Bred = B(p, q);
```

e. **Interaction Test**. Since the gasifier system is of a highly interactive nature, one may want to gauge how interactive it is. An interaction test—using Rosenbrock's row diagonal dominance [30], Gershgorin discs superimposed on the diagonal elements of the system frequency response—is shown in Figure 3.5. These plots indicate that the system is highly interactive over all frequencies, as observed from the increasing size of the discs in all rows.

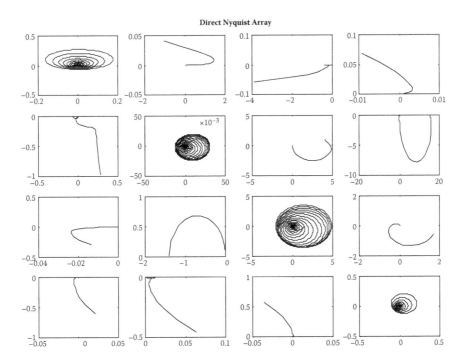

FIGURE 3.5 DNA of the 4 × 4 gasifier system.

```
      % DNA-Draw Gershogorin Disk
      % Applied initial scaling & permutation on output
ko=[0.00001 0 0 0;0 0.001 0 0; 0 0 0.001 0; 0 0 0 0.1];
k=[0 1 0 0;0 0 1 0;1 0 0 0;0 0 0 1];
sys=k*(ko*ss(A,B,C,D));
[Ass,Bss,Css,Dss]=ssdata(sys);
[G, num1,num2,num3,num4,den]=tfm(sys);
G11=G(1,1)
      % Transfer Function Matrix
  G44_OL=[G(1,1)  G(1,2)  G(1,3)  G(1,4);
          G(2,1)  G(2,2)  G(2,3)  G(2,4);
          G(3,1)  G(3,2)  G(3,3)  G(3,4);
          G(4,1)  G(4,2)  G(4,3)  G(4,4)];
  clear i n
  figure();box on
  title('Nyquist Diagram with Gershgorin Disk')
  subplot(4,4,1);
  sw=logspace(-2,10,100);
  hold on;
  G_ger_g11=G44_OL(1,1);
  [re_g11_1,im_g11_1]=nyquist(G_ger_g11,sw);

for n=1:length(sw)
  re_g11(n)=re_g11_1(:,:,n); im_g11(n)=im_g11_1(:,:,n);
  G_ger=evalfr(G44_OL,sw(n)*i);
  G11_ger=evalfr(G44_OL(1,1),sw(n)*i);
  G22_ger=evalfr(G44_OL(2,2),sw(n)*i);
  G33_ger=evalfr(G44_OL(3,3),sw(n)*i);
  G44_ger=evalfr(G44_OL(4,4),sw(n)*i);
  diag_block_ger=[G11_ger    0       0       0;
                  0       G22_ger    0       0;
                  0          0    G33_ger    0;
                  0          0       0    G44_ger];
  off_diag_block_ger=G_ger-diag_block_ger;
  radius_g11_1(n)=sum(abs(off_diag_block_ger(1,:)));
  radius_g22_2(n)=sum(abs(off_diag_block_ger(2,:)));
  radius_g33_3(n)=sum(abs(off_diag_block_ger(3,:)));
  radius_g44_4(n)=sum(abs(off_diag_block_ger(4,:)));
[lat_g11_1,lon_g11_1]=scircle1(re_g11_1(:,:,n),im_
g11_1(:,:,n),radius_g11_1(n));
    plot(lat_g11_1,lon_g11_1);
  end
plot(re_g11,im_g11)
```

As mentioned, the gasifier system is ill conditioned, that is, some combinations of the inputs have a strong effect on the outputs, whereas other combinations have a weak effect on the outputs. The condition number, which may be considered as the ratio between the gains in the strong and weak directions, may quantify this. The example below shows that the gasifier system represented by its transfer function evaluated at zero frequency is ill conditioned.

$$G(0) = \begin{bmatrix} 33069.40 & -36622 & 25380.90 & -40656.70 \\ -5806.63 & -1547.30 & 4318.95 & -738.44 \\ 505.47 & 13356.60 & 800.67 & -7768.50 \\ 1.45 & 21.14 & -20.36 & -46.66 \end{bmatrix} \quad (3.4)$$

The condition number of $G(0)$ is 1323(= 69711/52.7), which corresponds to the ratio of its maximum singular value to its minimum singular value at 0 rad/s. This could be seen from the Singular Value Decomposition (SVD) of $G(jw)$, shown in Figure 3.4, where its minimum singular values are much smaller than the others. The matrix $G(0)$ is nearly a rank three matrix instead of a full rank matrix, since $S = \{69711, 11015, 6779.4, 52.7\}$ S is the steady-state values of the system outputs. This implies that the system is very sensitive to change in the inputs. In other words, the steady-state outputs $y = |G(0)^{-1}u_1 - G(0)^{-1}u_2| = |G^{-1}(0)(u_1 - u_2)|$ will be very large for a different value of input u. For example, if $u_1 = [1\ 1\ 1\ 1]^T$ and $u_2 = [1\ 1\ 1\ 0.99]^T$, the outputs y will differ significantly by $y = [406.6\ 7.4\ 77.7\ 0.5]^T$ for a small change in the fourth input. This shows that the system is quite ill conditioned.

```
sys=ss(A,B,C,D);
G0=evalfr(sys,0);
svd_G0=svd(G0)
```

It is clearly desirable to precondition the system data prior to performing computation. The condition number of the system can be reduced if each of the state-space matrices are preconditioned using the norm reduction method discussed below.

f. **Osborne Preconditioning**. Before performing any numerical operation using the resulting state-space minimal realization matrices, a numerical preconditioning on these matrices was carried out, since the gasifier system is not well conditioned as mentioned before.

The Osborne preconditioning method [31], which uses the norm reduction method on a matrix that is pre- and postmultiplied by a diagonal matrix was employed, and resulted in an improvement in the condition number. The detailed derivations and steps involved in the Osborne preconditioning can be found in Osborne [31].

The condition number of matrix A was reduced from 5.2×10^{19} to 9.8×10^6 after the Osborne preconditioning. In comparing this approach with other norm methods, such as the one-norm, infinity-norm and two-norm, the following tests are applied. After all of its elements are divided by the respective norm, the resulting A matrix is denoted by A_n. The sum of the absolute error (SAE) of each element is then obtained by summing the absolute error between the original matrix A and A_n. These results are shown in Table 3.3.

TABLE 3.3

Comparison of Various Preconditioning Methods

Norm Methods	SAE	Maximum SAE
One-norm	4.13×10^{-15}	10×10^{-16}
Two-norm	1.1×10^{-15}	5.5×10^{-16}
Infinity-norm	4.2×10^{-16}	1.5×10^{-16}
Osborne	2.7×10^{-8}	2.1×10^{-9}

The condition number obtained with the Osborne method is the smallest and this implies that an inverse of A can be easily performed without significant numerical error if the Osborne preconditioning is used. Thus, it enables more reliable computations to be carried out in the latter stages of the chapter.

```
  % check error (1 norm)
n1=norm(A,1); % ninf=norm(A,inf) & n2=norm(A,2);
An1=A./n1;
[U1,d1]=eig(An1);
A1=U1*d1*inv(U1);
error1=abs(A1)-abs(An1);
one_norm_maxerror=max(max(error1));
one_norm_error=sum(sum(error1));
      % check error (Obsorne)
[mu,Ao,logD]=osborne(A);
Dss=diag(exp(logD));
[Uo,do]=eig(Ao);
Aoo=Uo*do*inv(Uo);
erroro=abs(Ao)-abs(Aoo);
osborne_maxerror=max(max(erroro));
osborne_error=sum(sum(erroro));
```

g. **Open-Loop TFM of the Gasifier**. With the resulting matrix A being a block upper triangular form and now well conditioned, the evaluation of the corresponding transfer-function matrix (TFM) [32], becomes simpler. Now, since

$$\mathbf{G(s)} = \mathbf{C_r} (s\mathbf{I} - \mathbf{A_r})^{-1} \mathbf{B_r} + \mathbf{D_r}$$

$$= \begin{bmatrix} \mathbf{C}_{11} & \mathbf{C}_{12} \end{bmatrix} \left(\begin{bmatrix} s & 0 \\ 0 & s \end{bmatrix} - \begin{bmatrix} \mathbf{A}_{11} & 0 \\ \mathbf{A}_{21} & \mathbf{A}_{22} \end{bmatrix} \right)^{-1} \begin{bmatrix} 0 \\ \mathbf{B}_{21} \end{bmatrix} + \mathbf{D_r}$$

$$= \begin{bmatrix} \mathbf{C}_{11} & \mathbf{C}_{12} \end{bmatrix} \begin{bmatrix} s - \mathbf{A}_{11} & 0 \\ -\mathbf{A}_{21} & \mathbf{A}_{22} \end{bmatrix}^{-1} \begin{bmatrix} 0 \\ \mathbf{B}_{21} \end{bmatrix} + \mathbf{D_r} \qquad (3.5)$$

by applying the matrix inversion lemma; namely,

$$
\begin{bmatrix} \mathbf{x} & \mathbf{0} \\ \mathbf{y} & \mathbf{z} \end{bmatrix}^{-1} = \begin{bmatrix} \mathbf{x}^{-1} & \mathbf{0} \\ -\mathbf{z}^{-1}\mathbf{y}\mathbf{x}^{-1} & \mathbf{z}^{-1} \end{bmatrix}
$$

the evaluation of $\mathbf{G}(s)$ becomes

$$
\mathbf{G}(s) = \begin{bmatrix} \mathbf{C}_{11} & \mathbf{C}_{12} \end{bmatrix} \begin{bmatrix} (s\mathbf{I} - \mathbf{A}_{11})^{-1} & \mathbf{0} \\ (s\mathbf{I} - \mathbf{A}_{22})^{-1}\mathbf{A}_{21}(s\mathbf{I} - \mathbf{A}_{11})^{-1} & (s\mathbf{I} - \mathbf{A}_{22})^{-1} \end{bmatrix} \begin{bmatrix} \mathbf{0} \\ \mathbf{B}_{21} \end{bmatrix} + \mathbf{D}_r
$$
(3.6)

$$
= \mathbf{C}_{12} \begin{bmatrix} s\mathbf{I} - \mathbf{A}_{22} \end{bmatrix}^{-1} \mathbf{B}_{21} + \mathbf{D}_r
$$

This allows the necessary inversion to be performed with fewer variables and thus gives a better numerical accuracy.

3.3 CONTROL STRUCTURE DESIGN

After achieving a better numerically conditioned system, the minimal-order 4×4 gasifier system at the 100% load condition is further examined. A very important part of a multivariable design is the input/output (I/O) pairing, which defines the so-called control structure design [33] problem. The section below shows the I/O pairing using Relative Gain Array (RGA) and other methods.

a. **RGA**. The RGA [33–34] of a complex nonsingular $n \times n$ matrix G, denoted $\Lambda(\mathbf{G})$, is a complex $n \times n$ matrix defined by

$$
\Lambda(\mathbf{G}) = \mathbf{G} * (\mathbf{G})^{-T}
$$
(3.7)

where the operation * denotes element-by-element multiplication (Hadamard or Schur product).

By applying the RGA to the 4×4 gasifier system, $\mathbf{G}_{100\%}$ the best I/O pairs to be used for control can be determined. Based on physical considerations, it was decided that maintaining the bed mass of the gasifier at a desired level was most important. Applying Equation (3.7) to the system at the steady-state condition, denoted by $\mathbf{G}_{100\%}$, we get

$$
\Lambda_{100\%}(\mathbf{0}) = \begin{bmatrix} 0.4133 & -0.0724 & 0.4187 & 0.2404 \\ \mathbf{0.5781} & -0.0302 & \mathbf{0.4246} & 0.0274 \\ 0.0150 & \mathbf{0.9066} & 0.0519 & 0.0265 \\ -0.0064 & 0.1959 & 0.1048 & \mathbf{0.7056} \end{bmatrix}
$$
(3.8)

```
sys=ss(Ar,Br,Cr,Dr); % use reduced system
G0=evalfr(sys,0);
rga_G0=G0.*(pinv(G0)).';
```

By examining the coefficients of the matrix $\Lambda_{100\%}(0)$, the first input (WCHR) should control the second output (MASS) as seen from the first column and second row of this matrix. The element (1,2) of the $\Lambda_{100\%}(0)$ is the closest to 1 in the column, which indicates that the aforementioned pairing is the most favorable. The remaining I/O pairs were paired similarly by examining the rest of the columns of $\Lambda_{100\%}(0)$. From the above, the MASS, PGAS, CVGAS, and TGAS are most suitably controlled by the WCHR, WAIR, WCOL, and WSTM, respectively. By reordering the system outputs using a corresponding row permutation matrix, the largest element in the resulting columns is now reflected by the diagonal entries.

Since the above RGA is applied at the steady state, or zero frequency, it is wise to examine the RGA at other frequencies. A term known as the *RGA number* [35] is used for this purpose and this has the form,

$$RGA-number = \|\Lambda(\mathbf{G}) - \mathbf{I}\|_{sum} \tag{3.9}$$

By evaluating the RGA number across the frequencies of interest for the $\mathbf{G}_{100\%}$ after row permutation, the plots obtained are shown in Figure 3.6. There is a slight decrease in the RGA number after ordering across the frequencies and therefore this implies that the selected I/O pairs will have an effect on the diagonal dominance of the system. Also, a decrease in the RGA number corresponds to a decrease in the condition number of the gasifier system, which is a "good" feature.

```
% Frequecy dependent RGA for full process
G=pck(Ar,Br,Cr,Dr);
w=logspace(-8,2,200);
Gjw = frsp(G,w);
rga = vrga(Gjw);
        % Select input structure
RGA1 = vrga( sel(Gjw, [1 2 3 4], ':') );
RGA2 = vrga( sel(Gjw, [2 3 1 4], ':') );
        % Find RGA-number
rn1 = veval('sum', veval('sum', vabs(msub(RGA1,eye(4)))));
rn2 = veval('sum', veval('sum', vabs(msub(RGA2,eye(4)))));
        % Plot VRGA
figure();
vplot('liv,d', rn1, '-', rn2,'-·')
axis([1e-8 1e2 0 70])
xlabel('Frequency')
title('RGA Numbers ');
legend('b/f ordering','a/f ordering')
drawnow
```

b. **Hankel Singular Values (HSVs).** Large HSVs are often preferred in a control system design. The diagonal entries of the grammians, which reflect the joint controllability and observability of the states resulting from a balanced realization of the system, are called the HSVs. The HSVs of the $\mathbf{G}_{100\%}$

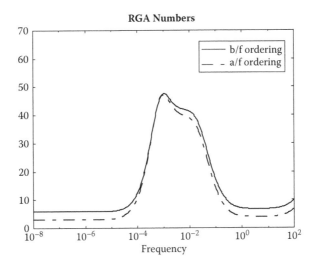

FIGURE 3.6 RGA number of gasifier plant.

model after row permutations are similar. The control structure selected does not affect the original system.

```
Cp1=[Cr(2,:);Cr(3,:);Cr(1,:);Cr(4,:)];
[HSV0,P,Q]=hksv(Ar,Br,Cr);
[HSV1,P,Q]=hksv(Ar,Br,Cp1);
HSV0-HSV1
```

c. **Niederlinski Index (Ni).** This is defined as the ratio of the determinant of G, evaluated at steady state, to the multiple of its diagonal elements.

$$\mathbf{Ni} = \frac{\det \mathbf{G}(0)}{\prod_{i=0}^{n} \mathbf{G}_{ii}(0)} \tag{3.10}$$

Equation (3.10) gives a sufficient condition for stability, if the Niederlinski index [35] is greater than zero. Note that this is a necessary and sufficient condition for two-input and two-output systems only. For the gasifier system, the **Ni** before and after the row permutations is 143.47 and 2.98, respectively. Though the **Ni** decreases after the I/O pairings, the system is still considered to be sufficiently stable.

```
        % Before row permutations
sys=ss(Ar,Br,Cr,Dr);
G0=evalfr(sys,0);
Ni_G0=abs(det(G0)/(G0(1,1)*G0(2,2)*G0(3,3)*G0(4,4)))
        % After row permutations
k=[0 1 0 0;0 0 1 0; 1 0 0 0; 0 0 0 1];
kG0=k*G0;
Ni_G0_af=abs(det(kG0)/(kG0(1,1)*kG0(2,2)*kG0(3,3)
*kG0(4,4)))
```

3.4   GASIFIER SYSTEM ANALYSIS

After the inherent properties tests and the control structure design are performed, the actions taken prior to gasifier control system design are discussed in this section.

a. **Proportional-Plus-Integral (PI) Controller Optimization.** The aim of using a Proportional-plus-Integral (PI) action controller on the first loop is to control the BED-MASS height directly using WCHR, as mentioned previously as a physical requirement. This implies that this 1×1 system has to be decoupled from the remaining 3×3 subsystem as shown in Figure 3.7.

Input and output scaling are employed simultaneously to further reduce the level of ill conditioning in the system and also to obtain the best diagonal dominance across the frequency range of interest, especially near the system bandwidth. The remaining 3×3 subsystem is then to be controlled by the few controllers, which is discussed in a subsequent section.

A simple optimization routine [36] was developed to minimize the time response error of the output using a PI controller: $PI = k_p + k_i / s$. This routine involved a line search of k_i in an increment to its left and right of the initial location and determined the minimum of the Integral Squared Error (ISE) at these three locations. The minimum ISE is obtained by simulating the time response and determines the error over the simulated time. The routine then performs a line search on k_p starting from the location of the minimum ISE obtained from the best k_i search. This procedure is repeated until the minimum ISE over the simulated time has reached a minimum value. The ISE convergence plot is shown in Figure 3.8, and it can be seen that the error converges to the minimum at two iteration steps.

After obtaining the initial PI controller parameters from this process, the controller is fine-tuned further to obtain an upper block triangular form of $\mathbf{G}(s)$, which would make it block diagonal dominant as shown in Figure 3.9.

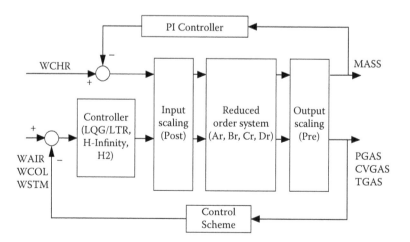

FIGURE 3.7   Final control structure.

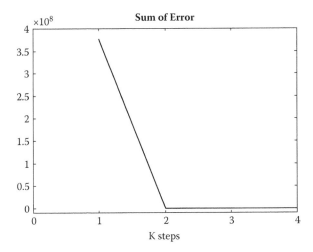

FIGURE 3.8 Convergent plot.

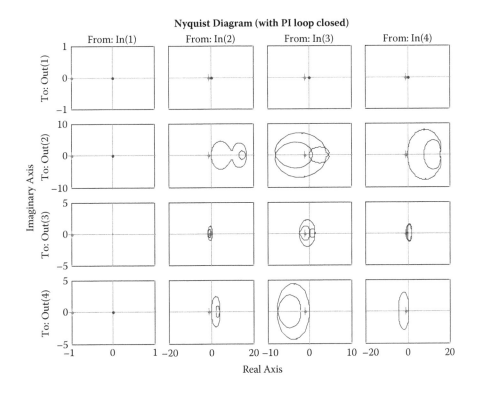

FIGURE 3.9 Nyquist Array.

```
% Closed first loop only & 3x3 subsystem is open loop
ko=[0.00001 0 0 0;0 0.001 0 0; 0 0 0.001 0; 0 0 0 0.1];
k=[0 1 0 0;0 0 1 0;1 0 0 0;0 0 0 1];
sys=k*(ko*ss(Ar,Br,Cr,Dr));
[Ass,Bss,Css,Dss]=ssdata(sys);
[G, num1,num2,num3,num4,den]=tfm(sys);kp=-0.159;ki=-0.001
;PI_num=[kp ki];
PI_den=[300 50];
controller=tf(PI_num,PI_den);
controller_G=[ controller 0 0 0;
                    0 1 0 0;
                    0 0 1 0;
                    0 0 0 1];
controller_G2=[ controller 0 0 0;
                    0 0 0 0;
                    0 0 0 0;
                    0 0 0 0];
cont_G= series(controller_G,G);
cont_G2= series(controller_G2,G);
cont_G2_fb=feedback(cont_G2,eye(4),-1);
G44_OL=[cont_G2_fb(1,1) cont_G2_fb(1,2) cont_G2_fb(1,3)
cont_G2_fb(1,4);
        cont_G2_fb(2,1) cont_G(2,2) cont_G(2,3)
        cont_G(2,4);
        cont_G2_fb(3,1) cont_G(3,2) cont_G(3,3)
        cont_G(3,4);
        cont_G2_fb(4,1) cont_G(4,2) cont_G(4,3)
        cont_G(4,4)];
figure()
nyquist(G44_OL);
title('Nyquist Diagram (with PI loop closed ) ')
```

b. **Char Off-Take Rate**. In addition, the input WCHR has to be checked for any violation of its maximum and minimum values and the rate limit given in the specifications. The closed-loop step response of the first loop is shown in Figure 3.10. It can be seen that the rate limit of the WCHR (dotted line) is still within the allowable value of 0.2 kg/s^2 after the PI loop is closed.

c. **Block Diagonal Dominance**. For the gasifer system, since the first loop is well decoupled from the remaining 3×3 subsystem, the TFM can be treated as a 2×2 block matrix where blocks $\mathbf{G}_{11}$ and $\mathbf{G}_{22}$ are both 1×1 and 3×3 TFM, respectively, as shown below.

$$\mathbf{G} = \begin{bmatrix} \mathbf{G}_{11} & \mathbf{G}_{12} \\ \mathbf{G}_{21} & \mathbf{G}_{22} \end{bmatrix} \tag{3.11}$$

The matrix $\mathbf{G}(s)$ is considered to be block diagonal dominant with this partitioning if the gain (minimum singular value) of the diagonal blocks dominates the gain (maximum singular value) of the off-diagonal blocks. This was found to be the case as shown in Figure 3.11.

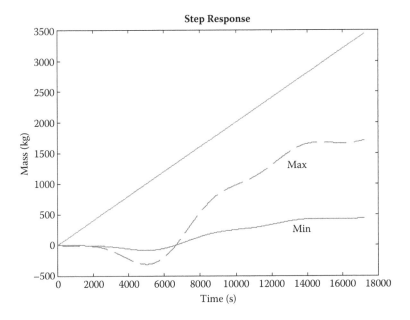

FIGURE 3.10 WCHR rate.

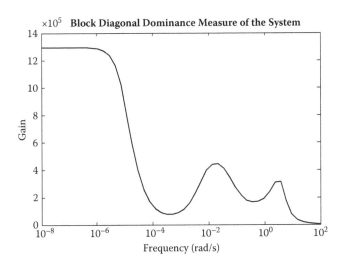

FIGURE 3.11 Block diagonal dominance measure.

```
% Min SVD of diagonal block / max svd of off-diagonal
  block--> Block diagonal dominance measure
s=logspace(-8,2);
clear i
rro=50;
G33_OL= cont_G([2 :4],[2 :4]) ;
```

```
for n=1:rro
    G_sv=evalfr(G44_OL,s(n)*i); % where i is complex number
    G11_sv=evalfr(cont_G(1,1),s(n)*i);
    G33_sv=evalfr(G33_OL,s(n)*i);
    diag_block=[G11_sv            0          0        0;
        0        G33_sv(1,1)  G33_sv(1,2)  G33_sv(1,3);
        0        G33_sv(2,1)  G33_sv(2,2)  G33_sv(2,3);
        0        G33_sv(3,1)  G33_sv(3,2)  G33_sv(3,3)];

    off_diag_block=G_sv-diag_block;
    sv_min=max(svd(abs(diag_block)));
    sv_max=max(svd(abs(off_diag_block)));
    ratio_sv(n)=sv_min/(sv_max);
end
figure();
semilogx(s,ratio_sv,'r');
title('Block Diagonal Dominance Measure of the System');
xlabel('Frequency (rad/s)');
ylabel('Gain');
drawnow
```

d. **System Bandwidth**. The bandwidth of the 3×3 subsystem of the gasifier can
   be obtained using the loop gains, $\mathbf{L}(jw)$, as shown in Figure 3.12. From this
   plot, it can be observed that the bandwidth, given by the minimum singular
   values, $\underline{\sigma}(\mathbf{L}(jw))$ of the system is approximately 0.005 rad/s, which is equivalent to a rise time of about 200 seconds $(= 1/w_B)$. This is quite acceptable for
   this system, since the gasifier is a physically large system that requires a time

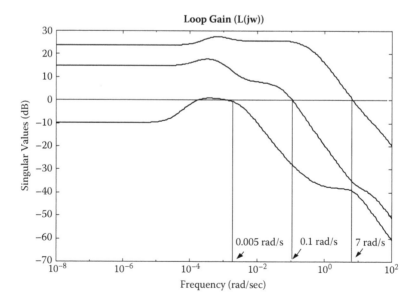

FIGURE 3.12   System bandwidth.

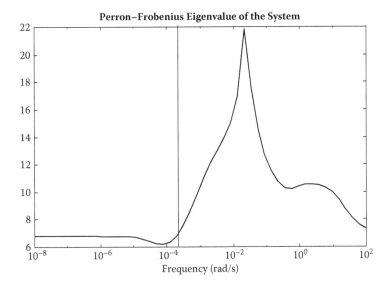

FIGURE 3.13 Perron–Frobenius eigenvalue plot.

of several hundred seconds to react. In addition, the plot also shows the middle and high bandwidths of the system, which are 0.1 and 7 rad/s, respectively.

```
w=logspace(-8,2,300);
figure();
sigma(G33_OL,w);
title (' Loop Gain (L(jw))')
```

e. **Worst Frequency**. The Perron–Frobenius eigenvalue plot is used to determine the frequency at which the highest interaction occurs. The Perron–Frobenius [37] eigenvalue over the frequencies of interest is defined as

$$\lambda_p \left| G(jw)G_{\text{diag}}^{-1}(jw) \right| \tag{3.12}$$

where λ_p is the eigenvalue of $|\cdot|$ and G_{diag} contains the diagonal elements of $G(jw)$. The location where the highest peak occurs is the frequency where interaction is mostly concentrated. From Figure 3.13, the 3×3 subsystem is unlikely to be made diagonal dominant by a simple diagonal constant scaling matrix, since all of its λ_p are greater than 6 dB, or 2, in absolute magnitude over the frequencies of interest.

```
    % Perron-Frobenius eigenvalue plot
rrr=50;
ccc=2;
table 1=cell(rrr,ccc);
p1=1;
s=logspace(-8,2,rrr);
```

```
for n=1:rrr
  G=abs(evalfr(G33_OL,s(n)*i));
  Gdiag=diag(diag(G));
  invGdiag=minv(Gdiag);
  GG=mmult(invGdiag,G);
  perGG=eig(GG);
  [pfval,vector_left,vector_right,gs]=speron(GG);
  vec_left1(n)=vector_left(1,1);
  vec_left2(n)=vector_left(2,2);
  vec_left3(n)=vector_left(3,3);
  maxperGG(n)=20*log10(max(perGG));
  table 1{p1,1}=s(n); % create cell to store maxperGG
  table 1{p1,2}=maxperGG(n);
  p1=p1+1;
 end
grid
figure();semilogx(s,maxperGG,'r');
title('Perron-Frobenius eigenvalue of the system');
xlabel('Frequency (rad/s)');
ylabel('Gain (dB) ');
```

f. **Design Scaling**. Next, systematic ways to make a system least interactive are given. Various design scaling methods such as Edmunds scaling, one-norm scaling [36] and Perron–Frobenius (PF) scaling [36] are compared. It is found that Edmunds scaling gives a more diagonal dominant system. This is reflected in less concentration of the Gershgorin discs at the origin of the plot, as shown (right-hand side) in Figure 3.14.

```
Edmund scaling (gs=post*g*pre)
w1=0.008i
Gw1=evalfr(G33_OL,w1);
[pre1e,post1e,gs1e,blocks1e,block1e]=normalise(Gw1);

sys1E=post1e*G33_OL*pre1e;
sys_feedback_G33e1=feedback(sys1E,zeros(3),-1);
gerdisk(sys1E,0,11)
```

This algorithm will give the same scaling independently of the ordering of the inputs and outputs and pre- and postcompensators that provide both scaling and I/O pairings as indicated by the resulting nonunity permutation matrices. From Figure 3.15, the condition number due to the Edmunds and one-norm scaling is the lowest as compared to the PF scaling.

By using the Edmunds scaling, the following I/O pairs for the remaining 3×3 system were obtained: WCOL-TGAS, WAIR-CVGAS, and WSTM-PGAS. The nonunity permutation matrices scaled at $w = 0.008i$ are as follows:

$$
\mathbf{Pre} = \begin{bmatrix} 0 & 0.395 & 0 \\ -0.432 & 0 & 0 \\ 0 & 0 & -0.423 \end{bmatrix} \quad \mathbf{Post} = \begin{bmatrix} 0 & 0 & 1.727 \\ 0 & 0.517 & 0 \\ 0.081 & 0 & 0 \end{bmatrix}
$$

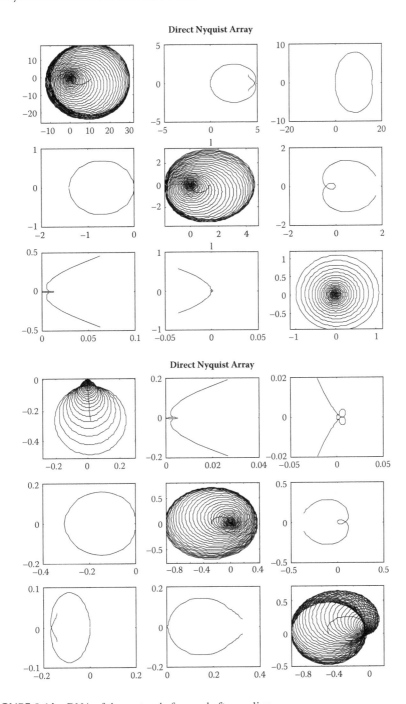

FIGURE 3.14 DNA of the system before and after scaling.

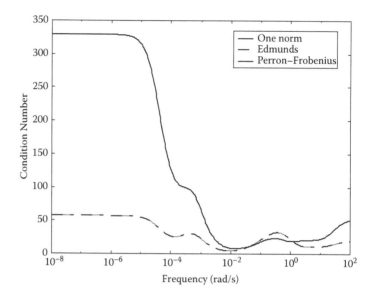

FIGURE 3.15   Condition number for various methods.

3.5   MODEL ORDER REDUCTION (MOR)

After the 3×3 subsystem is scaled, the next stage is to perform a MOR on the gasifier system (G) to determine a reduced order of the system ($\mathbf{G_r}$). For comparison
purposes, only two methods are considered; namely, modal truncation and Schur
balanced truncation, which uses the state-space approach. To judge the effectiveness
of these MOR methods, both time and frequency-response tests were used. This is to
check for any significant deviation of the reduced model from the full-order model.
In addition, the impulse response error matrix, and Hankel SV are used.

> **Time Response**. As can be seen in Figure 3.16, the Schur balanced trunca-
> tion preserves both the transient and the steady-state gain, while the modal
> truncation method on the other hand retains the gain at higher frequencies
> only. This implies $\mathbf{G_r}(0) \neq \mathbf{G}(0)$, $\mathbf{G_r}(\infty) = \mathbf{G}(\infty)$.

```
    % MOR using Schur Balanced Truncation, Modal
      Truncation and b/f reduction
sys=k*(ko*ss(Ar,Br,Cr,Dr));
[Ass,Bss,Css,Dss]=ssdata(sys);
Ass_m=Ass;
Bss_m=Bss(:,2:4);
Css_m=Css(2:4,:);
Dss_m=Dss(2:4,2:4);

model reduction using modal truncation
[Am,Bm,Cm,Dm]=modreal(Ass_m,Bss_m,Css_m,Dss_m,8);
```

```
      % model reduction using schur balanced truncation
[As,Bs,Cs,Ds]=schbal(Ass_m,Bss_m,Css_m,Dss_m,1,8);

      % model reduction using singular perturbation
sys_org=pck(Ass_m,Bss_m,Css_m,Dss_m);
sys_resd=sresid(sys_org,8);
[Ares,Bres,Cres,Dres]=unpck(sys_resd);
```

Error of the Impulse Response Matrix. A reduced-order model ($\mathbf{A_r}$, $\mathbf{B_r}$, $\mathbf{C_r}$, $\mathbf{D_r}$) of a system can be judged by the difference in its impluse response from that of the full order system. The error impulse response matrix is defined as $\mathbf{H_e}(t) = (\mathbf{Ce}^{At}\mathbf{B}+\mathbf{D})-(\mathbf{C_r}\mathbf{e}^{A_r t}\mathbf{B_r}+\mathbf{D_r})$ and characterizes the error of the reduced model. A reduced-order model is said to be "good" if the largest principal component of $\mathbf{H_e}(t)$ over $[0,\infty)$ is "small" compared to the smallest principal component of $\mathbf{Ce}^{At}\mathbf{B}$. By using the ratio of the 2 norm of the error signal, $\mathbf{H_e}(t)$ to the 2 norm signal of $\mathbf{H}(t)$, as shown in (3.13), a potential reduced model can be judged.

$$\left\|\left(\int_0^\infty \mathbf{H_e}(t)^2\,dt\right)^{1/2}\right\|_2 \left\|\left(\int_0^\infty \mathbf{H}(t)^2\,dt\right)^{-1/2}\right\|_2 \tag{3.13}$$

The relative error in the impulse response is lower in the Schur balanced truncation than its counterparts. This shows that the Schur balanced truncation provides a "good" fit, compared to the modal truncation.

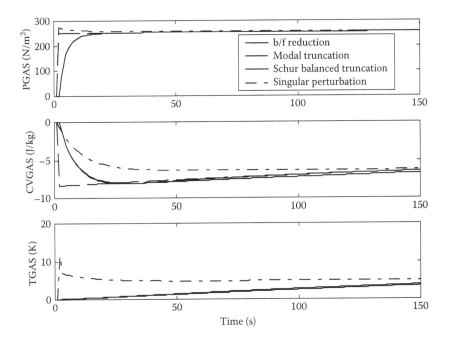

FIGURE 3.16   Steady-state time response.

Figure 3.17 shows how the relative error in the impulse response increases as the order of the system decreases.

Hankel SV. The Hankel singular value is used to measure the state controllability and observability of the system. The Hankel SVs for the Schur balanced truncation, shown in Figure 3.18, are larger than those of the modal truncation, which indicates that the former has better state controllability and observability properties than the latter.

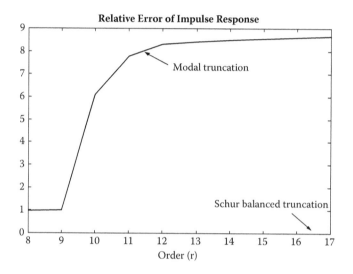

FIGURE 3.17   Relative error of impulse response.

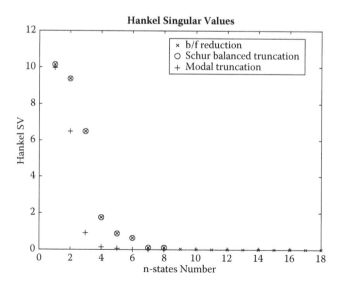

FIGURE 3.18   Hankel singular values of the gasifier system.

```
% Hankel Singular values
[hsvr,pr,qr]=hksv(Ass_m,Bss_m,Css_m);
[hsvs,ps,qs]=hksv(As,Bs,Cs);
[hsvm,pm,qm]=hksv(Am,Bm,Cm);
[hsvres,pres,qres]=hksv(Ares,Bres,Cres);
figure();
plot(1:18,hsvr,'x')
hold on
plot(1:8,hsvs,'o')
plot(1:8,hsvm,'+')
plot(1:8,hsvres,'^')
title('Hankel Singular Values');
xlabel('n-states number');
ylabel('Hankel SV ');
legend('b/f reduction','Modal Truncation','Schur Balanced
Truncation','Singular Perturbation')
```

Using these various criteria for judging the MOR methods employed on the gasifier system, it is found that the Schur balanced truncation gave a promising result.

System Poles. After the MOR is performed on the 3×3 subsystem, the system poles and zeros [30] are to be determined. These enable the properties of the gasifier system after the design scaling and MOR to be examined. For the gasifier system, the poles of the reduced-order system are shown in Table 3.4 below. The system after scaling and MOR remains open-loop stable as can be seen in their negative real parts.

Transmission Zeros. One way to calculate the transmission zeros and invariant zeros without involving much numerical operation is to use the Davison [38] approach. This basically uses the idea of the classical root locus technique. For a minimal system, as the gain of the output-feedback matrix $F_y\mathbf{I}_m \to \infty$, the closed-loop system poles approach the infinite zeros (transmission zeros). The matrix $F_y\mathbf{I}_m$ is the feedback gain matrix of the closed-loop system. The steps in the Davison method are shown below.

TABLE 3.4
Open-Loop Poles

Open-Loop Poles

Open-Loop Poles
−0.432
−0.161
−0.055
−0.016
$-8.054\times10^{-4} \pm 3.326\times10^{-4}$i
$-3.758\times10^{-4} \pm 1.033\times10^{-4}$i

- Generate an output feedback system with the feedback gain, F_y.
- Initialize the feedback gain, $F_y I_m$. Determine the closed-loop system eigenvalues (CLE) given by the

$$\det \begin{bmatrix} sI - A + BF_y & BF_y \\ -C & D \end{bmatrix} = 0 \tag{3.14}$$

for each $F_y I_m$.
- Repeat the previous step until some of the CLE (closed-loop eigenvalues) become invariant as $F_y I_m$ increases. Note as $F_y I_m \to \infty$, some of the closed-loop eigenvalues will approach the transmission zeros as shown in Table 3.5. These represent a physically unrealizable design, since in practice reaching infinite frequency is impossible.

The eigenvalues marked with an asterisk are clearly invariant with respect to F_y. These correspond to the transmission zeros of the system. As observed, the feedback gains F_y had to become as large as 10^{12} in order to obtain the transmission zeros of the system. This is due to the fact that the D term is so small ($\approx 10^{-11}$) and for the feedback gain to be dominant, this gain has to be very large. Hence, the F_y applied to the gasifier system should be at least 10^{11}, since the D term has a small magnitude of 10^{-11}.

In comparing these with the transmission zeros obtained from MATLAB, both resulted in almost identical transmission zeros. The only discrepancy is that there is one zero, which is different. This is due to the fact that the gain in F_y has to be increased further in order to reach this infinite zero, which in practice is impossible. The sum of absolute error is used to measure the error at each step change of the F_y. The reference transmission zeros used for this error calculation are the ones obtained in MATLAB. The convergence plot of the transmission zeros at each step size is shown in Figure 3.19.

The Davison method provides a simple and reliable way of determining the transmission zeros, since it uses the classical root locus concept instead of formulae, which are always limited by their assumptions.

TABLE 3.5

Eigenvalues of Output Feedback System

	Davison			
$F_y = 1 \times 10^2$		$F_y = 1 \times 10^8$	$F_y = 1 \times 10^{12}$	MATLAB
CLE	$-0.19 \pm 1.04i$	1.01×10^5	-3.22×10^{10}	
	0.16	1.01×10^5	$-5.33 \times 10^{8*}$	-5.207×10^8
	-0.013	0.12^*	0.12^*	0.12599
	-0.0046	-0.013^*	-0.013^*	-0.001372
	-0.0012	-0.0044^*	-0.0044^*	-0.000584
	-0.00057	-0.0012^*	-0.0122^*	-0.01375
	-0.00057^*	-0.00057^*	-0.00057^*	-0.00006

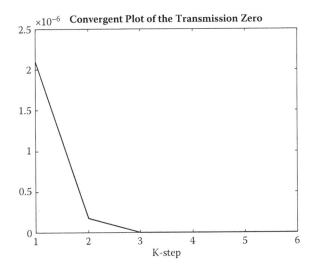

FIGURE 3.19   Convergent plot of transmission zeros.

In the next section, the various controller designs on the remaining 3×3 subsystem after scaling and MOR are discussed. The controller is designed at the 100% load condition and then tested at the 50% and 0% load condition for robustness. Various controller design methods such as the Linear Quadratic Regulator (LQR), the Linear Quadratic Gaussian (LQG) with Loop Transfer Recovery (LTR), the H_2 optimization, and the H_∞ optimization are demonstrated on the gasifier system. For each controller design method, its theory is explained followed by an outline of the design steps involved. Lastly, the performance tests using step and sinusoidal input disturbances to check the robustness of the controller at the 50% and 0% load condition are discussed.

3.6   LINEAR QUADRATIC REGULATOR (LQR)

The concept of the LQR [39–40], or commonly known as *optimal state-feedback design*, uses a set of weighting matrices to penalize the states and control inputs of the system. This requirement is set in the cost function to be minimized as shown below. The weighting matrices Q and R, which are applied to the states and inputs, respectively, are tuned iteratively until the desired performance is achieved. Inevitably, this becomes an *ad hoc* process.

3.6.1   LQR Theory

The linear quadratic regulator (LQR) problem aims to minimize the weighed sum of the energy of the states and control inputs in the form of the cost function:

$$J = \int_0^T [\mathbf{x}^T(t)\mathbf{Q}\mathbf{x}(t) + \mathbf{u}^T(t)\mathbf{R}\mathbf{u}(t)]dt \qquad (3.15)$$

with respect to the input vector $\mathbf{u}(t)$. For the optimization, Q must be a symmetric positive semidefinite matrix $(\mathbf{Q}^T = \mathbf{Q} \geq 0)$ and R must be a symmetric positive-definite matrix $(\mathbf{R}^T = \mathbf{R} > 0)$. Using for instance, the Pontryagin minimum principle, the solution is in the form of a time-varying control law:

$$\mathbf{u}(t) = -\mathbf{F}(t)\mathbf{x}(t) \tag{3.16}$$

where

$$\mathbf{F}(t) = \mathbf{R}^{-1}\mathbf{B}^T\mathbf{P}(t) \tag{3.17}$$

and $\mathbf{P}(t)$ is a solution of the Riccati Differential Equation (RDE):

$$\mathbf{A}^T\mathbf{P}(t) + \mathbf{P}(t)\mathbf{A} + \mathbf{Q} - \mathbf{P}(t)\mathbf{B}\mathbf{R}^{-1}\mathbf{B}^T\mathbf{P}(t) = -\frac{d}{dt}\mathbf{P}(t) \tag{3.18}$$

To implement the LQR controller in (3.17), all states $\mathbf{x}(t)$ are required to be measurable. If the time horizon is indefinite $(T \rightarrow \infty)$ and an optimal solution exists, then $\mathbf{P}(t)$ tends to a constant matrix P. The control law is then:

$$\mathbf{u}(t) = -\mathbf{F}\mathbf{x}(t) \tag{3.19}$$

where

$$\mathbf{F} = \mathbf{R}^{-1}\mathbf{B}^T\mathbf{P} \tag{3.20}$$

The matrix P is a solution of the Algebraic Riccati Equation (ARE):

$$\mathbf{A}^T\mathbf{P} + \mathbf{P}\mathbf{A} + \mathbf{Q} - \mathbf{P}\mathbf{B}\mathbf{R}^{-1}\mathbf{B}^T\mathbf{P} = 0 \tag{3.21}$$

Note the time dependent parameter becomes independent and the right-hand side of (3.18) becomes zero. For the positive-definite matrix P, the corresponding closed-loop system with LQR controller is asymptotically stable. A schematic diagram of the LQR design is shown in Figure 3.20.

LQR designed controllers ensure good stability margin and sensitivity properties. Using this approach, a minimum phase margin of 60 degrees and infinite gain margin can be achieved. By the Kalman identity, the return difference matrix, $\mathbf{I} + \mathbf{L}(s)$ is always greater or equal to 1 as shown below:

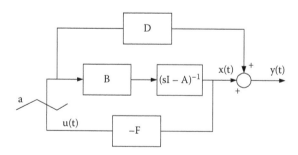

FIGURE 3.20   LQR design.

$$[I+F(sI-A)^{-1}B+D] \geq 1$$

$$[I+L(s)] \geq 1 \qquad (3.22)$$

where F is the optimal state gain matrix. This implies good disturbance rejection and tracking properties and hence robustness is guaranteed for the LQR approach. The data shown in Figures 3.21 and 3.22 show the robustness properties

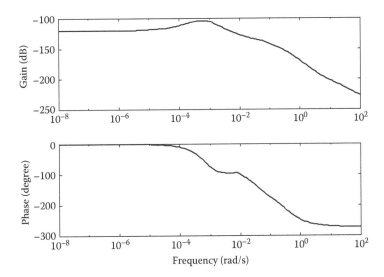

FIGURE 3.21 Bode plots of the LQR design.

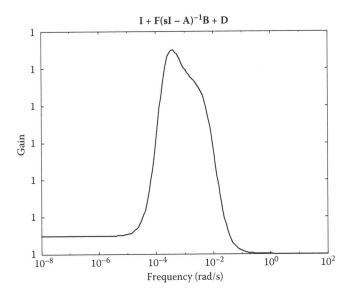

FIGURE 3.22 Return difference of the LQR design.

of the LQR-based controller designed for the gasifier. Instead of showing a 3×3 array of the element Bode diagrams, the minimum gain and phase plots are shown. As observed, the LQR has indeed good robustness properties. It has infinite gain margin and a $180°$ phase margin throughout the frequency range considered. This adequate phase margin will also be reflected in its closed-loop stability.

The frequency response of the return difference matrix, $[\mathbf{I} + \mathbf{L}(jw)]$, of the optimal closed-loop system, as shown in Figure 3.22, indicates that $\mathbf{L}(jw)$ always lies outside the unit circle centered at $(-1, 0i)$. This implies that the output of the system is insensitive to input disturbances.

```
%Design LQR
%Put disturbance column matrix back on to form 4 x 5
Bd1=[Br B(8:25,6)];
Dd1=[Dr D(:,6)];
        % Applied initial scaling & permutation on output
        on 4 x 5 plant
ko=[0.00001 0 0 0;0 0.001 0 0; 0 0 0.001 0; 0 0 0 0.1; 0
0 0 1]; % 5x4
k=[0 1 0 0 0;0 0 1 0 0;1 0 0 0 0;0 0 0 1 0]; % 4x5
sys=series(eye(5),k*(ko*ss(Ar,Bd1,Cr,Dd1))); % 4x5
        % Extract 3 x 3 subsystem from 4 x 5
[Ass,Bss,Css,Dss]=ssdata(sys);
Ass_m=Ass;
Bss_m=Bss(:,2:4);
Css_m=Css(2:4,:);
Dss_m=Dss(2:4,2:4);
syss=ss(Ass_m,Bss_m,Css_m,Dss_m);
        % Scale on 3x3 subsystem using Edmunds scaling
clear i;
w1=0.008i;
Gw1=evalfr(syss,w1);
[pre1e,post1e,gs1e,blocks1e,block1e]=normalise(Gw1);
sys_33=post1e*syss*pre1e;
[Alqr, Blqr, Clqr, Dlqr]=ssdata(sys_33);
        % LQR Design on 3 inputs and 3 outputs system 3x3
[len,width]=size(Blqr);
R=eye(width);
QQ=Clqr'*Clqr;
[K_state,SS,Eig_state]=lqr(Alqr,Blqr,QQ*0.001,R)
        %L(s)=F*(sI-A)^-1*B open loop gain
sysCL=ss(Alqr,Blqr,K_state,Dlqr);
ww=logspace(-8,2,200);
figure()
sv_CL=sigma(sysCL,ww,2);
semilogx(ww,min(sv_CL))
ylabel('Gain ')
xlabel('Frequency (rad/s)')
title('I+F(sI-A)^-^1B + D')
```

3.6.2   LQR DESIGN STEPS

To achieve the results presented above, the following design steps were taken. Note that the eighteenth-order model is used here instead of the eighth-order model obtained from the MOR steps. This helps to preserve the originality of the system and since the higher-order model caused no computational problems in the LQR designs or even in the LQG design considered later, it is therefore used. However, in the H_∞ and H_2 designs, the reduced order controller model is used.

1. Solve the **Algebraic Riccati Equation** for P using (3.21).
2. Determine the **optimal state-feedback gain**, F (LQR.m) using (3.20).
3. Weight selection.

    The weighting matrix Q is typically selected as $\mathbf{C}^T\mathbf{C}$ since this choice reflects the weighting of the states in the output since $\mathbf{y}^T\mathbf{y}=\mathbf{X}^T\mathbf{C}^T\mathbf{C}\mathbf{x}$. For the weighting matrix R, this is selected to be a diagonal matrix of equal weights on the control inputs. It was found that R equal to the identity matrix gave a good compromise result on the control input over the 50% and 0% load condition.

$$\mathbf{Q}=k_c\mathbf{C}^T\mathbf{C}, \quad \mathbf{R}=\mathbf{I}$$

    where k_c is a scalar used to decrease the states interaction. It has a value here of 0.0001.
4. Perform the simulation using Simulink.

    The simulation block diagram is formulated using the Simulink facility, as shown in Figure 3.23. Note that an initial scaling block is used in Figure 3.23 to reduce the size of the output. This applied to all subsequent controller designs.

3.6.3   PERFORMANCE TESTS ON LQR DESIGN

In order that a range of control system design techniques, which are introduced in this chapter, can be compared on a basis, which is as fair as possible, the following design criteria and test cases should be adhered to:

1. The controller design is to be undertaken for the 100% load condition.
2. Apply a pressure step disturbance PSINK of −0.2 bar to the system in Figure 3.23 at 30 seconds:
    a. Run the simulation for 300 seconds and calculate the integral of absolute error (IAE) for the outputs CVGAS and PGAS.
    b. Observe any input and output performance requirement violations.
3. Apply a sine wave pressure disturbance of amplitude 0.2 bar and frequency of 0.04 Hz at 0 seconds:
    a. Over a 300 second run and calculate the integral of absolute error as before.
    b. Observe any input and outputs performance violations.
4. Repeat steps 2 and 3 for the 50% and 0% load conditions.

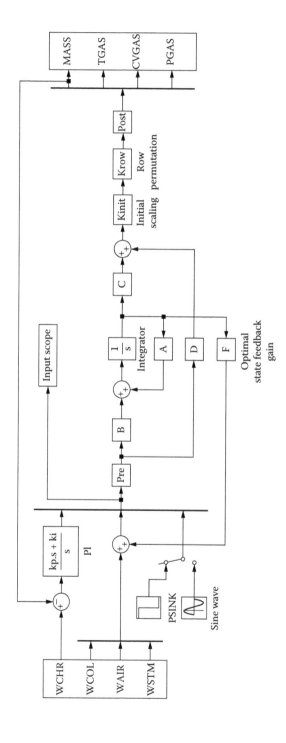

FIGURE 3.23   Simulation block diagram of LQR.

The results of the simulation and the resulting performance test tables are shown in Appendix A2. In these tables, the numbers given in bold print are the values that do not satisfy the specifications. This applied to all the controllers designed in this chapter. It can be observed from these results that the input constraint imposed on WCHR is violated at all loading conditions. In particular, for the 100% load condition, using the step function as the disturbance input, it is observed to have a nonminimum phase response. This is mainly due to the RHP zeros of the system. It occurs during the transient part of the response since the RHP zeros are situated very near to the origin of the s-plane. Simply stated, these RHP zeros dominate the response. For the rest of the loading conditions, a decrement in the WCHR response is observed. This is mainly due to the small negative value, of order 10^{-11}, in the feed-forward matrix of the system.

3.7 LINEAR QUADRATIC GAUSSIAN (LQG)

Since in practice, not all the states are available for feedback, the use of a state estimator becomes inevitable. A new controller called the *LQG design* [39–40], arises due to the optimal state feedback and state estimator being used. The task of finding this controller is to minimize the stochastic cost function below. The stochastic framework is used due to the presence of noise that is inherent during the state estimation process. To establish the solution of the optimal estimation problem, the following illustrates the derivation and the application of the LQG design on our gasifier.

3.7.1 LQG THEORY

Suppose that the gasifier plant state-space model is given in stochastic form as

$$\dot{\mathbf{x}}(t) = \mathbf{A}\mathbf{x}(t) + \mathbf{B}\mathbf{u}(t) + \mathbf{\Gamma}\mathbf{w}(t)$$
$$\mathbf{y}(t) = \mathbf{C}\mathbf{x}(t) + \mathbf{D}\mathbf{u}(t) + \mathbf{v}(t)$$

(3.23)

where $\mathbf{w}(t)$ and $\mathbf{v}(t)$ represent statistical knowledge about plant disturbances matrices:

$$\mathbf{E}\{\mathbf{w}(t)\mathbf{w}^T(t)\} = \mathbf{W} \geq 0$$
$$\mathbf{E}\{\mathbf{v}(t)\mathbf{v}^T(t)\} = \mathbf{V} \geq 0$$

(3.24)

and uncorrelated:

$$\mathbf{E}\{\mathbf{v}(t)\mathbf{w}^T(t+\tau)\} = 0$$

(3.25)

for all t and τ. $\mathbf{E}\{.\}$ is the statistical expectation operator. The LQG control problem is to devise a control law that minimizes the cost function

$$J = \lim_{T \to \infty} \mathbf{E}\left\{\int_0^T [\mathbf{x}^T(t)\mathbf{Q}\mathbf{x}(t) + \mathbf{u}^T(t)\mathbf{R}\mathbf{u}(t)]dt\right\}$$

(3.26)

Using the separation principle, the LQG control problem can be decomposed into two subproblems. First, obtaining an optimal estimate $\hat{\mathbf{x}}(t)$ of the state $\mathbf{x}(t)$ is done by minimizing

$$E\{[\mathbf{x}(t) - \hat{\mathbf{x}}(t)]^T[\mathbf{x}(t) - \hat{\mathbf{x}}(t)]\} \tag{3.27}$$

Using the Kalman filter theory, the optimal estimator becomes

$$\dot{\hat{\mathbf{x}}}(t) = \mathbf{A}\hat{\mathbf{x}}(t) + \mathbf{B}\mathbf{u}(t) + \mathbf{F}_e[\mathbf{y}(t) - \mathbf{C}\hat{\mathbf{x}}(t) + \mathbf{D}\mathbf{F}\mathbf{x}(t) - \mathbf{D}\mathbf{F}\hat{\mathbf{x}}(t)] \tag{3.28}$$

where the filter gain $\mathbf{F}_e$ is

$$\mathbf{F}_e = \mathbf{P}\mathbf{C}^T\mathbf{V}^{-1} \tag{3.29}$$

The estimation error variance $\mathbf{P}_e$ is a solution of the Filter Algebraic Riccati Equation (FARE):

$$\mathbf{A}\mathbf{P}_e + \mathbf{P}_e\mathbf{A}^T + \mathbf{\Gamma}\mathbf{W}\mathbf{\Gamma}^T - \mathbf{P}_e\mathbf{C}^T\mathbf{V}^{-1}\mathbf{C}\mathbf{P}_e = 0 \tag{3.30}$$

Second, the application of the standard deterministic LQR control law with the Kalman state estimate $\hat{\mathbf{x}}(t)$ instead of the state $\mathbf{x}(t)$ becomes

$$\mathbf{u}(t) = \mathbf{P}\mathbf{C}^T\mathbf{V}^{-1}\hat{\mathbf{x}}(t) \tag{3.31}$$

where

$$\mathbf{F}_e = \mathbf{P}\mathbf{C}^T\mathbf{V}^{-1} \tag{3.32}$$

A schematic block diagram of the LQG control is presented in Figure 3.24.

3.7.2 Loop Transfer Recovery (LTR)

In the previous section, the LQG design, which uses an optimal state estimator, was introduced. Unfortunately, the state estimation process used destroyed the robustness margin that is achieved in the LQR case. However, with the use of the LTR [39–42] technique, the robustness properties can be recovered. When the closed-loop system in Figure 3.20 is broken at the point a, the open-loop transfer function is

$$\mathbf{F}(s\mathbf{I} - \mathbf{A})^{-1}\mathbf{B} + \mathbf{D} \tag{3.33}$$

The corresponding open-loop transfer function for the LQG control problem is

$$\mathbf{F}(s\mathbf{I} - \mathbf{A} - \mathbf{B}\mathbf{K} - \mathbf{F}_e\mathbf{C} - \mathbf{F}_e\mathbf{F}\mathbf{D})^{-1}\,\mathbf{F}_e\,[\mathbf{C}(s\mathbf{I} - \mathbf{A})^{-1}\mathbf{B} + \mathbf{D}] \tag{3.34}$$

When the plant is a minimum phase system and the cost function J weights are chosen so that $\mathbf{R} = \mathbf{I}$ and $\mathbf{Q} = q^2\mathbf{B}\mathbf{B}^T$, it can be shown that for $q \to \infty$ the open-loop transfer function for the LQG problem approaches that for the LQR problem:

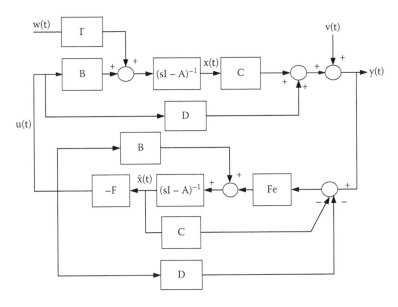

FIGURE 3.24   LQG schematic block diagram.

$$\lim_{q \to \infty} \left\{ \mathbf{F}(s\mathbf{I} - \mathbf{A} - \mathbf{BK} - \mathbf{F_e C} - \mathbf{F_e FD})^{-1} \mathbf{F_e} \ [\mathbf{C}(s\mathbf{I} - \mathbf{A})^{-1} \mathbf{B} + \mathbf{D}] \right\}$$

$$= \mathbf{F}(s\mathbf{I} - \mathbf{A})^{-1} \mathbf{B} + \mathbf{D}$$

(3.35)

This suggests a method of control system design, which is known as the *LQG with LTR*.

```
        % LQG Design on 3 inputs and 3 outputs system
r_set=20;
re_set=0.5;
Qe_set=12;
sigma_set=0;
[len,width]=size(Blqr);
Q=Clqr'*Clqr;
r=r_set*eye(width);
f=lqr(Alqr,Blqr,Q,r)
[len2,width2]=size(Clqr');
Qe=Qe_set*eye(len)+(sigma_set*eye(len)* Blqr*Blqr');
re=re_set*eye(width2);
fe=lqr(Alqr',Clqr',Qe,re);
fe=fe'
        % Form Klqg
[Ac,Bc,Cc,Dc]=reg(Alqr,Blqr,Clqr,Dlqr,f,fe);
sys_LQG=ss(Ac,Bc,Cc,Dc);
```

3.7.3   LQG/LTR Design Steps

The following steps are adopted for the LQG design.

1. Solve the **ARE** for P using (3.21).
2. Determine the **optimal state-feedback gain**, F using (3.20).
3. Solve the **FARE** for P_e using (3.30).
4. Determine the **optimal state-estimator gain**, F_e (LQE.m) using (3.32).
5. Weight selection.

   Again, the Q matrix is typically selected as C^TC, which reflects the weighting of the states on the outputs since $y^Ty = x^TC^TCx$. The matrix Q_e was selected to be a diagonal scalar matrix of different weights on each state. For the R and R_e matrices, these were selected to be diagonal matrices of different weights on each control input. The following are the weighting matrices used.

$$Q = C^TC,$$
$$Q_e = diag\ \{0.5, 0.2, 0.1, 0.2, 0.3, 0.1, 0.3, 0.1, 0.9, 0.1, 0.6, 0.8, 0.2, 0.7, 0.01,$$
$$0.02, 0.01, 0.1\}$$

$$R = \begin{bmatrix} 35 & 0 & 0 \\ 0 & 25 & 0 \\ 0 & 0 & 21 \end{bmatrix}, \quad R_e = \begin{bmatrix} 0.5 & 0 & 0 \\ 0 & 0.4 & 0 \\ 0 & 0 & 1 \end{bmatrix}$$

For the LQG/LTR method, the weight Q_{el} is

$$Q_{el} = 0.005\,Q_e + q^2BB^T \quad \text{where } q = 10$$

For example, the effects of the weight Q_e and R_e on the outputs can be determined using the following methodology. The effect of increasing the weighting matrices Q_e and R_e on each output, as shown in Figure 3.25, is used as a guide to adjusting the weights. After several iterations, a suitable choice of the weighting matrices Q_e and R_e were determined. It should be observed that equal weighting should not be applied on each state and instead a different weighting for each state is used.

By having different weight Q_e on each state, the corresponding output can be shaped. It can also be observed that the weights Q_{10}, Q_{12}, Q_{15}, Q_{17},

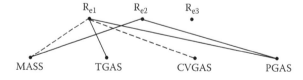

FIGURE 3.25   Effect of Q_e and R_e on the system outputs.

and Q_{18} have a fairly high influence on the output MASS, while the Q_7, Q_9, Q_{10}, Q_{12}, Q_{15}, and Q_{17}, on the other hand, have a substantial effect on TGAS, and so on. Interestingly, the Q_1, Q_2, Q_5, Q_6, Q_{11}, Q_{13}, Q_{14}, and Q_{16} have no influence on any outputs.

As for the weighting matrix $\mathbf{R_e}$, R_{e1} has an influence on all the outputs while R_{e2} has an effect on MASS and PGAS, whereas, R_{e3} has no influence on any outputs. Similarly, the same analysis can be performed on the weighting matrix R. From these interpretations, the system is indeed highly interactive. Note that this procedure is used as a guide and still requires further adjustment to meet some compromise between all the load conditions.

6. Perform simulation using Simulink.

The simulation block diagram is formulated using the Simulink facility, as shown in Figure 3.26.

Note that simulation block diagram for the LQG/LTR is identical to that used above except that the weightings ($\mathbf{Q_{e1}}$) used are different.

3.7.4   PERFORMANCE TESTS ON LQG/LTR

The performance tests for the LQG/LTR controller design using the specified step and sinusoidal disturbance input were performed. The results, which are summarized in Appendix A3 and Appendix A4, were obtained by only offsetting the steady-state values provided from the inputs. The graphical results, on the other hand, are presented for the 100%, 50% and 0% load conditions for the step disturbance input. As required in the ALSTOM challenge, the step disturbance was applied at $t = 30s$, whereas the sinusoidal disturbance signal was injected at $t = 0s$. The simulation was run until $t = 300s$ in both cases.

With the LQG/LTR controller design, all the input and output constraints are met at all load conditions. The input rates observed at all load conditions are very small and hence limit the outputs within the specifications. As far as the integral of the absolute error (IAE) associated with the CVGAS is concerned, it increases as the load condition decreases from 100% to 0%. This is not an unexpected trend, since the primary design was undertaken for the 100% load condition and the controller parameters were tuned for this particular load condition. On the other hand, with PGAS this does not seem to be the case, and this reduces progressively from the 100% load condition, through the 50% load condition, to the 0% load condition.

3.8   H-INFINITY OPTIMIZATION

H_∞ optimization [43–46], as its name implies, is a method used to optimize some form of cost function to meet the given control objectives. It was introduced to overcome the shortcomings of LQG control. The poor robustness properties of LQG could be attributed to the integral criterion in terms of the H_2 norm, and also the representation of disturbances by white noise processes is often unrealistic. Although,

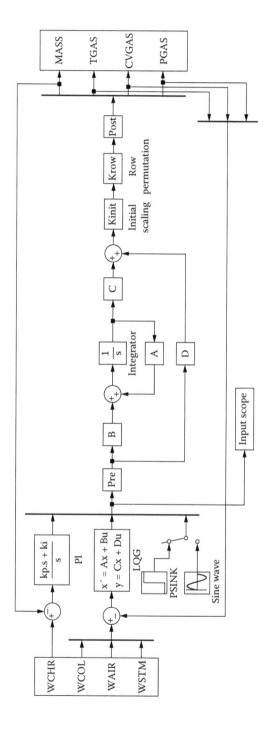

FIGURE 3.26 Simulation-block diagram of LQG/LTR.

LQG/LTR gives fairly good robustness properties in terms of changes in operating conditions, the H_∞ method is introduced here for comparison.

Before advancing to the controller design, the theory required for H_∞ design has to be examined. Basically, it involves the casting of the problem into a generalized plant configuration, the assumptions used in H_∞ design, the description of its optimization routine, the formulation of a mixed sensitivity problem, and lastly the selection of the weighing functions to shape the sensitivity of the closed-loop system.

3.8.1 GENERALIZED PLANT

There are many ways in which feedback design problems can be treated as H_∞ optimization problems. It is therefore useful to have a standard problem formulation into which any particular problem may be manipulated. Such a general configuration is shown in Figure 3.27.

The system shown in Figure 3.27 above is described by

$$\begin{bmatrix} \mathbf{Z} \\ \mathbf{V} \end{bmatrix} = \mathbf{P(s)} \begin{bmatrix} \mathbf{w} \\ \mathbf{u} \end{bmatrix} = \begin{bmatrix} \mathbf{P}_{11}(s) & \mathbf{P}_{12}(s) \\ \mathbf{P}_{21}(s) & \mathbf{P}_{22}(s) \end{bmatrix} \begin{bmatrix} \mathbf{w} \\ \mathbf{u} \end{bmatrix} \quad (3.36)$$

$$\mathbf{u} = \mathbf{K(s)V}$$

with a state-space realization of the generalized plant or open-loop system transfer function P given by

$$\mathbf{P} = \begin{bmatrix} \mathbf{A} & \mathbf{B}_1 & \mathbf{B}_2 \\ \mathbf{C}_1 & \mathbf{D}_{11} & \mathbf{D}_{12} \\ \mathbf{C}_2 & \mathbf{D}_{21} & \mathbf{D}_{22} \end{bmatrix} \quad (3.37)$$

Note that the matrix P is partitioned into four blocks $\mathbf{P}_{11}$, $\mathbf{P}_{12}$, $\mathbf{P}_{21}$, and $\mathbf{P}_{22}$. However, the general feedback configuration above has the controller K as a separate block. In order to analyze the closed-loop performance and minimize the H_∞ norm

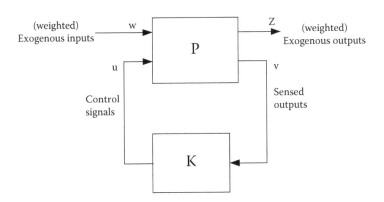

FIGURE 3.27 A generalized plant.

FIGURE 3.28   Closed-loop transfer function.

of the resulting closed-loop transfer function, the controller K may be absorbed into the interconnection structure as shown in Figure 3.27 and results in a system called *the lower fractional transformation* (LFT) in Figure 3.28. The lower means that the controller K is situated at the bottom of the generalized plant while the w to Z loop remains on top.

Let

$$Z = H(s)V \tag{3.38}$$

where the LFT denoted by $\mathbf{H}(s)$ is

$$\mathbf{H}(s) = \mathbf{P}_{11} + \mathbf{P}_{12}\mathbf{K}(\mathbf{I} - \mathbf{P}_{22}\mathbf{K})^{-1}\,\mathbf{P}_{21} \tag{3.39}$$

For completeness, the upper LFT is given as

$$\mathbf{H}_u(s) = \mathbf{P}_{22} + \mathbf{P}_{21}\mathbf{K}(\mathbf{I} - \mathbf{P}_{11}\mathbf{K})^{-1}\,\mathbf{P}_{12} \tag{3.40}$$

After obtaining the general closed-loop representation of the system in Figure 3.28, the next stage is to minimize the H_∞ norm of the LFT. In short, it is to minimize the H_∞ norm of the transfer function from w to Z. The control problem is then to find a controller K that, based on the information in V, generates a control signal u, which counteracts the influence of w on Z, thereby minimizing the closed-loop norm from w to Z.

3.8.2   H-INFINITY DESIGN ASSUMPTIONS

The following assumptions are typically made in H_∞ design problems:

A1 $(\mathbf{A}, \mathbf{B}_2, \mathbf{C}_2)$ is stabilizable and detectable.

A2 $\mathbf{D}_{12}$ and $\mathbf{D}_{12}$ have full rank.

A3 $\begin{bmatrix} \mathbf{A} - jw\mathbf{I} & \mathbf{B}_2 \\ \mathbf{C}_1 & \mathbf{D}_{12} \end{bmatrix}$ has full column rank for all w (3.41)

A4 $\begin{bmatrix} \mathbf{A} - jw\mathbf{I} & \mathbf{B}_1 \\ \mathbf{C}_2 & \mathbf{D}_{21} \end{bmatrix}$ has full row rank for all w (3.42)

A5 $\mathbf{D}_{12} = 0$ and $\mathbf{D}_{22} = 0$

Assumption A1 is required for the existence of stabilizing controllers K, and assumption A2 is sufficient to ensure that these controllers are proper and hence realizable. Assumption A3 and A4 ensure that the optimal controller does not try to cancel poles or zeros on the imaginary axis, which would result in closed-loop instability.

3.8.3 H_∞ Optimization Routine

H_∞ optimization is to minimize the infinity-norm of the closed-loop transfer function, H. For a stable scalar closed-loop transfer function $\mathbf{H}(s)$:

$$\|\mathbf{H}\|_\infty = \sup_{w \in R} |\mathbf{H}(jw)| \tag{3.43}$$

For a MIMO system, the minimization becomes

$$\|\mathbf{H}\|_\infty = \sup_{w \in R} \|\mathbf{H}(jw)\|_2$$

$$= \sup_{w \in R} \bar{\sigma}(\mathbf{H}(jw)) \tag{3.44}$$

$$= \sup_{w \in R} \sqrt{\lambda_{\max}(\mathbf{H}(jw)^H \mathbf{H}(jw))}$$

where the 2-norm $\|\mathbf{H}(jw)\|_2 = \left(\int_{-\infty}^{\infty} |\mathbf{H}(jw)|^2 \, dw \right)^{1/2}$, $\bar{\sigma}$ is the maximum singular value and $\lambda_{\max}$ is the maximum eigenvalue. It should be noted that the use of "max" (the maximum value) instead of "sup" (the supremum, the least upper bound) should be avoided. This is because the maximum may only be approached as $w \to \infty$ and may therefore not actually be achieved. However, for engineering purposes there is no difference between "sup" and "max".

3.8.4 Mixed Sensitivity Problem Formulation

Typically, the H_∞ control is formulated using mixed sensitivity. Mixed-sensitivity H_∞ control is the name given to transfer function shaping problems in which the sensitivity function $\mathbf{S} = (\mathbf{I} + \mathbf{GK})^{-1}$, for the case of negative feedback, is shaped together with some other closed-loop transfer function such as input sensitivity ($\mathbf{KS}$) and complementary sensitivity (T). S is the transfer function between an output disturbance and the output, $\mathbf{KS}$ is the transfer function between the input disturbance and the output. T is the transfer function between the command input and output (Figure 3.29).

To have a clearer picture, Table 3.6 below shows the relationship between the functions $\mathbf{S}$, $\mathbf{KS}$, and T.

As desired, the maximum singular value of the sensitivity function S is to be small in order to reject disturbances. The input sensitivity $\mathbf{KS}$ is used to limit the size and bandwidth of the controller, and hence the control energy used. The size of $\mathbf{KS}$ is also important for robust stability. In summary, the purposes of each function can be tabulated as shown in Table 3.7 below.

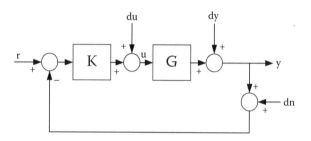

FIGURE 3.29   Feedback system with disturbance input.

TABLE 3.6
Sensitivity Function Relationship

Type	Transfer Functions
Sensitivity (**S**)	$(I + GK)^{-1}$
Input sensitivity (**KS**)	$K(I + GK)^{-1} = (I + KG)^{-1} K$
Complementary sensitivity (**T**) or (**GKS**)	$GK(I + GK)^{-1} = G(I + KG)^{-1} K$

TABLE 3.7
Purpose of Each Sensitivity Function

Sensitivity Function	Purpose
S	Use for performance measure
KS	Use to ensure robustness and to avoid sensitivity to noise
T	Use to ensure robust stability and to penalize large input

From the above discussion of the S, **KS** and T transfer functions, it can be observed that there is some interrelation between them. Among these, S is the most suitable function to manipulate, since it appears in each function. Typically, the function **S** should be small at low frequencies and not greater than 0 dB at higher frequencies, while the function T is its opposite. These requirements may be combined into a stacked H_∞ problem, where the function to be minimized is as follows

$$\min_{K} \left\| \begin{matrix} \mathbf{W_u KS} \\ \mathbf{W_T T} \\ \mathbf{W_p S} \end{matrix} \right\|_\infty \tag{3.45}$$

where K is a stabilizing controller and the $\mathbf{W_u}$, $\mathbf{W_T}$, $\mathbf{W_p}$ are the sets of weighting functions or matrices. The resulting H_∞ norm should be less than 1 in order to achieve robust stability. Note that any negative signs have no effect when evaluating $\|\cdot\|_\infty$. From the above discussion, it can be seen that the H_∞ control is a synthesis process of selecting the weighting functions to shape the sensitivity and its complementary function in the relevant frequency region, in order to achieve performance and robustness.

The main disadvantage of this method is the high-order controller obtained from the design. Nevertheless, a MOR method can be applied to reduce the order of the controller.

3.8.5 SELECTION OF WEIGHTING FUNCTION

The weighting function is selected to shape the sensitivity and the input sensitivity of the system to the required bandwidth and the upper (indicate the percent of disturbances rejection) and lower bounds. To illustrate this process, a SISO system is used.

$$\mathbf{L} = \mathbf{GK} = \frac{w_n^2}{s(s+2\xi w_n)} = \frac{4}{s(s+2.8)}; \quad \xi = 0.7, \quad w_n = 2 \ rad \ / \ s \qquad (3.46)$$

Note that the same principle can be applied to the MIMO system as well.

Sensitivity (S) Shaping. In order to select a weighting function to penalize any large S, the following weighting functions are used, such that

$$|\mathbf{S}(s)| \leq \left| \frac{s + w_b e}{s / M_s + w_b} \right|, \quad s = jw, \forall w$$

$$\Leftrightarrow |\mathbf{W_s S}| \leq 1, \quad \mathbf{W_s} = \frac{s / M_s + w_b}{s + w_b e_s} \qquad (3.47)$$

Figure 3.30 shows a typical sensitivity function S, which may be encountered in practice and has low sensitivity at low frequency and increasingly higher sensitivity at high frequency.

Input Sensitivity (KS) Shaping. A typical $\mathbf{KS}$ and the corresponding inverse weighting function are shown in Figure 3.31. The control weighting function $\mathbf{W_u}$ for the input sensitivity shaping can be determined as follows:

$$\mathbf{W_u} = \frac{s + w_{bc} / M_u}{e_u s + w_{bc}} \qquad (3.48)$$

Complementary Sensitivity (T) Shaping. A typical T and the corresponding inverse weighting function are shown in Figure 3.32. In order to

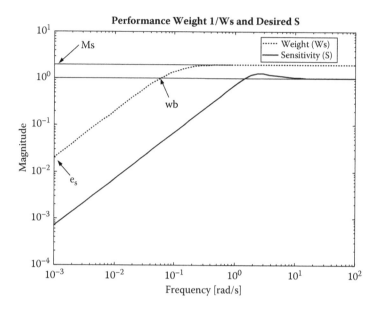

FIGURE 3.30 Sensitivity plot.

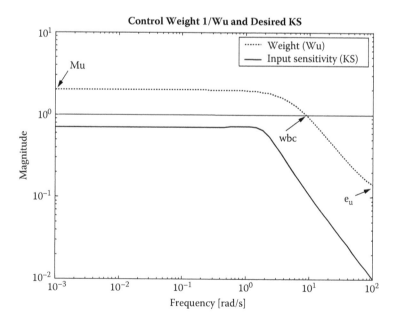

FIGURE 3.31 Input sensitivity plot.

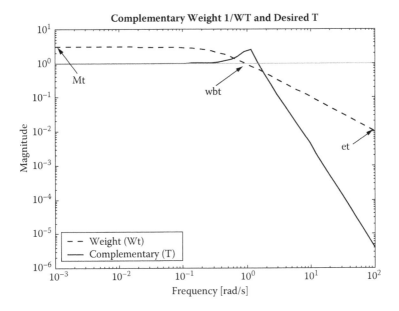

FIGURE 3.32   Complementary sensitivity plot.

select a weighting function, W_T to penalize the T, the following weighting function for complementary sensitivity shaping is used, such that

$$\mathbf{W_T} = \frac{s + w_{bt} \, / \, M_t}{e_t s + w_{bt}} \tag{3.49}$$

It has the same form as the input sensitivity weighting function. Note that as S is decreased, the T has to be increased since $\mathbf{S} + \mathbf{T} = 1$. Therefore, a compromise has to be sought in shaping the S and T.

Both weighting functions used for sensitivity shaping are able to meet the specification in terms of bandwidth and upper and lower bound limits. Another way to examine whether the complementary sensitivity function or any other function is well shaped is to plot $|\mathbf{W_T T}|$ instead of $1/\mathbf{W_T}$ directly, as shown in Figure 3.33.

Since there is an overshoot beyond the 1 (= 0 dB) level, the criteria of $|\mathbf{W_T T}| < 1$ is not met. This can be seen similarly in the plot of $1/\mathbf{W_T}$ in Figure 3.32.

Alternatively, using the MATLAB function MAGSHAPE.m enables an appropriate shaping filter to be designed in a more systematic way.

3.8.6   H-INFINITY DESIGN STEPS

The procedure for designing an H_∞ controller for the gasifier plant is shown below.

1. Identify the exogenous input signal $w(t)$ and the exogenous output $Z(t)$ of the plant.

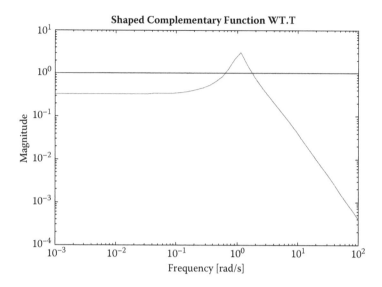

FIGURE 3.33   Shaped complementary plot.

2.

$$w = \begin{bmatrix} \mathbf{r} \\ \mathbf{v}_1 \\ \mathbf{u} \end{bmatrix}; \quad \mathbf{Z} = \begin{bmatrix} \mathbf{z}_1 \\ \mathbf{z}_2 \\ \mathbf{z}_3 \end{bmatrix} \tag{3.50}$$

3. Formulate the generalized plant, $\mathbf{P}(s)$ that represents the open-loop transfer function between the $\mathbf{Z}(t)$ and $\mathbf{w}(t)$. Note that the matrix $\mathbf{G_p}$ refers to the transfer function between the input disturbances and the output.

   From the block diagram shown in Figure 3.34, the following set of equations can be obtained:

$$\mathbf{z}_1 = \mathbf{W}_1\mathbf{W}_4\mathbf{r} - \mathbf{W}_1\mathbf{G}\mathbf{u} - \mathbf{W}_1\mathbf{G_p}\mathbf{W}_5\mathbf{v}_1$$

$$\mathbf{z}_2 = \mathbf{W}_2\mathbf{u}$$

$$\mathbf{z}_3 = \mathbf{W}_3\mathbf{G}\mathbf{u} \tag{3.51}$$

$$\mathbf{V} = \mathbf{W}_4\mathbf{r} - \mathbf{G}\mathbf{u} - \mathbf{G_p}\mathbf{W}_5\mathbf{v}_1$$

So the generalized plant P from $\begin{bmatrix} \mathbf{r} & \mathbf{v}_1 & \mathbf{u} \end{bmatrix}^T$ to $\begin{bmatrix} \mathbf{z}_1 & \mathbf{z}_2 & \mathbf{z}_3 & \mathbf{V} \end{bmatrix}^T$ is

$$P = \begin{bmatrix} \mathbf{W}_1\mathbf{W}_4 & -\mathbf{W}_1\mathbf{G_P}\mathbf{W}_5 & -\mathbf{W}_1\mathbf{G} \\ 0 & 0 & \mathbf{W}_2 \\ 0 & 0 & \mathbf{W}_3\mathbf{G} \\ \mathbf{W}_4 & -\mathbf{G_P}\mathbf{W}_5 & -\mathbf{G} \end{bmatrix} \tag{3.52}$$

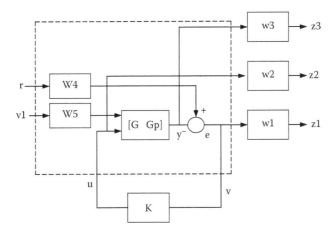

FIGURE 3.34 H$_\infty$ generalized plant.

P is then partitioned as

$$\mathbf{P} = \begin{bmatrix} \mathbf{P}_{11} & \mathbf{P}_{12} \\ \mathbf{P}_{21} & \mathbf{P}_{22} \end{bmatrix} \tag{3.53}$$

such that its parts are compatible with the signals $\mathbf{w}, \mathbf{Z}, \mathbf{u}, \mathbf{V}$ in the generalized control configuration, where

$$\mathbf{Z} = \mathbf{P}_{11}\mathbf{w} + \mathbf{P}_{12}\mathbf{u}$$
$$\mathbf{V} = \mathbf{P}_{21}\mathbf{w} + \mathbf{P}_{22}\mathbf{u} \tag{3.54}$$

From this,

$$\mathbf{P}_{11} = \begin{bmatrix} \mathbf{W}_1\mathbf{W}_4 & -\mathbf{W}_1\mathbf{G}_P\mathbf{W}_5 \\ 0 & 0 \\ 0 & 0 \end{bmatrix}, \quad \mathbf{P}_{12} = \begin{bmatrix} -\mathbf{W}_1\mathbf{G} \\ \mathbf{W}_2 \\ \mathbf{W}_3\mathbf{G} \end{bmatrix} \tag{3.55}$$

$$\mathbf{P}_{21} = \begin{bmatrix} \mathbf{W}_4 & -\mathbf{G}_P\mathbf{W}_5 \end{bmatrix}, \quad \mathbf{P}_{22} = -\mathbf{G}$$

The generalized plant is then converted into the state-space realization (PCK.m, MKSYS.m) in order to be usable in the MATLAB routine, HINF.m.

$$\mathbf{P}^s = \begin{bmatrix} \mathbf{A} & \mathbf{B}_1 & \mathbf{B}_2 \\ \mathbf{C}_1 & \mathbf{D}_{11} & \mathbf{D}_{12} \\ \mathbf{C}_2 & \mathbf{D}_{21} & \mathbf{D}_{22} \end{bmatrix} \tag{3.56}$$

4. Determine the closed-loop transfer function to be minimized. The cost function is obtained from the closed-loop derivation of the generalized plant. Using the LFT of P with K as the parameter, the LFT is obtained as follows:

$$H = P_{11} + P_{12}K(I - P_{22}K)^{-1}P_{21}$$

$$= \begin{bmatrix} W_1 W_4 & -W_1 G_P W_5 \\ 0 & 0 \\ 0 & 0 \end{bmatrix} + \begin{bmatrix} -W_1 G \\ W_2 \\ W_3 G \end{bmatrix} \frac{K}{I + GK} \begin{bmatrix} W_4 & -G_P W_5 \end{bmatrix}$$

$$= \begin{bmatrix} W_1 W_4 & -W_1 G_P W_5 \\ 0 & 0 \\ 0 & 0 \end{bmatrix} + \begin{bmatrix} -W_1 W_4 G & W_1 G G_P W_5 \\ W_2 W_4 & -W_2 G_P W_5 \\ W_3 W_4 W & -W_3 G G_P W_5 \end{bmatrix} \frac{K}{I + GK} \qquad (3.57)$$

$$= \begin{bmatrix} W_1 S W_4 & -W_1 S G_P W_5 \\ W_2 K S W_4 & -W_2 K S G_P W_5 \\ W_3 T W_4 & -W_3 T G_P W_5 \end{bmatrix}$$

where

$$S = \frac{I}{I + GK}, \quad T = \frac{GK}{I + GK}$$

5. Select the weighting matrices used as shown in Figure 3.34.

The weighting function is used to shape the sensitivity of the closed-loop system to the desired level. A typical type of performance function used is the low-pass filter, high-pass filter and a constant weight. The tuning of the weighting function, irrespective of the type used, is performed iteratively. The sequence is to tune the first entry of the weighting function so that the nominal performance is still enforced. Retain this value for the first entry and repeat the procedure for the second entry, and so on.

For a high- or low-pass filter, the tuning parameter is naturally the bandwidth and the maximum and minimum amplitude, as well as the scalar gain accompanying it. Usually the weightings are chosen to be simple functions of first order to avoid the total order of the resulting closed-loop system becoming too large. In the multivariable case, the weighting matrices are often chosen to be diagonal. From the closed-loop transfer function of the system, the weighting function W_4 can be treated as a scalar matrix. This leads to the use of only one weighting function instead of using two filters for shaping:

$$W_1 \rightarrow S, W_2 \rightarrow KS, W_3 \rightarrow T, W_4 \rightarrow const, W_5 \rightarrow SG_p$$

TABLE 3.8

H$_\infty$ Performance Function Specification

Weighting Functions	Specifications
W$_1$	w$_b$ = 0.008 rad/s M$_s$ = 0.1 e$_s$ = 0.01
W$_2$	w$_{bc}$ = 1 rad/s M$_u$ = 0.1 e$_u$ = 0.01
W$_3$	w$_{bt}$ = 0.008 rad/s M$_t$ = 0.1 e$_t$ = 0.01
W$_4$	Uses a constant weighting
W$_5$	w$_b$ = 0.008 rad/s M$_s$ = 0.2 e$_s$ = 0.01

Before the types of weighting functions used are explained, the shaping specifications for each weight related to its corresponding sensitivity function are shown in Table 3.8.

Note that besides using the specifications in Table 3.8, an extra scalar term is applied to each of the entries of the weighting function to increase the percentage of the disturbance rejection. This is shown in the selection of these weights below.

Selection of W$_1$. The weight W$_1$ was chosen as a high-pass filter in order to achieve the nominal output performance and also shape the sensitivity of the closed-loop system.

$$\mathbf{W_1} = \begin{bmatrix} 0.2 & 0 & 0 \\ 0 & 0.8 & 0 \\ 0 & 0 & 0.3 \end{bmatrix} \frac{10s + 0.008}{s + 0.00008} \tag{3.58}$$

Selection of W$_2$. A low-pass filter was selected to shape the input sensitivity of the closed-loop system. A similar first-order, low-pass filter was used in each channel with a corner frequency of 1 rad/s, in order to limit the input magnitudes at high frequencies and thereby limit the closed-loop bandwidth. Different gains for each entry were selected, sequentially.

$$\mathbf{W_2} = \begin{bmatrix} 0.01 & 0 & 0 \\ 0 & 0.01 & 0 \\ 0 & 0 & 0.02 \end{bmatrix} \frac{s + 10}{0.01s + 1} \tag{3.59}$$

Selection of W$_3$. Again a low-pass filter was designed for controlling the output of the gasifier, and hence shaping the closed-loop system. It limits the size of the output specified.

$$\mathbf{W_3} = \begin{bmatrix} 0.1 & 0 & 0 \\ 0 & 0.1 & 0 \\ 0 & 0 & 0.09 \end{bmatrix} \frac{s + 0.08}{0.01s + 0.008} \tag{3.60}$$

Selection of W$_4$. The weight here was chosen to be sufficiently small in order to prevent the appearance of some badly damped modes in the closed-loop system. Any slight increase in the gain of this filter causes the closed-loop system to be unstable.

$$\mathbf{W_4} = \begin{bmatrix} 0.002 & 0 & 0 \\ 0 & 0.001 & 0 \\ 0 & 0 & 0.005 \end{bmatrix} \tag{3.61}$$

Selection of $\mathbf{W_5}$. $\mathbf{W_5}$ is a scalar low-pass filter with a crossover frequency approximately equal to that of the desired closed-loop bandwidth. With such a weight, the maximum disturbance amplification in the low to middle frequencies should not exceed 0.2. The crossover frequency was adjusted iteratively to speed up the output time responses, while maintaining the control effort within the specified limits.

$$\mathbf{W_5} = 0.1\frac{s+0.04}{0.01s+0.008} \tag{3.62}$$

6. Minimize the H_∞ norm of the closed-loop transfer function to obtain a stabilizing controller, K (HINF.m).
7. Simulate the system using the block diagram shown in Figure 3.35. All the weightings used have dimensions of 4×4 where the position {1,1} is a unity weight for the first loop.

```
        %Design H_Infinity Controller
plant=k*(ko*ss(Ar,Bd1,Cr,Dd1));
POST=[ 1 0          0            0;            % 4x4
       0 post1e(1,1)  post1e(1,2)  post1e(1,3);
       0 post1e(2,1)  post1e(2,2)  post1e(2,3);
       0 post1e(3,1)  post1e(3,2)  post1e(3,3)];
PRE= [ 1 0          0            0            0; % 5x5
       0 pre1e(1,1)  pre1e(1,2)  pre1e(1,3)  0;
       0 pre1e(2,1)  pre1e(2,2)  pre1e(2,3)  0;
       0 pre1e(3,1)  pre1e(3,2)  pre1e(3,3)  0;
       1 1          1            1            1];
plant_tot=POST*plant*PRE;
       % W1
n11=0.2*[1/0.1 0.008]; d11=[1 0.008*0.01];
n12=0.8*[1/0.1 0.008]; d12=[1 0.008*0.01];
n13=0.3*[1/0.1 0.008]; d13=[1 0.008*0.01];
[aw11,bw11,cw11,dw11]=tf2ss(n11,d11);
[aw12,bw12,cw12,dw12]=tf2ss(n12,d12);
[aw13,bw13,cw13,dw13]=tf2ss(n13,d13);
aw1=daug(aw11,aw12,aw13);
bw1=daug(bw11,bw12,bw13);
cw1=daug(cw11,cw12,cw13);
dw1=daug(dw11,dw12,dw13);
       % Using the same method for other weighting-
       W2 to W5

       % Form Generalized Plant G
```

```
WW1=ss(aw1,bw1,cw1,dw1);
WW2=ss(aw2,bw2,cw2,dw2);
WW3=ss(aw3,bw3,cw3,dw3);
WW4=ss(aw4,bw4,cw4,dw4);
WW5=ss(aw51,bw51,cw51,dw51);
Gp=ss(Ar,B(8:25,6),Cr(2:4,:),D(2:4,6));
G33=syss;
GGG=[WW1*WW4  -WW1*Gp*WW5  -WW1*G33;
     zeros(3)  zeros(3,1)   WW2;
     zeros(3)  zeros(3,1)   WW3*G33;
     WW4       -Gp*WW5      -G33];
        %Separate the input w and reference input r from
        control input u
  [AA,BB,CC,DD]=ssdata(GGG);
  BB1=BB(:,1:4);
  BB2=BB(:,5:7);
  CC1=CC(1:9,:);
  CC2=CC(10:12,:);
  DD11=DD(1:9,1:4);
  DD12=DD(1:9,5:7);
  DD21=DD(10:12,1:4);
  DD22=DD(10:12,5:7);
        %Form state space realization of Generalized Plant
  pp=pck(AA,[BB1 BB2],[CC1;CC2],[DD11 DD12;DD21 DD22]);
  PH22 = rct2lti(mksys(AA,BB1,BB2,CC1,CC2,DD11,DD12,DD21,D
  D22,'tss'));
        % H- Infinity Controller
  [ss_cp,ss_cl]=hinf(PH22);
  [Ainf,Binf,Cinf,Dinf]=ssdata(ss_cp);
  ss_hinf=ss(Ainf,Binf,Cinf,Dinf);
        % Model order reduction on H-inf controller
        (originally = 98 states)
[Ainfs,Binfs,Cinfs,Dinfs]=balmr(Ainf,Binf,Cinf,Dinf,1,70);
ss_cps=ss(Ainfs,Binfs,Cinfs,Dinfs);
```

8. Analyze the resulting **S**, **KS**, and *T* plots.

   As shown in the *S* plots, the weighting function used has indeed shaped the sensitivity of the system below the 0-dB margin. Since the singular value of the sensitivity function is low, it shows that the closed-loop system with a disturbance signal has less influence on the system output. Conversely, a large value of *S* means the system has a poor stability margin (Figure 3.36).

   Examining the *T* plots of the system, the gain over the frequencies concerned is less than the 0 dB (= 1) margin. This indicates that the system output is small for any large input. This is reflected in the simulation results shown in Appendix A6 that indicate that the responses are indeed very small (Figure 3.37).

   From the KSG_p plot, it can be seen the effect of the input disturbances is quite negligible on the gasifier output. The KSG_p after applying the shaping function is well below the margin of 0 dB (Figure 3.38).

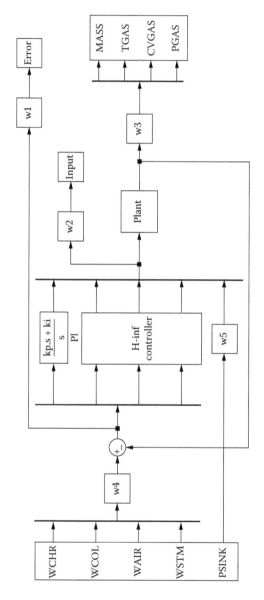

FIGURE 3.35   Simulation block diagram for H_∞ design.

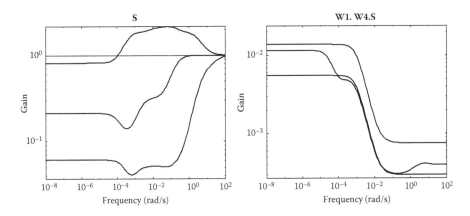

FIGURE 3.36   *S* plots for H$_\infty$ design.

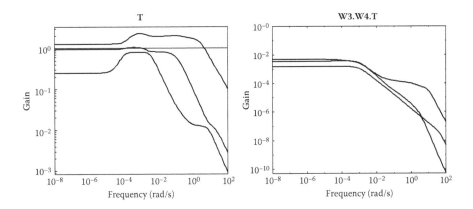

FIGURE 3.37   *T* plots for H$_\infty$ design.

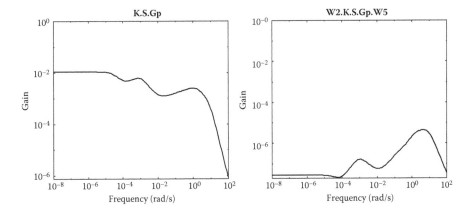

FIGURE 3.38   **KSG$_p$** plots for H$_\infty$ design.

```
% Analyze the resulting S, KS and T plots.
% Put PI & Hinf controller together--> 4 x 4
kp=-0.159;ki=-0.001;
PI_num=[kp ki];PI_den=[1 0];
[Api,Bpi,Cpi,Dpi]=tf2ss(PI_num,PI_den);
controller=ss(Api,Bpi,Cpi,Dpi);
K_PI_HInfinity=[controller zeros(1,3) ;
                zeros(3,1) ss_cps ];
ko1=[0.00001 0 0 0;0 0.001 0 0; 0 0 0.001 0; 0 0 0 0.1]; %
4x4
k1=[0 1 0 0 ;0 0 1 0 ;1 0 0 0 ;0 0 0 1 ]; % 4x4
sys_tot_OL=(POST*k1*(ko1*ss(Ar,Br,Cr,Dr))*PRE(1:4,1:4)
)*K_PI_HInfinity;
sys_tot_cl=feedback(sys_tot_OL,eye(4,4),-1);
        % Output Robustness- ( 1+(GK)^-1)>1
ww=logspace(-8,2,100);
figure 4);
out_robust=sigma(sys_tot_OL,ww,3);
loglog(ww,(out_robust));
hold on
loglog(ww,1);
title(' Robust stability at the Plant output')
ylabel('Gain')
xlabel('Frequency (rad/s)')
        % Input Robustness- ( 1+(KG)^-1)>1
sys_tot_OL2=K_PI_LQG*(POST*k1*(ko1*ss(Ar,Br,Cr,Dr))*
PRE(1:4,1:4));
xlabel('Frequency (rad/s)')

figure();
inp_robust=sigma(sys_tot_OL2,ww,3);
loglog(ww,min(inp_robust));
hold on
loglog(ww,1);
title(' Robust stability at the Plant input')
ylabel('Gain')
xlabel('Frequency (rad/s)')
        % Sensitivity min_sigma[1/(1+GK)]<1
sens_100=sigma(eye(4)+sys_tot_OL,ww,1);
figure();
loglog(ww,(sens_100));
hold on
loglog(ww,1);
title(' Output Sensitivity ')
ylabel('Gain')
xlabel('Frequency (rad/s)')
        % Asymptotically stability min_sigma[1+GK]>0
asym_100=sigma(eye(4)+sys_tot_OL,ww,1);
figure();
loglog(ww,max(asym_100));
```

```
hold on
loglog(ww,1);
title(' Asymptotical Plant stability ')
ylabel('Gain')
xlabel('Frequency (rad/s)')
```

3.8.7 PERFORMANCE TESTS ON H-INFINITY DESIGN

The performance tests for the H_∞ design were performed using the same methods as used in the previous section. The results of the simulation and the performance test tables are shown in Appendix A6. For the H_∞ control schemes, all the input and output constraints are met at all loading conditions using the sine disturbance input. For the case of the step disturbance input, all the specifications are met except for the maximum limit of WCHR at the 0% load condition, which is about 3% greater than allowed. Also, the maximum CVGAS exceeded the output limit of 10 kJ/kg by 30 kJ/kg at this condition. The input rates at all load conditions are very small as observed in the peak values. As far as the integral of the absolute error associated with CVGAS and PGAS are concerned, this increases as the load condition decreases from 100% to 0%. This is an expected trend, since the primary design and weight tuning is undertaken for the 100% load condition.

3.9 H₂ OPTIMIZATION

The H_2 formulation [47] replaces the stochastic minimum least-squares interpretation of LQG optimization with the minimization of the 2-norm of the closed-loop system. One considerable advantage of this approach is that there is no need to interpret parameters such as the intensity of the various white noise processes that enter into the LQG problem as stochastic data, whose values need to be determined by modeling or identification experiments. These can simply be taken as tuning parameters for the design process. In LQG practice, this is, of course, common procedure. A pedagogical advantage of the H_2 formulation is that there is no need to go into the intricacies of white noise.

For a stochastic system described by (3.23), to minimize notation, the time-dependent term is omitted. The controlled output Z and the measured output y are given by

$$\mathbf{Z} = \mathbf{Fx}$$

$$\mathbf{y} = \mathbf{Cx} + \mathbf{Du} + \mathbf{v}$$

(3.63)

where the white noise v represents sensor noise. Assuming that the controlled output Z and the input u are suitably scaled, LQG optimization amounts to finding a feedback compensator K that stabilizes the system and minimizes the stochastic cost function given in (3.26).

As mentioned in the previous section, the Kalman filter and the state feedback gains can be obtained from the solution of two algebraic Riccati equations, namely

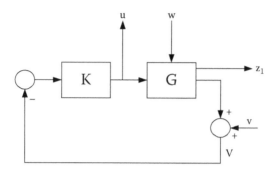

FIGURE 3.39   Generalized LQG feedback system.

the ARE and FARE. The generalized feedback system for LQG optimization can be shown as in Figure 3.39 below.

It can easily be seen that the configuration in Figure 3.39 resembled the generalized plant shown in Figure 3.27 by redefining Z and V as

$$Z = \begin{bmatrix} z_1 \\ u \end{bmatrix}, \quad V = \begin{bmatrix} w \\ v \end{bmatrix} \tag{3.64}$$

The LQG problem now amounts to the minimization of the steady-state value of $E(Z^TZ)$. In the generalized plant, the output Z may be expressed in terms of Laplace transforms as

$$Z = H(s)V \tag{3.65}$$

where $H(s)$ is the LFT defined as

$$H(s) = P_{11} + P_{12}K(I - P_{22}K)^{-1}P_{21} \tag{3.66}$$

If V is white noise with intensity matrix I, $Q = R = I$, and the closed-loop system is stable, then

$$J = \lim_{t \to \infty} E[z_1^Tz_1 + u^Tu] = \frac{1}{2\pi} tr \int_{-\infty}^{\infty} H^T(-jw)H(jw)dw \tag{3.67}$$

The above uses Parseval's identity; namely, $\dfrac{1}{2\pi}\displaystyle\int_{-\infty}^{\infty} \Phi_u(w)dw = \int_{-\infty}^{\infty} z_1(t)z_1^T(t)dt$. In

(3.67) E denotes the mathematical expectation $E[.] = \lim_{N \to \infty} \dfrac{1}{N}\displaystyle\int_{0}^{N} [.]dt$. The $\frac{1}{2\pi}$ is

used due to working in the frequency domain while tr is used for measuring the signal size of the matrix. The expression on the right-hand side is the square of the 2-norm

$$\|\mathbf{H}\|_2 = \left(\frac{1}{2\pi} tr \int_{-\infty}^{\infty} \mathbf{H}^T(-jw)\mathbf{H}(jw)dw \right)^{1/2} \tag{3.68}$$

of the stable transfer function matrix H. Hence, the optimization is tantamount to minimization of the 2-norm of the closed-loop system, H. This minimization problem is the celebrated H_2 problem.

Hence, solution of the LQG problem amounts to minimization of the 2-norm, $\|\mathbf{H}\|_2$, of the transfer function H from the white noise input to the output. The problem has changed from a signal interpretation to a systems interpretation. The norm interpretation is nonstochastic.

3.9.1 H_2 Design Steps

The design steps are basically as follows.

1. Identify the exogenous input signal, $w(t)$, and the exogenous output $\mathbf{Z}(t)$ of the plant shown in Figure 3.40.

$$\mathbf{w} = \begin{bmatrix} \mathbf{v}_1 & \mathbf{v}_2 & \mathbf{r} & \mathbf{u} \end{bmatrix}^T, \quad \mathbf{Z} = \begin{bmatrix} \mathbf{z}_1 & \mathbf{z}_2 & \mathbf{z}_3 \end{bmatrix}^T \tag{3.69}$$

2. Formulate the generalized plant, $\mathbf{P(s)}$, that represents the open-loop transfer function between the $\mathbf{Z}(t)$ and $\mathbf{w}(t)$ is shown in Figure 3.40.

From the block diagram above, the following set of equations can be obtained:

$$\mathbf{z}_1 = \mathbf{W}_1\mathbf{W}_3\mathbf{v}_1 + \mathbf{W}_1\mathbf{Gu}$$

$$\mathbf{z}_2 = \mathbf{W}_2\mathbf{u} \tag{3.70}$$

$$\mathbf{V} = \mathbf{r} - \mathbf{W}_4\mathbf{v}_2 - \mathbf{W}_3\mathbf{v}_1 - \mathbf{Gu}$$

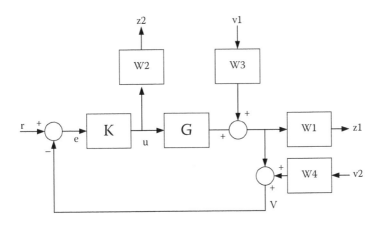

FIGURE 3.40 H_2 generalized plant.

So, the generalized plant P from $\begin{bmatrix} \mathbf{v}_1 & \mathbf{v}_2 & \mathbf{r} & \mathbf{u} \end{bmatrix}^T$ to $\begin{bmatrix} \mathbf{z}_1 & \mathbf{z}_2 & \mathbf{V} \end{bmatrix}^T$ is

$$\mathbf{P} = \begin{bmatrix} \mathbf{W}_1\mathbf{W}_3 & \mathbf{0} & \mathbf{0} & \mathbf{W}_1\mathbf{G} \\ \mathbf{0} & \mathbf{0} & \mathbf{0} & \mathbf{W}_2 \\ -\mathbf{W}_3 & -\mathbf{W}_4 & \mathbf{I} & -\mathbf{G} \end{bmatrix} \qquad (3.71)$$

Partitioning the P into four parts gives

$$\mathbf{P}_{11} = \begin{bmatrix} \mathbf{W}_1\mathbf{W}_3 & \mathbf{0} & \mathbf{0} \\ \mathbf{0} & \mathbf{0} & \mathbf{0} \end{bmatrix}, \quad \mathbf{P}_{12} = \begin{bmatrix} \mathbf{W}_1\mathbf{G} \\ \mathbf{W}_2 \end{bmatrix}$$

$$(3.72)$$

$$\mathbf{P}_{21} = \begin{bmatrix} -\mathbf{W}_3 & -\mathbf{W}_4 & \mathbf{I} \end{bmatrix}, \quad \mathbf{P}_{22} = -\mathbf{G}$$

3. Determine the closed-loop transfer function to be minimized. The cost function is obtained from the closed-loop derivation of the generalized plant using the LFT, as

$$\mathbf{H} = \mathbf{P}_{11} + \mathbf{P}_{12}\mathbf{K}(\mathbf{I} - \mathbf{P}_{22}\mathbf{K})^{-1}\mathbf{P}_{21}$$

$$= \begin{bmatrix} \mathbf{W}_1\mathbf{W}_3 & \mathbf{0} & \mathbf{0} \\ \mathbf{0} & \mathbf{0} & \mathbf{0} \end{bmatrix} + \begin{bmatrix} \mathbf{W}_1\mathbf{G} \\ \mathbf{W}_2 \end{bmatrix} \frac{\mathbf{K}}{\mathbf{I}+\mathbf{GK}} \begin{bmatrix} -\mathbf{W}_3 & -\mathbf{W}_4 & \mathbf{I} \end{bmatrix}$$

$$= \begin{bmatrix} \mathbf{W}_1\mathbf{W}_3 & \mathbf{0} & \mathbf{0} \\ \mathbf{0} & \mathbf{0} & \mathbf{0} \end{bmatrix} + \begin{bmatrix} -\mathbf{W}_1\mathbf{W}_3\mathbf{G} & -\mathbf{W}_1\mathbf{W}_4\mathbf{G} & \mathbf{W}_1\mathbf{G} \\ -\mathbf{W}_2\mathbf{W}_3 & -\mathbf{W}_2\mathbf{W}_4 & \mathbf{W}_2 \end{bmatrix} \frac{\mathbf{K}}{\mathbf{I}+\mathbf{GK}} \qquad (3.73)$$

$$= \begin{bmatrix} \mathbf{W}_1\mathbf{W}_3\mathbf{S} & -\mathbf{W}_1\mathbf{W}_4\mathbf{T} & \mathbf{W}_1\mathbf{T} \\ -\mathbf{W}_2\mathbf{W}_3\mathbf{KS} & -\mathbf{W}_2\mathbf{W}_4\mathbf{KS} & \mathbf{W}_2\mathbf{KS} \end{bmatrix}$$

where S and T are as before.

4. Select the weighting functions to be used.

   The weighting function, or the performance weighting, is used to shape the sensitivity of the closed-loop system to the desired level. The weighting functions have to be selected such that only one weight is used to shape the desired sensitivity. This is done in order to have a more convenient control over the sensitivity. The following are the weights used to shape the sensitivity.

$$\mathbf{W}_1 \to \mathbf{T}, \ \mathbf{W}_2 \to \mathbf{KS}, \ \mathbf{W}_3 \to \mathbf{S}, \ \mathbf{W}_4 \to const$$

The performance specification for each weighting function that is used to shape the corresponding sensitivity functions are shown in Table 3.9.

Note that besides using the specifications in Table 3.9, an extra scalar term is applied to each entry of the weighting function to increase the percentage of the disturbance rejection. This is shown in the selection of these weights below.

TABLE 3.9

H$_2$ Performance Function Specifications

Weighting Functions	Specifications
W$_1$	$w_{bt} = 0.008$ rad/s M$_t$ = 0.1 e$_t$ = 0.01
W$_2$	$w_{bc} = 1$ rad/s M$_u$ = 0.1 e$_u$ = 0.01
W$_3$	Uses a constant weighting
W$_4$	Uses a constant weighting

Selection of W$_1$. The weight **W$_1$** was used to shape the nominal output response of the closed-loop system. It was a low-pass filter with –20 dB at low frequency and a cut-off frequency at 0.008 rad/s. The constant weighing that was iteratively adjusted was used throughout the entry.

$$\mathbf{W_1} = \begin{bmatrix} 0.1 & 0 & 0 \\ 0 & 0.05 & 0 \\ 0 & 0 & 0.005 \end{bmatrix} \frac{s + 0.08}{0.01s + 0.008} \tag{3.74}$$

Selection of W$_2$. Again, a low-pass filter was used. This weight aims to shape the nominal closed-loop control effort and hence it is used to limit the maximum amplification between the disturbance and the controlled inputs. The weight was chosen such that the maximum disturbance amplification should not exceed 0.1. The cut-off frequency of **W$_2$** was adjusted iteratively to speed up the output time responses while maintaining the control effort within the desired limits.

$$\mathbf{W_2} = \begin{bmatrix} 10 & 0 & 0 \\ 0 & 7 & 0 \\ 0 & 0 & 12 \end{bmatrix} \frac{s + 0.08}{0.01s + 0.008} \tag{3.75}$$

Selection of W$_3$. **W$_3$** was a constant weighting function applied to all channels. Instead of using a high-pass filter to shape the sensitivity of the closed-loop system, a constant weighting was chosen to allow **W$_1$** to enhance the sensitivity as well as the nominal closed-loop system response.

$$\mathbf{W_3} = \begin{bmatrix} 0.08 & 0 & 0 \\ 0 & 0.8 & 0 \\ 0 & 0 & 0.08 \end{bmatrix} \tag{3.76}$$

Selection of W$_4$. A constant weighting function was again used to provide a certain weighting effect on the sensitivity, in particular the input and

closed-loop transfer functions. The gain applied to each entry was iteratively adjusted to reduce the effect of noise on the system.

$$\mathbf{W}_4 = \begin{bmatrix} 1 & 0 & 0 \\ 0 & 0.5 & 0 \\ 0 & 0 & 0.3 \end{bmatrix} \tag{3.77}$$

5. Minimize the H_∞ norm of the stacked cost functions to obtain a stabilizing
   controller K (H2LQG.m).
6. Simulate the H_2 design using the block diagram shown in Figure 3.41. All
   the weightings used have dimensions of 4×4, where the position $\{1,1\}$ is a
   unity weight for the first control loop.

```
% Design H_2 Controller
% W1
n11=0.1*[1 0.008/0.1]; d11=[0.01*1 0.008];
n12=0.05*[1 0.008/0.1];d12=[0.01*1 0.008];
n13=0.005*[1 0.08/0.1]; d13=[0.01*1 0.008];
[aw11,bw11,cw11,dw11]=tf2ss(n11,d11);
[aw12,bw12,cw12,dw12]=tf2ss(n12,d12);
[aw13,bw13,cw13,dw13]=tf2ss(n13,d13);
aw1=daug(aw11,aw12,aw13);
bw1=daug(bw11,bw12,bw13);
cw1=daug(cw11,cw12,cw13);
dw1=daug(dw11,dw12,dw13);
        % Using the same method for other weighting- W2
          to W4
        % Form Generalized Plant G
WW1=ss(aw1,bw1,cw1,dw1);
WW2=ss(aw2,bw2,cw2,dw2);
WW3=ss(aw3,bw3,cw3,dw3);
WW4=ss(aw4,bw4,cw4,dw4);
G33=sys_33;
GGG=[WW1*WW3 zeros(3) zeros(3) WW1*G33;
      zeros(3) zeros(3) zeros(3) WW2;
      -WW3 -WW4 eye(3) -G33];
        % Separate the input w and reference input r from
          control input u
  [AA,BB,CC,DD]=ssdata(GGG);
  BB1=BB(:,1:9);
  BB2=BB(:,10:12);
  CC1=CC(1:6,:);
  CC2=CC(7:9,:);
  DD11=DD(1:6,1:9);
  DD12=DD(1:6,10:12);
  DD21=DD(7:9,1:9);
  DD22=DD(7:9,10:12);
        % Form state space realization of Generalized Plant
  pp=pck(AA,[BB1 BB2],[CC1;CC2],[DD11 DD12;DD21 DD22]);
```

```
PH22 = rct2lti(mksys(AA,BB1,BB2,CC1,CC2,DD11,DD12,DD21,
DD22,'tss'));
     % H2-LQG control
[ss_cp,ss_cl]=h2lqg(PH22);
[Ah2,Bh2,Ch2,Dh2]=ssdata(ss_cp);
```

7. Analyze the resulting S, **KS**, and T of the system after applying the respective weighting functions. From the S plots below, the high sensitivity to input at low frequencies is improved after the weighting functions were used. The gain is well below 0 dB (= 1) margin over the frequencies (Figure 3.42).

The T has improved over the low and middle frequencies as reflected from the modified T. The system output is very small for any large input as seen from the small gains in the resulting T plots (Figure 3.43).

From the **KS** plots, it can be seen the effect of the input disturbances is quite negligible on the gasifier output. The **KG** after applying the shaping function is well below the margin of 0 dB (Figure 3.44).

```
% Analyze the resulting S, KS and T plots.
ss_cp2=ss(Ah2,Bh2,Ch2,Dh2);
clear sigma
ww=logspace(-8,2,200);
     % S=(1+GK)^-1
figure()
I_GK=eye(3)+G33*ss_cp2;
sv_I_GK=sigma(I_GK,ww,1);
loglog(ww,sv_I_GK);
title(' Sensitivity (S) ')
xlabel('Frequency (rad/s)');
ylabel('Gain')
axis([min(ww) max(ww) min(min(sv_I_GK))-0.01
max(max(sv_I_GK))+1])
     % KS=k(1+GK)^-1
KI_GK=ss_cp2*inv(eye(3)+G33*ss_cp2);
[sv_KI_GK]=sigma(KI_GK,ww);
figure()
loglog(ww,sv_KI_GK,'r-');
hold on
loglog(ww,1,':');
title(' Input Sensitivity (KS)')
xlabel('Frequency (rad/s)');
ylabel('Gain')
axis([min(ww) max(ww) min(min(sv_KI_GK))-0.01 max(max(sv_
KI_GK))+1])
     % T=(1+GK)^-1*GK
KI_GK=inv(eye(3)+G33*ss_cp2)*(G33*ss_cp2);
[sv_GKI_GK]=sigma(KI_GK,ww);
figure()
loglog(ww,sv_GKI_GK,'r-');
hold on
loglog(ww,1,':');
```

```
title(' Complementary Sensitivity (T)')
xlabel('Frequency (rad/s)');
ylabel('Gain')
axis([min(ww) max(ww) min(min(sv_GKI_GK))-0.01
max(max(sv_GKI_GK))+1])

        % original S of the system
G33_I=eye(3)+G33;
[sv_org]=sigma(G33_I,ww);
figure 8)
loglog(ww,sv_org,'r-');
hold on;
loglog(ww,1,':');
title(' Sensitivity (S)-original ')
xlabel('Frequency (rad/s)');
ylabel('Gain')
axis([min(ww) max(ww) min(min(sv_org))-0.01 max(max(sv_
org))+1])
hold off;

        % original T of the system-->(1+G)^-1*G
I_G=inv(eye(3)+G33)*G33;
[sv_IG]=sigma(I_G,ww);
figure 9)
loglog(ww,sv_IG,'r-');
hold on
loglog(ww,1,':');
title(' Complementary Sensitivity (T)-original')
xlabel('Frequency (rad/s)');
ylabel('Gain')
axis([min(ww) max(ww) min(min(sv_IG))-0.01 max(max(sv_
IG))+1])
hold off

        % W1*W3*S
WS=sigma(WW1*WW3*inv(I_GK),ww);
figure 10)
loglog(ww,WS);
hold on
loglog(ww,1,':')
title(' W1.W3.S ')
xlabel('Frequency (rad/s)');
ylabel('Gain')
axis([min(ww) max(ww) min(min(WS))-0.01 max(max(WS))+10])
hold off

        % W1*W4*T
WS=sigma(WW1*WW4*G33*ss_cp2*inv(I_GK),ww);
figure 11)
loglog(ww,WS);
hold on
loglog(ww,1,':')
title(' W1.W4.T ')
```

```
xlabel('Frequency (rad/s)');
ylabel('Gain')
axis([min(ww) max(ww) min(min(WS))-0.01 max(max(WS))+10])
hold off

       % W1*T
WS=sigma(WW1*G33*ss_cp2*inv(I_GK),ww);
figure 12)
loglog(ww,WS);
hold on
loglog(ww,1,':')
title(' W1.T ')
xlabel('Frequency (rad/s)');
ylabel('Gain')
axis([min(ww) max(ww) min(min(WS))-0.01 max(max(WS))+10])
hold off

       % W2*W3*KS
WS=sigma(WW2*WW3*ss_cp2*inv(I_GK),ww);
figure 13)
loglog(ww,WS);
hold on
loglog(ww,1,':')
title(' W2.W3.KS ')
xlabel('Frequency (rad/s)');
ylabel('Gain')
axis([min(ww) max(ww) min(min(WS))-0.01 max(max(WS))+10])
hold off

       % W2*W4*KS
WS=sigma(WW2*WW4*ss_cp2*inv(I_GK),ww);
figure 14)
loglog(ww,WS);
hold on
loglog(ww,1,':')
title(' W2.W4.KS ')
xlabel('Frequency (rad/s)');
ylabel('Gain')
axis([min(ww) max(ww) min(min(WS))-0.01 max(max(WS))+10])
hold off

       % W2*KS
WS=sigma(WW2*ss_cp2*inv(I_GK),ww);
figure 15)
loglog(ww,WS);
hold on
loglog(ww,1,':')
title(' W2.KS ')
xlabel('Frequency (rad/s)');
ylabel('Gain')
axis([min(ww) max(ww) min(min(WS))-0.01 max(max(WS))+10])
hold off
```

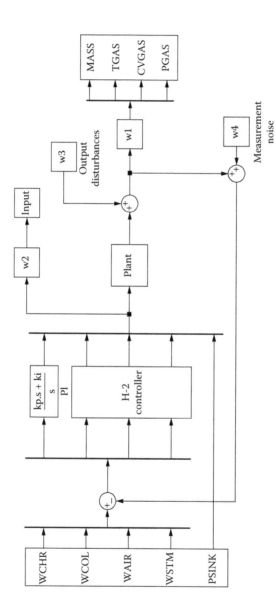

FIGURE 3.41 H$_2$ simulation block diagram.

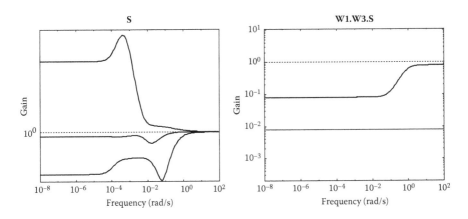

FIGURE 3.42 *S* plots for H$_2$ design.

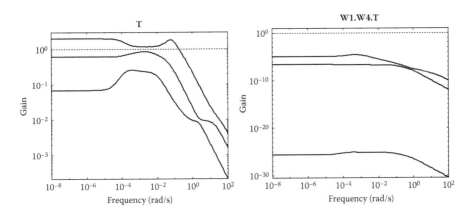

FIGURE 3.43 *T* plots for H$_2$ design.

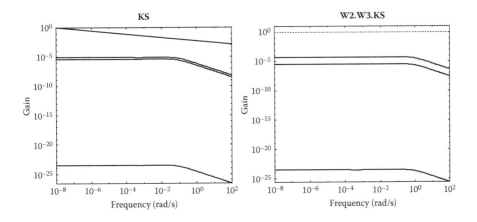

FIGURE 3.44 **KS** plots for H$_2$ design.

3.9.2 PERFORMANCE TESTS ON H_2 DESIGN

The performance tests for the H_∞ control scheme were performed using the same methods as those used in previous section. The results of the simulation and the performance test tables are shown in Appendix A5. For the H_2 control, a decrement response exists at the input WSTM for all load conditions. This is mainly due to the presence of a small negative term in the feed forward matrix, D. The input rates at all load conditions are very small as can be seen in the peak values.

3.10 COMPARISON OF CONTROLLERS

In this section, methods used to compare the various controllers designed are discussed. These methods involved the used of the sensitivity function, the robust stability of the system, the MIMO system asympototical stability and so on. These characteristics are each explained followed by their application to the gasifier system. Additionally, other controllers designed using (LQR, LQG, H_2 optimization, and H_∞ optimization design), but not presented here, which are used in this comparison can be found in Chin [36].

3.10.1 SENSITIVITY (S)

The sensitivity [48] of a system, denoted by $S(jw) = [I+GK(jw)]^{-1}$, is used to measure the effects of a disturbance at the input or output on the output of the feedback system. For a MIMO system, the maximum singular value of the sensitivity function evaluated over the frequencies concerned should be less than one;

$$\bar{\sigma}[S(jw)] \leq 1 \qquad (3.78)$$

The following two subsections consider the sensitivity measure of input and output disturbances on the feedback system.

1. **Output Sensitivity**. Insensitivity to output disturbances is essentially achieved if the system is strongly stable. If $G(s)$ is allowed to vary by $\Delta G(s)$ and the controller K remains unchanged, and the variation $\Delta R(s)$ in the closed-loop transfer function $R(s)$ is considered, then it can be shown that

$$\frac{\left\|\Delta R(jw)R(jw)^{-1}\right\|}{\left\|\Delta G(jw)G(jw)^{-1}\right\|} \leq \left\|[I+GK(jw)]^{-1}\right\| \qquad (3.79)$$

This is equivalent to $\underline{\sigma}[1+GK(jw)] \geq 1$, since the 2-norm of the inverse of a matrix is the smallest singular value of the matrix, and the left-hand side of (3.79) is equal to 1.
2. **Input Sensitivity**. Insensitivity to input disturbances, as shown in Figure 3.45, is achieved if $\bar{\sigma}[(I+GK)^{-1}G] \leq 1$. This is because the contribution at the plant output due to a disturbance input, d_u, is $y = (I+GK)^{-1}Gd_u$ assuming $r = d_y = 0$. Note that $(I+GK)^{-1}G = G(I+KG)^{-1}$.

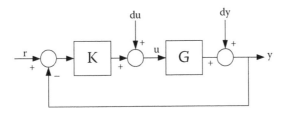

FIGURE 3.45 Typical feedback system.

So it can be seen that the sensitivity is small if $\underline{\sigma}[\mathbf{GK}]$ is large; that is, the fact that large loop gains (tight control) correspond to good performance is a well-known adage. There are, of course, limitations on the amount of gain that one can introduce into a system. Large gains correspond to large energy requirements that may breach the control specifications. So the amount of gain that can be applied without causing any problem can be gauged using the robust stability concept discussed in the next section.

3.10.2 ROBUST STABILITY (RS)

A control system is robust if it is insensitive to a difference between the actual system and the model of the system that is used to design the controller. The system is robustly stable [49] at the plant outputs if

$$\underline{\sigma}[\mathbf{I}+(\mathbf{GK})^{-1}] \geq 1 \tag{3.80}$$

Robust stability at the plant inputs is measured by simply swapping the G and K matrices, which becomes

$$\underline{\sigma}[\mathbf{I}+(\mathbf{KG})^{-1}] \geq 1 \tag{3.81}$$

The reason for using $\underline{\sigma}$ is the interpretation of the smallest singular value of a matrix as the distance between the matrix and the nearest singular matrix, since this is precisely the concept needed to determine the nearness of a stable transfer function to an unstable one. Hence, its use as a measure of stability robustness is natural.

3.10.3 MIMO SYSTEM ASYMPTOTIC STABILITY (MIMO AS)

Besides using the asymptotic stability concept on a SISO system, it can also be applied to a MIMO system. In general, for a MIMO system to be asymptotically stable, the minimum singular value of the return difference function should be greater than zero. Note that this is similar to the expression used for the sensitivity measure of the system. Therefore, if the system is stable, the system should be insensitive to parameter change and vice versa; that is,

$$\underline{\sigma}[\mathbf{I}+\mathbf{GK}(jw)] > 0 \tag{3.82}$$

3.10.4   NYQUIST TYPE CRITERION (NTC)

Using the Nyquist type of stability criteria, for a SISO system, a sufficient condition
for stability under perturbation is that

$$\left| \mathbf{GK}(jw) - \mathbf{G}_0\mathbf{K}(jw) \right| < \left| \mathbf{I} + \mathbf{G}_0\mathbf{K}(jw) \right| \quad w \in \Re \qquad (3.83)$$

where $\mathbf{G}_0\mathbf{K}(jw)$ and $\mathbf{GK}(jw)$ are the original and perturbed open-loop gain of the
plant, respectively. This can be easily extended to MIMO systems by considering the
singular values instead of the modulus as before, to give

$$\left\| \Delta\mathbf{GK}(jw) \right\|_\infty \le \left\| \mathbf{I} + \mathbf{GK}(jw) \right\|_\infty$$

$$\overline{\sigma}[\Delta\mathbf{GK}(jw)\overline{\sigma}[\mathbf{I} + \mathbf{GK}(jw)] \qquad (3.84)$$

If the condition above is satisfied, it implies that the gasifier system designed at
the 100% load condition (original plant), when operating at the 50% and 0% load
condition (perturbed plants), remains stable. Ideally, the maximum singular value of
the 50% and 0% load condition should be less than the maximum singular value of
the 100% load condition in order to satisfy (3.84).

3.10.5   INTERNAL STABILITY (IS)

In adition to considering the *input-output stability* of a feedback system using one
closed-loop TFM, which is assumed to have no RHP pole-zero cancellations between
the controller and the plant, the use of internal stability could be used.

A system is internally stable [35] if none of its components contain hidden unstable modes, and the injection of bounded external signals at any point in the system
results in bounded output signals measured anywhere in the system. Instead of performing these measurements at each point in the system, the equation relating these
inputs to the corresponding outputs can be established, as shown below. Consider the
SISO system in Figure 3.45, where signals are injected and measured at both locations between the two components, G and K. The system equations in matrix form
can be established as follows.

$$\begin{bmatrix} \mathbf{u} \\ \mathbf{y} \end{bmatrix} = \begin{bmatrix} (\mathbf{I}+\mathbf{KG})^{-1} & -\mathbf{K}(\mathbf{I}+\mathbf{GK})^{-1} \\ \mathbf{G}(\mathbf{I}+\mathbf{KG})^{-1} & (\mathbf{I}+\mathbf{GK})^{-1} \end{bmatrix} \begin{bmatrix} \mathbf{d}_\mathbf{u} \\ \mathbf{d}_\mathbf{y} \end{bmatrix} \qquad (3.85)$$

With further simplification, Equation (3.85) gives

$$\begin{bmatrix} \mathbf{u} \\ \mathbf{y} \end{bmatrix} = \begin{bmatrix} \mathbf{I} & -\mathbf{K} \\ \mathbf{G} & \mathbf{I} \end{bmatrix}^{-1} \begin{bmatrix} \mathbf{d}_\mathbf{u} \\ \mathbf{d}_\mathbf{y} \end{bmatrix}$$

$$= \mathbf{M(s)} \begin{bmatrix} \mathbf{d}_\mathbf{u} \\ \mathbf{d}_\mathbf{y} \end{bmatrix} \qquad (3.86)$$

By assuming that there are no RHP pole-zeros cancellations between G and K; that is, all RHP-poles in G and K are contained in the minimal realizations of $\mathbf{KG}$ or $\mathbf{GK}$; the feedback system in Figure 3.45 is internally stable if and only if one of the four closed-loop transfer function matrices in (3.85) is stable.

Hence, for an internally stable SISO system, the matrix $\mathbf{M(s)}$ has to be stable. In other words, $(\mathbf{I+GK})^{-1}$or $(\mathbf{I+KG})^{-1}$ has to be small in order not to amplify the input signal. For a MIMO system, the same concept is applied except that the maximum singular value (worst case measure) is used. This means that $\overline{\sigma}[(\mathbf{I+GK})^{-1}]<1$ or $\overline{\sigma}[(\mathbf{I+KG})^{-1}]<1$.

3.10.6   INSTANTANEOUS ERROR (ISE)

The concept of the sensitivity matrix relates the output errors due to the parameter variations in a feedback system to the output errors due to the parameter variations in a corresponding open-loop system. Using the integrated square of the error (ISE) as a performance index, the performance of the feedback system can be judged. The performance index using the ISE is defined as

$$\int_0^t \mathbf{e}^T(t)\mathbf{e}(t)dt \tag{3.87}$$

where $e(t)$ is either the error of the open-loop or the closed-loop system, and the time, t, is any positive finite quantity. In a practical case, a meaningful choice for t might be about four to five times the largest "time constant" of the system. For a feedback system to be better than a corresponding open-loop system, the following inequality must be satisfied:

$$\int_0^t \mathbf{e_c}^T(t)\mathbf{e_c}(t)dt < \int_0^t \mathbf{e_o^T}(t)\mathbf{e_o}(t)dt \tag{3.88}$$

where the subscript c and o refer to the closed-loop and open-loop system, respectively.

3.10.7   FINAL VALUE THEOREM (FVT)

As in classical control, the asymptotical error of the closed-loop system, $\mathbf{G_c}$ can be determined by using the Final Value Thorem. It is defined as:

$$\mathbf{e_\infty} = \lim_{t\to\infty}\mathbf{e}(t) = \lim_{\mathbf{s}\to 0}\mathbf{sE(s)}$$

$$= \lim_{\mathbf{s}\to 0}\mathbf{s}\,(\mathbf{I-G_c}) \tag{3.89}$$

$$= \lim_{\mathbf{s}\to 0}\mathbf{s\,S(s)}$$

Instead of evaluting this at $t \to \infty$, which is not possible in practice (only in theory), the author determined the final value of the error as time tended to be very large; that is, $[0, \infty)$.

3.10.8 CONTROLLER ORDER (CO)

The order of the resulting controller is also used to compare the various types of controller designs. This criteria is unlikely to favor the H_2 and H_∞ design techniques, which give high-order controllers. Usually, a lower-order controller is desired, since it is easier to implement and handle.

3.10.9 CONDITION NUMBER (CN)

The condition number was also applied to check the numerical conditioning of the closed-loop system after the feedback system is designed. The condition number, κ, for the closed-loop system should be smaller or equal to that of the open-loop system model used.

$$\kappa(\mathbf{G}_c) \leq \kappa(\mathbf{G}_o) \tag{3.90}$$

The subscripts c and o used in Equation (3.90) refer to the closed-loop and open-loop systems, respectively.

3.11 COMPARISON OF ALL CONTROLLERS

Table 3.10 gives a full overview of the criteria mentioned above. Note that this section is not targeted toward choosing the best controller for the gasifier system. Its purpose is simply to compare the various controllers designed using the criteria described above. A detailed explanation of the various controllers designed for the gasifier system using the criteria above is given in Chin [36]. This section gives the overview of the criteria used for controller comparison.

The dash indicates criteria that are satisfied over the frequency range from 10^{-8} to 100 rad/s, while the bold print refers to a particular condition that is unsatisfactory. The **O** and **I** refer to output and input, respectively. The value of the **ISE**, the **FVT**, and the condition number of the closed-loop configurations are tabulated, and it can be observed that the H_∞ optimization gives the *highest-order* controller which exceeds that of the eighteenth-order gasifier plant itself. Also, when compared to its counterpart determined using the H_2 optimization, or any other controller design method, the computational burden appears to be excessive. While some of the above criticisms may be valid in certain situations, it cannot be denied that the H_∞ approach fits robust control and stability like a glove.

The H_2 design, on the other hand, is numerically unsatisfactory, as can be seen in the large condition number. This is due to the presence of uncertainties, which means that the H_2 norm cannot serve as a suitable measure of goodness for robust control, although the norm used (quadratic performance criteria) is a more natural norm for system performance than that of the H_∞ design.

TABLE 3.10
Summary of the Criteria Used for Each Controller Design

	S		RS		MIMO				FVT	CO	CN	RHP Zeros
	O	I	O	I	AS	NTC	IS	ISE				
LQR	1	—	1	10	1	—	—	5×10^7	$\{-1, -733,\ 361, -14\}$	19	1.5×10^8	2
LQG	0.01	—	0.008	1×10^{-4}	0.01	—	—	1×10^5	$\{0, 47, 21, 2\}$	19	1×10^9	1
LQG/LTR	0.01	—	0.008	1×10^{-4}	0.01	—	—	9×10^4	$\{0, 0, 2, 20, 4\}$	19	3×10^9	2
H_∞	—	—	0.01	0.01	—	—	—	4×10^3	$\{0.006, -0.08, -1, 2\}$	19	3×10^9	4
h_2	—	—	0.01	0.01	—	—	—	2×10^6	$\{-3, 25, 18, 3\}$	19	4×10^{17}	8

The H_∞ design gives the smallest *ISE* as compared to the H_2 and other controller designs. For the LQG and LQG/LTR design, *robust stability* at the gasifier input within the bandwidth ($w_B = 0.005$ rad/s) of interest could be met. Therefore, these design are robustly stable for both plant output and input disturbances, when operating at the 50% and 0% load condition. All controllers designed seem to satisfy the *internal stability* criteria. Hence, any signal injected at any point in the closed-loop system of the gasifier plant would result in a stable or bounded output at any other point.

The *Nyquist type criterion* indicates that most controller designs are quite robustly stable except for the LQR and even H_2 design. This was observed in the simulation results for the LQR and H_2 designs, which show some violations in the input values and their rate limits.

The *input and output sensitivity* are met for all controller designs. This shows that the feedback system designed using any of the controllers is insensitive to a disturbance input, such as a step or sinusoidal function, and the disturbance outputs such as noise incurred during the measurement process.

With these comments, this section is concluded with a *ranking* of the controllers based on the criteria above. This is shown in Table 3.11 below. Note that this ranking is only with respect to the gasifier system. As observed, the LQR design is ranked quite low due to the violation of the robustness test. On the other hand, the H_2 design which uses a more natural norm such as a quadratic performance function, or H_2 norm, should be able to fair better than the H_∞ design. However, it is ranked the last due its inability to meet some of the constraints in the performance specifications as well as the higher-order, closed-loop system and controller obtained from its formulation. In addition, the high condition number obtained from this design is taken into account as well.

The H_∞ design is ranked after the LQG and LQG/LTR due to its high-order controller and plant resulting from its formulation. Merits such as the ability to produce a stabilized controller and a good measure on the robustness aspects are taken into account. The H_∞ controller complexity is deemed to be less attractive as compared with its robust competitor the LQG/LTR, which is simpler to implement in practice and gives moderately good robustness margins.

Lastly, a graphical user interface (GUI) in MATLAB was used. Why use a GUI in MATLAB? The main reason GUIs are used is because it makes things simple for the end-users of the program. If GUIs were not used, users have to work

---

TABLE 3.11

Ranking of the Controllers

Type of Controllers
LQG/LTR and LQG
H_∞
LQR and H_2

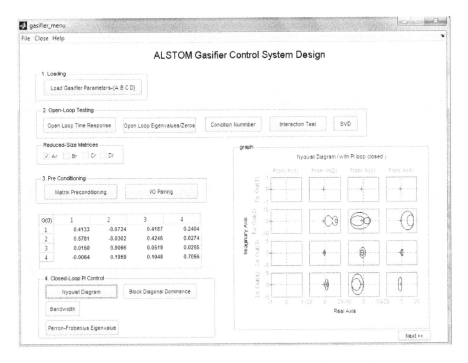

FIGURE 3.46 The first page in the GUI for the ALSTOM gasifier control system design.

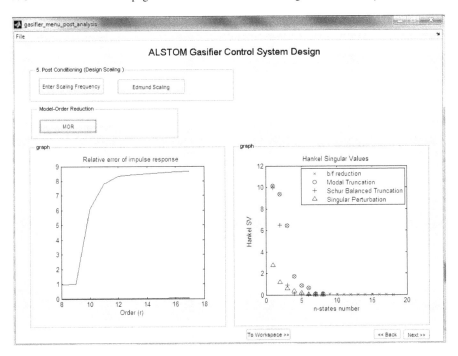

FIGURE 3.47 The second page in the GUI for the ALSTOM gasifier control system design.

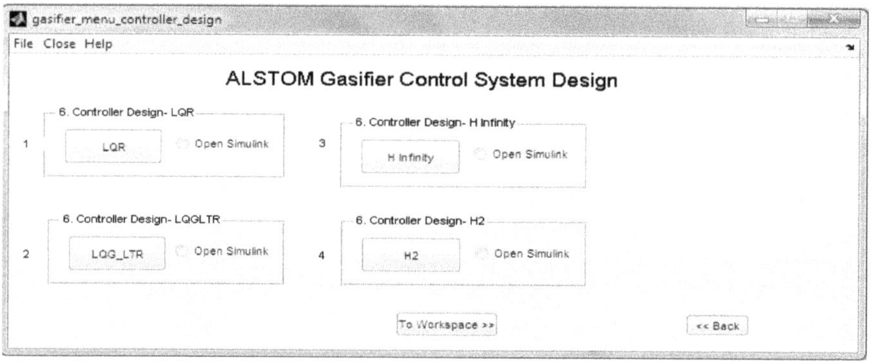

FIGURE 3.48   The third page in the GUI for the ALSTOM gasifier control system design.

from the command line interface or run multiple m-files, which can be extremely frustrating and mistakes may occur such as running a wrong file. A few GUIs for the ALSTOM gasifier control system design are shown in Figures 3.46 through 3.48. It allows the users to click on the buttons and display the results as shown in the previous sections.

4 Modeling of a Remotely Operated Vehicle

4.1 BACKGROUND OF THE URV

In the last four decades, the use of the URV (underwater robotic vehicle) has experienced tremendous growth. Many are used for underwater inspection of subsea cables and oil and gas installations like Christmas trees, structures, and pipelines. They are essential at depths where the use of human divers is impractical. The deployments of URVs present several difficulties, as they are difficult to control remotely and autonomously. Besides, the control systems design is quite challenging, as the vehicle dynamics are highly nonlinear, uncertain and vulnerable to unknown disturbances.

Traditionally, URVs can be broadly classified as a remotely operated vehicle (ROV) and an autonomous underwater vehicle (AUV), depending on their designed tasks and modes of operations. A new class of URV called the Hybrid ROV-URV (HROV), which has both the capabilities of a ROV and an AUV, has been proposed and experimented with. Nevertheless, ROVs are best suited for work that involves operating from a stationary point or cruising at relatively slow speeds such as pipeline inspection. Furthermore, for any tasks involving manipulation and requiring maneuverability, they are the most cost-effective platform. The following shows some of the common URVs present in the literature.

For example, the ROV JASON II/MEDEA system is designed by the Woods Hole Oceanographic Institution's Deep Submergence Laboratory (DSL) for scientific investigation of the deep ocean and seafloor. The ROV system includes tether management by the MEDEA vehicle that decouples Jason from wave motion. SIRENE is an AUV designed and built by the French Research Institute for Exploration of the Sea (IFREMER). It has an open frame structure with a dry weight of 4000 kg and a maximum depth of 4000 m. It is used for environmental survey and monitoring. It is equipped with two back thrusters for surge and yaw motion control in the horizontal plane, and one vertical thruster for heave control. SWIMMER-PHENIX is a HROV developed by IFREMER in collaboration with a French partner, CYBERNETIX. It consists of the SWIMMER AUV, which carries and delivers a standard work-class ROV PHENIX from the surface to a subsea docking station installed close to the equipment. The subsea docking station is connected to the surface production facilities through a preinstalled power or control umbilical. The ROV has a short 200 m umbilical that is wound round a self-contained winch mounted on the SWIMMER. Generally, the focus of URV research has been centered on an AUV or HROV, but recent experiences with commercial AUVs have highlighted some of their limitations—lost AUVs and their lack of real-time communication with the surface vessels. In addition, current AUV designs have not been able to replace ROVs for nearly four decades, in many operations such as pipeline inspection for the oil and gas industry, where the ROV hardwire link is still required.

4.2 BASIC DESIGN OF A ROV AND TASKS UNDERTAKEN

In this section, the basic design of the ROV and its task are shown. The ROV designed by Robotics Research Centre (RRC) in Nanyang Technological University (NTU) is known as *RRC ROV* (see Figure 4.1). It is used to perform underwater pipeline inspections such as locating pipe leakages or cracks. Since pipelines carry hydrocarbons such as natural gas or oil at relatively high pressures, a crack in the line can result in an explosion as well as the accidental discharge of oil and/or gas with adverse environmental consequences. It is therefore desirable to inspect the pipelines periodically and to perform the tasks efficiently. This can be done with the operator focusing on inspection while the ROV automatically tracks the pipeline.

The twin *eye-ball* ROV depicted in Figure 4.1 has an open-frame structure and is 1 m long, 0.9 m wide, and 0.9 m high. It has a dry weight of 115 kg and a current operating depth of 100 m. Its designed tasks include inspections of underwater pipelines. The RRC ROV is underactuated as it has four thruster inputs for six degrees of freedom (DOFs) (that is, surge, sway, heave, roll, pitch, and yaw velocity) with a high degree of cross coupling between them. The vehicle is equipped with two lateral thrusters mainly for surge, sway and yaw motion control in the horizontal plane, two vertical thrusters for heave in the vertical plane and a suite of sensors for position and velocity measurements (see Table 4.1). Roll and pitch motions are passive as the metacentric height (that is, the distance between the center of gravity of a ROV and its metacenter) is sufficiently large to provide adequate static stability. A brief description of the component layout of the RRC ROV is given.

1. Four thrusters, each providing up to 70 N of thrust
2. Two cylindrical floats with four balancing steel weights

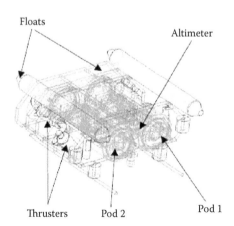

FIGURE 4.1   Schematic of RRC ROV.

TABLE 4.1

Sensors Used in RRC ROV

Sensors	Purpose	Outputs	Functioning	Locations
2" Micro-CTD	Depth	z	Yes	External
Magnetic Compass-TCM2	Heading	ψ	Yes	Pod 2
Argonaut DVL	Velocity	u, v, w	Yes	External
Tritech PA500 - Digital Precision Altimeters	Distance	x	Yes	External

3. Main pod (Pod 1) and sensors with navigational pod (Pod 2)
4. Two halogen lamps and an external sensor such as an altimeter (see Table 4.1).

The predefined underwater pipeline in subsea operation is shown in Figure 4.2. In an actual vehicle deployment, the predefined pipeline locations are usually provided by the company and the vehicle uses external cameras onboard the ROV and a navigation system to track it.

With the operational tasks defined, a system and control design can be undertaken. Upon completing the preliminary design, MATLAB scripting language and graphical block diagram model, Simulink, together with Virtual Reality Modeling Language (VRML) visualization were used to model the ROV dynamics, to refine controller designs, and to examine the feasibility of the overall control system design through execution of an underwater pipeline tracking mission given in Figure 4.3a. The Simulink model consists of the following: (a) pipeline tracking planner; (b) navigation sensors; (c) control system design, and (d) ROV dynamic block diagram.

The *pipeline-tracking planner* is the high level decision-maker onboard the ROV as shown in Figure 4.3b. It monitors all system states and issue commands to the controllers. The planner observes the states as well as flags from simulated emergency events such as collision avoidance (not included in the simulation) and system commands. Mainly, the planner decides the locations and the speed to go. During the simulation trials, the planner was set to command the ROV to descend to certain depth and then move to the first location of the underwater pipeline in a horizontal and vertical plane manner. The planner also issues a flag, which stops the simulation

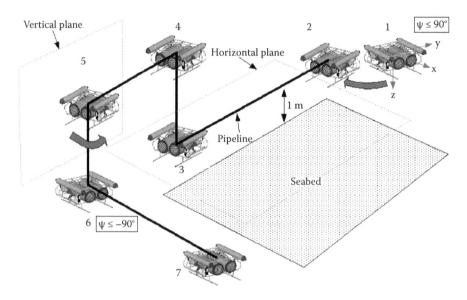

FIGURE 4.2   Proposed underwater pipeline tracking profile for RRC ROV.

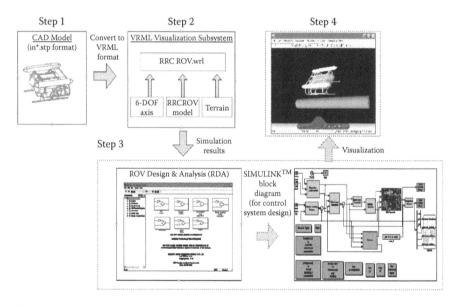

FIGURE 4.3a   RDA with VRML visualization integration and flowchart for ROV simulation.

when the ROV completes the designated pipeline-tracking mission. Then, the ROV ascends to the surface that is within sufficient capture range.

The *navigation sensor* block simulates the navigation system consisting of external sensors for estimating the location and pose of the ROV with respect to its body-fixed frame. On the other hand, Euler's transformation maps the body-fixed frame to

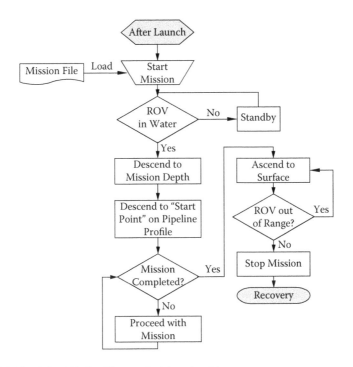

FIGURE 4.3b ROV Mission Planner decision algorithm.

the earth-fixed frame. From this block, the simulated system states are output to the pipeline tracking planner and controllers.

The six-DOFs *ROV dynamic model* is identified as the *ROV Dynamic*. It is described by a set of differential equations derived using Newton's second law of motion. Modeling the dynamic equations of the ROV is usually the first step in developing an accurate simulation. To create a more realistic model, perturbations due to uncertain hydrodynamic forces such as damping and added mass on the ROV are included in the ROV model. These values were numerically determined by a hydrodynamic software package such as Wave Analysis MIT (WAMIT) and ANSYS Computation Fluid Dynamic (CFD) package.

The *controller design* is proposed to enable the ROV to achieve the given operating specifications. For example, a control algorithm that can be used in the RRC ROV is shown in Figure 4.4.

4.3 NEED FOR ROV CONTROL

Similar to other marine vehicles such as a ship [50–51] and the AUV [52–53], remotely operated vehicle (ROV) [54–56] dynamics are nonlinear, highly coupled in motion and susceptible to hydrodynamic uncertainties. In addition, the RRC ROV is underactuated (i.e., the number of DOFs exceeds the number of actuators). The performance can suffer significantly due to lack of thrusters for the unactuated DOFs when the vehicle is performing a pipeline inspection that involves both station keeping and pipeline tracking at points on the pipeline. This deteriorating performance can be observed when the ROV

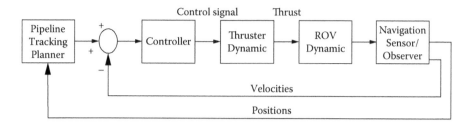

FIGURE 4.4   Block diagram of a control system.

is operating in the open-loop condition. This causes the ROV to exhibit a whirling motion whereby the linear velocities (and its corresponding linear positions) are not asymptotically stable during stabilization. The problem of whirling ROV motion is due to the added mass components in the Coriolis and centripetal term creating a cross-coupling effect on the ROV's motion—the Munk moments [57]—and arises from the change in the directions of the fluid as the ROV moves with small pitch and roll angles.

Hence, a robust control system to control both the velocities and position is required. There are many control strategies adopted for the ROV control; just to name a few, an example of simultaneous stabilization and tracking of ROV (the ODIN ROV) using output feedback with backstepping method is given [58–59]. However, a restriction of the ROV velocity [58–59] can be conservative and the controller design becomes complex, as a solution to partial differential equations are required in the derivation of the control law. Moreover, their approach solves the problem in the two-dimensional space only. The measurement noises in the output measurements and the quadratic cross terms of the unmeasured velocities appearing in the Coriolis and centripetal matrix make the estimation of the velocities complicated. To avoid this, a coordinate transformation [58–59] is used to cancel the quadratic cross terms to simply design of the observer. Motivated by the above nonlinear dynamic, a robust system for both tracking and stabilization under the hydrodynamic uncertainties on the ROV operating in two and higher dimensional spaces is required. As shown in the subsequent chapters on control systems design, few controllers were designed for comparisons. However, the dynamic equations of the ROV need to be derived.

In conjunction with the modeling, computer numerical models were developed. The computer models for the ROV were developed in MATLAB/Simulink and compared with experimental and computational fluid dynamics results. The models are used to determine the value of design parameters and to test the control algorithms before actual implementation.

4.4   DYNAMIC EQUATION USING THE NEWTONIAN METHOD

The following assumptions can be made when deriving the general ROV dynamic equation in order to simplify the effort in modeling. They are namely:

a. An ROV is a rigid body and is fully submerged once in water
b. Water is assumed to be the ideal fluid that is incompressible, inviscid (frictionless) and irrotational

c. An ROV is slow moving for operations such as pipeline inspection

d. The earth-fixed frame of reference is inertial

e. Disturbance due to wave is neglected as it is fully submerged

f. Tether dynamics attached to the ROV are not modeled.

The standard Society of Naval Architects & Marine Engineers (SNAME) notations used for the marine vehicle are shown in Table 4.2. A rigid body ROV dynamic equation is commonly expressed in the body-fixed frame since the control forces and measurement devices are easily and intuitively related to this body frame of reference. Using the Newtonian approach, the motion of a rigid body with respect to the body-fixed reference frame at the origin (see Figure 4.5) is given by the following set of equations [60]:

$$\mathbf{M}_{mass}[\dot{\mathbf{v}}_1 + \mathbf{v}_2 \times \mathbf{v}_1 + \dot{\mathbf{v}}_2 \times \mathbf{r}_G + \mathbf{v}_2 \times (\mathbf{v}_2 \times \mathbf{r}_G)] = \tau_1 \qquad (4.1)$$

TABLE 4.2
Notations Used in ROV

DOF	Motion Descriptions	Positions and Orientations	Linear and Angular Velocities
1	Motions in the x- direction (surge)	X	u
2	Motions in the y- direction (sway)	y	v
3	Motions in the z- direction (heave)	z	w
4	Rotations about the x- axis (roll)	φ	p
5	Rotations about the y- axis (pitch)	θ	q
6	Rotations about the z- axis (yaw)	ψ	r

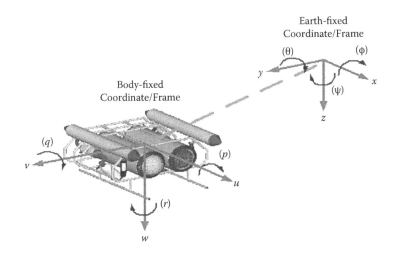

FIGURE 4.5   Coordinate systems used in ROV.

$$\mathbf{I}\,\dot{\mathbf{v}}_2 + \mathbf{v}_2 \times (\mathbf{I}\ \mathbf{v}_2) + \mathbf{M}_{mass}\mathbf{r}_G \times (\dot{\mathbf{v}}_1 + \mathbf{v}_2 \times \mathbf{v}_1) = \tau_2 \qquad (4.2)$$

where $\mathbf{r}_G = [x_G\ y_G\ z_G]^T$ is the location of the center of gravity, $\tau_1 \in \mathfrak{R}^3$ and $\tau_2 \in \mathfrak{R}^3$ are the external force and moment vector, respectively; $\mathbf{v}_1 = [u\ v\ w]^T \in \mathfrak{R}^3$ is the linear velocity vector and $\mathbf{v}_2 = [p\ q\ r]^T \in \mathfrak{R}^3$ is the angular velocity vector. $M_{mass} \in \mathfrak{R}^{3\times3}$ is the ROV mass matrix:

$$\mathbf{M}_{mass} = \begin{bmatrix} m & 0 & 0 \\ 0 & m & 0 \\ 0 & 0 & m \end{bmatrix} = m\mathbf{I}_{3\times3} \qquad (4.3a)$$

$\mathbf{I}_{3\times3}$ is the identity matrix and $\mathbf{I} \in \mathfrak{R}^{3\times3}$ is the constant inertia tensor:

$$\mathbf{I} = \begin{bmatrix} I_x & -I_{xy} & -I_{xz} \\ -I_{yx} & I_y & -I_{yz} \\ -I_{zx} & -I_{zy} & I_z \end{bmatrix}; \quad \mathbf{I} = \mathbf{I}^T > 0, \quad \dot{\mathbf{I}} = 0 \qquad (4.3b)$$

I_x, I_y and I_z are the moments of inertia about the X, Y and Z axes, respectively. As $\mathbf{I} = \mathbf{I}^T > 0$, the cross products of inertia $I_{xy} = I_{yx}$, $I_{xz} = I_{zx}$ and $I_{yz} = I_{zy}$.

The rigid body equation consisting of the inertia forces and the Coriolis and centrifugal forces can be expressed in matrix form:

$$\mathbf{M}_{RB}\dot{\mathbf{v}} + \mathbf{C}_{RB}(\mathbf{v}) = \tau \qquad (4.4)$$

where $\mathbf{M}_{RB} \in \mathfrak{R}^{6\times6}$ is the mass-inertia matrix, $\mathbf{C}_{RB}(\mathbf{v}) \in \mathfrak{R}^{6\times6}$ is the Coriolis and centripetal matrix, $\tau = [\tau_1\ \tau_2]^T \in \mathfrak{R}^6$ is a vector of external forces and moments and $\mathbf{v} = [\mathbf{v}_1\ \mathbf{v}_2]^T \in \mathfrak{R}^6$ is the linear and angular velocity vector.

The rigid body inertia matrix $\mathbf{M}_{RB}$ can be uniquely determined from (4.1) and (4.2) as:

$$\mathbf{M}_{RB}\dot{\mathbf{v}} = \begin{bmatrix} \mathbf{M}_{mass}\dot{\mathbf{v}}_1 + m\dot{\mathbf{v}}_2 \times \mathbf{r}_G \\ \mathbf{I}\dot{\mathbf{v}}_2 + m\mathbf{r}_G \times \dot{\mathbf{v}}_1 \end{bmatrix} \qquad (4.5)$$

From the above, the positive $\mathbf{M}_{RB} = \mathbf{M}_{RB}^T > 0$ can be separated and defined as:

$$\mathbf{M}_{RB} = \begin{bmatrix} \mathbf{M}_{mass} & -m\mathbf{S}(\mathbf{r}_G) \\ m\mathbf{S}(\mathbf{r}_G) & \mathbf{I} \end{bmatrix} \qquad (4.6)$$

where $\mathbf{S}(\mathbf{r_G}) = \begin{bmatrix} 0 & -z_G & y_G \\ z_G & 0 & -x_G \\ -y_G & x_G & 0 \end{bmatrix}$. Expanding the individual terms in the inertia matrix in (4.6) gives:

$$\mathbf{M_{RB}} = \begin{bmatrix} m & 0 & 0 & 0 & mz_G & -my_G \\ 0 & m & 0 & -mz_G & 0 & mx_G \\ 0 & 0 & m & my_G & -mx_G & 0 \\ 0 & -mz_G & my_G & I_x & -I_{xy} & -I_{xz} \\ mz_G & 0 & -mx_G & -I_{yx} & I_y & -I_{yz} \\ -my_G & mx_G & 0 & -I_{zx} & -I_{zy} & I_z \end{bmatrix} \qquad (4.7)$$

Similarly, the Coriolis and centripetal terms, describing the angular motion of the ROV in (4.1) and (4.2), can be rewritten as:

$$\mathbf{C_{RB}}(\mathbf{v})\mathbf{v} = \begin{bmatrix} m[\mathbf{v_2} \times \mathbf{v_1} + \mathbf{v_2} \times (\mathbf{v_2} \times \mathbf{r_G})] \\ \mathbf{v_2} \times (\mathbf{Iv_2}) + m\mathbf{r_G} \times (\mathbf{v_2} \times \mathbf{v_1}) \end{bmatrix} \qquad (4.8)$$

The Coriolis and centripetal matrix created using Simulink is shown in Figure 4.6. Using the results $\mathbf{a} \times \mathbf{b} \times \mathbf{c} = \mathbf{S}(\mathbf{a})\mathbf{S}(\mathbf{b})\mathbf{c}$ and $\mathbf{S}(\mathbf{v_1})\mathbf{v_1} = 0$, (4.8) can be manipulated to give:

$$\mathbf{C_{RB}}(\mathbf{v})\mathbf{v} = \begin{bmatrix} -m\mathbf{S}(\mathbf{v_1})\mathbf{v_2} - m\mathbf{S}(\mathbf{v_2})\mathbf{S}(\mathbf{r_G})\mathbf{v_2} \\ -m\mathbf{S}(\mathbf{v_1})\mathbf{v_1} + m\mathbf{S}(\mathbf{r_G})\mathbf{S}(\mathbf{v_2})\mathbf{v_1} - \mathbf{S}(\mathbf{Iv_2})\mathbf{v_2} \end{bmatrix} \qquad (4.9)$$

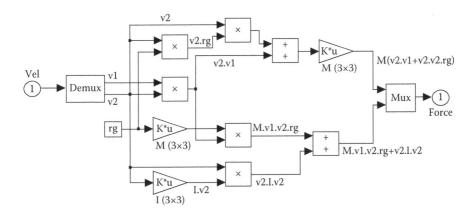

FIGURE 4.6 Coriolis and centripetal matrix in Simulink.

Separating $\mathbf{v}_1$ and $\mathbf{v}_2$ yields:

$$\mathbf{C}_{RB}(\mathbf{v}) = \begin{bmatrix} \mathbf{0}_{3\times 3} & -m\mathbf{S}(\mathbf{v}_1) - m\mathbf{S}(\mathbf{v}_2)\mathbf{S}(\mathbf{r}_G) \\ -m\mathbf{S}(\mathbf{v}_1) + m\mathbf{S}(\mathbf{r}_G)\mathbf{S}(\mathbf{v}_2) & -\mathbf{S}(\mathbf{I}\mathbf{v}_2) \end{bmatrix} \quad (4.10)$$

Expanding the terms on (4.10) gives:

$$\mathbf{C}_{RB}(\mathbf{v}) = \begin{bmatrix} \mathbf{0}_{3\times 3} & \mathbf{C}_{12}(\mathbf{v}) \\ -\mathbf{C}_{12}^{T}(\mathbf{v}) & \mathbf{C}_{22}(\mathbf{v}) \end{bmatrix} \quad (4.11)$$

with

$$\mathbf{C}_{12}(\mathbf{v}) = \begin{bmatrix} m(y_G q + z_G r) & -m(x_G q - w) & -m(x_G r + v) \\ -m(y_G p + w) & m(z_G r + x_G p) & -m(y_G r - u) \\ -m(z_G p - v) & -m(z_G q + u) & m(x_G p + y_G q) \end{bmatrix} \quad (4.12)$$

$$\mathbf{C}_{22}(\mathbf{v}) = \begin{bmatrix} 0 & -I_{yz}q - I_{xz}p + I_z r & I_{yz}r + I_{xy}p - I_y q \\ I_{yz}q + I_{xz}p - I_z r & 0 & -I_{xz}r - I_{xy}q + I_x p \\ -I_{yz}r - I_{xy}p + I_y q & I_{xz}r + I_{xy}q - I_x p & 0 \end{bmatrix} \quad (4.13)$$

In (4.4), the external force and moment vector τ includes the hydrodynamic forces and moments due to damping and inertia of surrounding fluid known as *added mass*, and the *restoring force* and *moment*. These forces and moments tend to oppose the motion of the ROV and, other than the restoring force, are dependent on the velocities and accelerations of the vehicle and are hence expressed in the body-fixed frame. The open-loop nonlinear ROV dynamic equations can be expressed according to:

$$\tau = \tau_A + \tau_H = \tau_A - \mathbf{M}_A - \mathbf{C}_A(\mathbf{v}) - \mathbf{D}(\mathbf{v})\mathbf{v} - \mathbf{G}_f(\eta_2)$$

$$\mathbf{M}\dot{\mathbf{v}} + \mathbf{C}(\mathbf{v})\mathbf{v} + \mathbf{D}(\mathbf{v})\mathbf{v} + \mathbf{G}_f(\eta_2) = \tau_A \quad (4.14)$$

where $\mathbf{v} = [u\ v\ w\ p\ q\ r]^T$ is the body-fixed velocity vector and $\eta [\eta_1\ \eta_2]^T$ is the earth-fixed vector, comprising the position vector $\eta_1 = [x\ y\ z]^T \in \mathfrak{R}^3$ and the orientation vector of Euler angles, $\eta = [\phi\ \theta\ \psi]^T \in S^3$ (a torus of three dimensions); the angles are defined on the interval $[0, 2\pi)$. $\mathbf{M} = \mathbf{M}_{RB} + \mathbf{M}_A \in \mathfrak{R}^{6\times 6}$ is the sum of the rigid body inertia mass and added fluid inertia mass matrix, $\mathbf{C}(\mathbf{v}) = \mathbf{C}_{RB}(\mathbf{v}) + \mathbf{C}_A(\mathbf{v}) \in \mathfrak{R}^{6\times 6}$ is the sum of Coriolis and centripetal and the added mass forces and moments matrix, and $\mathbf{D}(\mathbf{v}) \in \mathfrak{R}^{6\times 6}$ is the damping matrix due to the surrounding fluid. The input force and moment vector $\tau_A = \mathbf{T}\mathbf{u} \in \mathfrak{R}^6$ relates the thrust output vector $\mathbf{u} = \mathbf{F}_T\bar{\mathbf{u}} \in \mathfrak{R}^4$

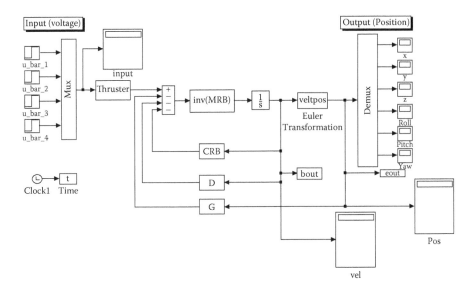

FIGURE 4.7   Overall view of ROV model in Simulink.

with the thruster configuration matrix $\mathbf{T} \in \Re^{6 \times 4}$, defined in a later section; $\mathbf{F_T} \in \Re^{4 \times 4}$ is the dynamics of each thruster and converts the input voltage command $\mathbf{\bar{u}} \in \Re^4$ into thrusts to propel the vehicle as seen in Section 4.6.5.

The following are the assumptions used in the RRC ROV modeling. The ROV is designed to move very slowly and under the assumption of an ideal fluid that is a nonviscous liquid and the hydrodynamic forces are determined under the assumption of potential flow. Main fluid-body interactions are the hydrodynamic forces like added mass and damping. As the vehicle is operating in a least coupled in motion, the linear and diagonal hydrodynamic damping is used. Besides, with an additional balancing mass, the ROV is made to be neutrally buoyant with the X-Y coordinate of the Center of Gravity (CG) coinciding with the Center of Buoyancy (CB). The overall view of the ROV model created in Simulink is shown in Figure 4.7.

4.5   KINEMATICS EQUATIONS AND EARTH-FIXED FRAME EQUATION

The transformation using the Euler angles provides an important transformation between the dynamics expressed in the body-fixed frame in (4.14) to the earth-fixed frame as seen in Figure 4.5. As the accelerations of a point on the surface of the Earth can be neglected for slow-moving marine vehicles, the earth-fixed frame can be considered to be an inertial frame. The kinematics equations [60] that represent Euler's transformation can be written as:

$$\dot{\eta} = \mathbf{J}(\eta_2)\mathbf{v} \tag{4.15}$$

Alternatively, Euler's transformation between the body and earth-fixed frames $(\boldsymbol{\eta}, \mathbf{v})$ to $(\boldsymbol{\eta}, \dot{\boldsymbol{\eta}})$ can be written as:

$$
\begin{bmatrix} \boldsymbol{\eta} \\ \dot{\boldsymbol{\eta}} \end{bmatrix} = \begin{bmatrix} \mathbf{I}_{6\times 6} & \mathbf{0}_{6\times 6} \\ \mathbf{0}_{6\times 6} & \mathbf{J}(\boldsymbol{\eta}_2) \end{bmatrix} \begin{bmatrix} \boldsymbol{\eta} \\ \mathbf{v} \end{bmatrix}
\tag{4.16}
$$

where the Euler transformation matrix, $\mathbf{J}(\boldsymbol{\eta}_2)$ is derived by successive rotation of the Euler angles $(\boldsymbol{\eta}_2 = [\phi \ \theta \ \psi]^T)$ about Z, Y and X axes to give:

$$
\mathbf{J}(\boldsymbol{\eta}_2) = \begin{bmatrix} \mathbf{J}_1(\boldsymbol{\eta}_2) & 0 \\ 0 & \mathbf{J}_2(\boldsymbol{\eta}_2) \end{bmatrix}
\tag{4.17}
$$

where

$$
\mathbf{J}_1(\boldsymbol{\eta}_2) = \begin{bmatrix} c(\psi)c(\theta) & -s(\psi)c(\phi)+c(\psi)s(\theta)s(\phi) & s(\psi)s(\phi)+c(\psi)c(\phi)s(\theta) \\ s(\psi)c(\theta) & c(\psi)c(\phi)+s(\phi)s(\theta)s(\psi) & -c(\psi)s(\phi)+s(\theta)s(\psi)c(\theta) \\ -s(\theta) & c(\theta)s(\phi) & c(\theta)c(\phi) \end{bmatrix}
$$

$$
\mathbf{J}_2(\boldsymbol{\eta}_2) = \begin{bmatrix} 1 & s(\phi)t(\theta) & c(\phi)t(\theta) \\ 0 & s(\phi) & -s(\phi) \\ 0 & \dfrac{s(\phi)}{c(\theta)} & \dfrac{c(\phi)}{c(\theta)} \end{bmatrix}
$$

and $s = \sin(.)$, $c = \cos(.)$, $t = \tan(.)$. The transformation is undefined for $\theta = \pm 90°$. To overcome this singularity, a quaternion approach must be considered. However, for the RRC ROV, this problem does not exist because the vehicle is neither designed nor required to operate at $\theta = \pm 90°$. The Euler transformation matrix in Simulink is shown in Figure 4.8.

Differentiating (4.16) and rearranging the equation gives:

$$
\begin{bmatrix} \mathbf{v} \\ \ddot{\boldsymbol{\eta}} \end{bmatrix} = \begin{bmatrix} \mathbf{J}^{-1}(\boldsymbol{\eta}_2) & \mathbf{0}_{6\times 6} \\ \dot{\mathbf{J}}(\boldsymbol{\eta}_2)\mathbf{J}^{-1}(\boldsymbol{\eta}_2) & \mathbf{J}(\boldsymbol{\eta}_2) \end{bmatrix} \begin{bmatrix} \dot{\boldsymbol{\eta}} \\ \dot{\mathbf{v}} \end{bmatrix}
\tag{4.18}
$$

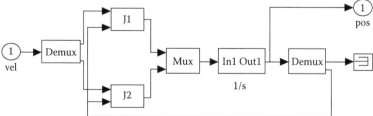

FIGURE 4.8a Euler's transformation matrix in Simulink.

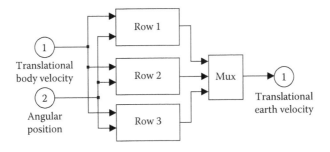

FIGURE 4.8b $J_1(\eta_2)$ matrix in Simulink.

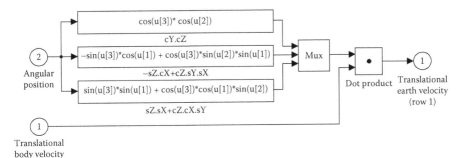

FIGURE 4.8c Row 1 of $J_1(\eta_2)$ matrix in Simulink.

where the body-fixed dynamic can be transformed into the earth-fixed frame (or vice versa) by substituting the $\mathbf{v}$ and $\dot{\mathbf{v}}$ as:

$$\mathbf{v} = \mathbf{J}^{-1}(\eta)\dot{\eta} \tag{4.19}$$

$$\dot{\mathbf{v}} = \mathbf{J}^{-1}(\eta)\ddot{\eta} - \mathbf{J}^{-2}(\eta)\dot{\mathbf{J}}(\eta)\dot{\eta}$$
$$= \mathbf{J}^{-1}(\eta)[\ddot{\eta} - \dot{\mathbf{J}}(\eta)\mathbf{J}^{-1}(\eta)\dot{\eta}] \tag{4.20}$$

into the body-fixed dynamic in (4.14), and this gives the ROV's dynamic in the earth-fixed frame:

$$\mathbf{M}_\eta(\eta)\ddot{\eta} + \mathbf{C}_\eta(\mathbf{v},\eta)\dot{\eta} + \mathbf{D}_\eta(\mathbf{v},\eta)\dot{\eta} + \mathbf{G}_\eta(\eta) = \tau_\eta \tag{4.21}$$

where
$$\mathbf{M}_\eta(\eta) = \mathbf{J}^{-T}(\eta)\mathbf{M}\mathbf{J}^{-1}(\eta)$$
$$\mathbf{C}_\eta(\mathbf{v},\eta) = \mathbf{J}^{-T}(\eta)[\mathbf{C}(\mathbf{v}) - \mathbf{M}\mathbf{J}^{-1}(\eta)\dot{\mathbf{J}}(\eta)]\mathbf{J}^{-1}(\eta)$$
$$\mathbf{D}_\eta(\mathbf{v},\eta) = \mathbf{J}(\eta)^{-T}\mathbf{D}\mathbf{J}(\eta)^{-1}$$

$$G_\eta(\eta) = J^{-T}(\eta)G_f(\eta)$$
$$\tau_\eta = J^{-T}(\eta)\tau_A$$

4.6 RRC ROV MODEL

The RRC ROV dynamics and kinematics equations are:

$$M\dot{v} + C(v) + D(v)v + G_f(\eta) = \tau_A \qquad (4.22)$$

$$\dot{\eta} = J(\eta)v \qquad (4.23)$$

Equation (4.22) and (4.23) are expressed in state-space form to facilitate the analysis and design of model-based controllers.

$$\dot{x} = f(x,t) + g(\bar{u},t), \quad x(0) = x_0 \qquad (4.24a)$$

The full-order state-space functions in (4.24a) are as follows:

$$x = [\eta \quad v]^T$$

$$f(x,t) = \begin{bmatrix} 0_{6\times 6} & J(\eta) \\ 0_{6\times 6} & -M^{-1}[C(v) + D(v)] \end{bmatrix} \begin{bmatrix} \eta \\ v \end{bmatrix} + \begin{bmatrix} 0_{6\times 1} \\ -M^{-1}G_f(\eta) \end{bmatrix}$$

$$g(\bar{u},t) = \begin{bmatrix} 0_{6\times 1} \\ M^{-1}TF_T\bar{u} \end{bmatrix}$$

The output equation can be expressed as:

$$y = h(x,t) \qquad (4.24b)$$

where

$$h(x,t) = \begin{bmatrix} I_{6\times 6} & 0_{6\times 6} \\ 0_{6\times 6} & 0_{6\times 6} \end{bmatrix} x$$

The state vector $f(x,t)$ is piecewise continuous in t and locally Lipschitz in x, $g(\bar{u},t)$ is piecewise continuous in t and locally Lipschitz in $\bar{u}$, $h(x,t)$ is piecewise continuous in t and continuous in x.

The parameters used in $f(x,t)$ are determined using CAD tools and experiments. They are observed to have the following properties expressed by Property 4.1 and are consistent with [60].

Property 4.1   Property of Body-Fixed Matrices

The matrices in (4.22) satisfy the following property:

$$\text{a) } \mathbf{M} = \mathbf{M}^T > 0, \quad \dot{\mathbf{M}} = 0$$

where

$$\mathbf{M} = \mathbf{M}_{RB} + \mathbf{M}_A$$

$$\text{b) } \mathbf{C}(\mathbf{v}) = -\mathbf{C}^T(\mathbf{v}), \quad \mathbf{v} \in \Re^6$$

where

$$\mathbf{C}(\mathbf{v}) = \mathbf{C}_{RB}(\mathbf{v}) + \mathbf{C}_A(\mathbf{v})$$

$$\text{c) } s^T[\dot{\mathbf{M}} - 2\mathbf{C}(\mathbf{v})]s = 0, \quad \forall s \in \Re^6$$

$$\text{d) } \mathbf{D}(\mathbf{v}) > 0$$

4.6.1   RIGID-BODY MASS AND CORIOLIS AND CENTRIPETAL MATRIX

The CAD software Pro/ENGINEER (see Figure 4.9) is used to determine the rigid-body mass and Coriolis and centrifugal force parameters. The principal components as shown in Figure 4.10 were included in the complete RRC ROV geometric model and using the density and mass values available, the rigid-body mass and Coriolis and centripetal properties with respect to the ROV's CG can be determined.

A simple form of the $\mathbf{M}_{RB}$ can be obtained as the ROV body axes (origin O) coincide with the CG of the ROV. This is achieved by adding balancing weights at designated locations in the ROV. This means that the locations of the CG, $\mathbf{r}_G$ become zero. Hence, the rigid-body mass matrix in (4.7) can be simplified to:

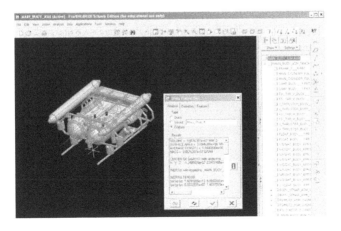

FIGURE 4.9   CAD software Pro/ENGINEER for ROV.

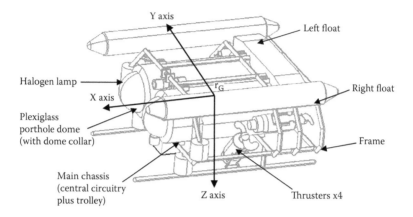

FIGURE 4.10 ROV Components.

$$
\mathbf{M_{RB}} =
\begin{bmatrix}
115.0000 & 0 & 0 & 0 & 0 & 0 \\
0 & 115.0000 & 0 & 0 & 0 & 0 \\
0 & 0 & 115.0000 & 0 & 0 & 0 \\
0 & 0 & 0 & 6.1000 & -0.00016 & -0.1850 \\
0 & 0 & 0 & -0.00016 & 5.9800 & 0.0006 \\
0 & 0 & 0 & -0.1850 & 0.0006 & 5.5170
\end{bmatrix}
$$

(4.25)

where the parameters used in (4.7) are as follows

$$m = 115.00 \ \text{kg}$$

$$I_x = 6.1000 \ \text{kg} \cdot \text{m}^2$$

$$I_y = 5.9800 \ \text{kg} \cdot \text{m}^2$$

$$I_z = 5.5170 \ \text{kg} \cdot \text{m}^2$$

(4.26)

$$I_{xy} = I_{yx} = -0.0002 \ \text{kg} \cdot \text{m}^2$$

$$I_{xz} = I_{zx} = -0.1850 \ \text{kg} \cdot \text{m}^2$$

$$I_{yz} = I_{zy} = 0.0006 \ \text{kg} \cdot \text{m}^2$$

Recall in (4.11), the structure of the rigid-body Coriolis and centripetal matrix can be rewritten as:

$$
\mathbf{C_{RB}}(\mathbf{v}) =
\begin{bmatrix}
\mathbf{0}_{3\times 3} & \mathbf{C_{12}}(\mathbf{v}) \\
-\mathbf{C_{12}^T}(\mathbf{v}) & \mathbf{C_{22}}(\mathbf{v})
\end{bmatrix}
$$

(4.27)

where $\mathbf{C}_{12}^{\mathrm{T}}(\mathbf{v})$ and $\mathbf{C}_{22}(\mathbf{v})$ were substituted by the numerical values in (4.26). The submatrices become:

$$\mathbf{C}_{12}(\mathbf{v}) = \begin{bmatrix} 0 & 115.00w & -115.00v \\ -115.00w & 0 & 115.00u \\ 115.00v & -115.00u & 0 \end{bmatrix}$$

$$\mathbf{C}_{22}(\mathbf{v}) = \begin{bmatrix} 0 & -0.0006q + 0.1850p + 5.5170r \\ 0.0006q - 0.1850p - 5.5170r & 0 \\ -0.0006r + 0.0002p + 5.9800q & -0.1850r - 0.0002q - 6.1000p \end{bmatrix}$$

$$\begin{bmatrix} 0.0006r - 0.0002p - 5.9800q \\ 0.1850r + 0.0002q + 6.1000p \\ 0 \end{bmatrix}$$

The subsequent information, simulation, experimental and pool test results on the hydrodynamic added mass and damping coefficients were directly obtained from the report [61] by Mr You Hong Tan and our joint effort paper [62] from my PhD works at NTU (in Singapore) under Dr Michael Lau Wai Shing supervision.

4.6.2 HYDRODYNAMIC ADDED MASS FORCES

As shown in this section, the hydrodynamic added mass [61-62] (or sometimes called *derivatives* or *virtual mass*) coefficients were obtained using MultiSurf® and WAMIT software. Prior to the application on the ROV [61–62], a few studies on the empirical results of a sphere and a cylinder were performed to verify the program setup and parameters. An experiment was conducted on a scaled ROV and later, laws of Similitude [63] were used to predict the actual ROV added mass coefficients from the scaled ROV results.

The hydrodynamic added forces and moments come about from the acceleration that the fluid particles experience when they encounter the vehicle. The motion of the surrounding body of fluid in response to the ROV motion manifests itself as the hydrodynamic forces and moments resisting the vehicle motion. The effect appears to be like "added" mass and inertia. For a fully submerged vehicle, the added mass and inertia are independent of the wave circular frequency.

The relationship between the hydrodynamic forces and moments and accelerations can be represented by the added mass or sometimes called *hydrodynamics derivatives*. For example, if there is acceleration $\dot{u}$ in the X-direction, the hydrodynamic force X_A arising from that motion can be given as:

$$X_A = X_{\dot{u}}\dot{u} \tag{4.28}$$

where the hydrodynamic derivative $X_{\dot{u}} = \dfrac{\partial X}{\partial \dot{u}}$. The single DOF equation of motion to describe the ROV moving in the X-direction can be written as: $(m - X_{\dot{u}})\dot{u} = \tau_x + X_u u$; m is the mass, X_u is the linear damping and τ_x is the thrust. In this form, the hydrodynamic derivative $X_{\dot{u}}$ can be considered as an added mass term on the left-hand side of the expression.

The added mass inertia and Coriolis and centripetal matrices in (4.22) are defined. The hydrodynamic added mass matrix is written as:

$$
\mathbf{M_A} =
\begin{bmatrix}
X_{\dot{u}} & X_{\dot{v}} & X_{\dot{w}} & X_{\dot{p}} & X_{\dot{q}} & X_{\dot{r}} \\
Y_{\dot{u}} & Y_{\dot{v}} & Y_{\dot{w}} & Y_{\dot{p}} & Y_{\dot{q}} & Y_{\dot{r}} \\
Z_{\dot{u}} & Z_{\dot{v}} & Z_{\dot{w}} & Z_{\dot{p}} & Z_{\dot{q}} & Z_{\dot{r}} \\
K_{\dot{u}} & K_{\dot{v}} & K_{\dot{w}} & K_{\dot{p}} & K_{\dot{q}} & K_{\dot{r}} \\
M_{\dot{u}} & M_{\dot{v}} & M_{\dot{w}} & M_{\dot{p}} & M_{\dot{q}} & M_{\dot{r}} \\
N_{\dot{u}} & N_{\dot{v}} & N_{\dot{w}} & N_{\dot{p}} & N_{\dot{q}} & N_{\dot{r}}
\end{bmatrix}
\tag{4.29}
$$

On the other hand, the hydrodynamic added Coriolis and centripetal matrix $\mathbf{C_A}(\mathbf{v}) = -\mathbf{C_A^T}(\mathbf{v})$ is given by:

$$
\mathbf{C_A}(\mathbf{v}) =
\begin{bmatrix}
0 & 0 & 0 & 0 & -a_3 & a_2 \\
0 & 0 & 0 & a_3 & 0 & -a_1 \\
0 & 0 & 0 & -a_2 & a_1 & 0 \\
0 & -a_3 & a_2 & 0 & -b_3 & b_2 \\
a_3 & 0 & -a_1 & b_3 & 0 & -b_1 \\
-a_2 & a_1 & 0 & -b_2 & b_1 & 0
\end{bmatrix}
\tag{4.30}
$$

where

$$
a_1 = X_{\dot{u}}u + X_{\dot{v}}v + X_{\dot{w}}w + X_{\dot{p}}p + X_{\dot{q}}q + X_{\dot{r}}r
$$
$$
a_2 = X_{\dot{v}}u + Y_{\dot{v}}v + Y_{\dot{w}}w + Y_{\dot{p}}p + Y_{\dot{q}}q + Y_{\dot{r}}r
$$
$$
a_3 = X_{\dot{w}}u + Y_{\dot{w}}v + Z_{\dot{w}}w + Z_{\dot{p}}p + Z_{\dot{q}}q + Z_{\dot{r}}r
$$
$$
b_1 = X_{\dot{p}}u + Y_{\dot{p}}v + Z_{\dot{p}}w + K_{\dot{p}}p + K_{\dot{q}}q + K_{\dot{r}}r
\tag{4.31}
$$
$$
b_2 = X_{\dot{q}}u + Y_{\dot{q}}v + Z_{\dot{q}}w + K_{\dot{q}}p + M_{\dot{q}}q + M_{\dot{r}}r
$$
$$
b_3 = X_{\dot{r}}u + Y_{\dot{r}}v + Z_{\dot{r}}w + K_{\dot{r}}p + M_{\dot{r}}q + N_{\dot{r}}r
$$

The hydrodynamic added mass matrix was computed through the use of the following assumptions commonly stated in the literature [60]:

ASSUMPTIONS 4.1 HYDRODYNAMIC ADDED MASS MATRIX

1. To simplify the problem, the ideal fluid is used. It refers to fluid that is incompressible, inviscid (frictionless) and irrotational (fluid particles are not rotating). For a rigid body moving very slowly, the hydrodynamic system inertia matrix, $\mathbf{M_A}$ is positive and constant [64]. As shown in Wendel [65],

the numerical values of the added mass derivatives in a real fluid are usually in good agreement with those obtained from ideal theory. Hence $\mathbf{M}_A > 0$ is a good approximation.

2. The effects of the off-diagonal elements in $\mathbf{M}_A$ on an underwater vehicle are small compared to the diagonal elements. For most low speed underwater vehicles, the off-diagonal terms are often neglected. This approximation is found to hold true for many applications. Hence, $\mathbf{M}_A$ has the simplified diagonal form as follows:

$$\mathbf{M}_A = -\text{diag}\{X_{\dot{u}}, Y_{\dot{v}}, Z_{\dot{w}}, K_{\dot{p}}, M_{\dot{q}}, N_{\dot{r}}\} \tag{4.32}$$

3. As the off-diagonal elements in $\mathbf{M}_A$ are neglected, the Coriolis and centripetal added mass matrix $\mathbf{C}_A(\mathbf{v})$ is simplified to:

$$\mathbf{C}_A(\mathbf{v}) = \begin{bmatrix} 0 & 0 & 0 & 0 & -Z_{\dot{w}}w & Y_{\dot{v}}v \\ 0 & 0 & 0 & Z_{\dot{w}}w & 0 & -X_{\dot{u}}u \\ 0 & 0 & 0 & -Y_{\dot{v}}v & X_{\dot{u}}u & 0 \\ 0 & -Z_{\dot{w}}w & Y_{\dot{v}}v & 0 & -N_{\dot{r}}r & M_{\dot{q}}q \\ Z_{\dot{w}}w & 0 & -X_{\dot{u}}u & N_{\dot{r}}r & 0 & -K_{\dot{p}}p \\ -Y_{\dot{v}}v & X_{\dot{u}}u & 0 & -M_{\dot{q}}q & K_{\dot{p}}p & 0 \end{bmatrix} \tag{4.33}$$

RESULT 4.1 HYDRODYNAMIC ADDED MASS MATRIX

In this section, the added mass coefficients of the ROV are obtained as shown in Figure 4.11. The added mass matrix for the ROV was computed using the CAD software, MultiSurf, that converts the 3D geometric model into finite surface panels. The geometry from MultiSurf was imported to WAMIT and the problem was solved

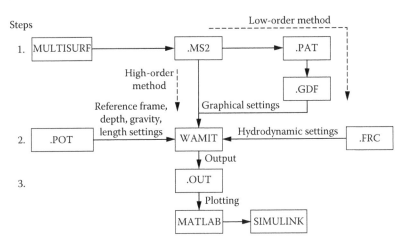

FIGURE 4.11 Programs flowchart.

using the high-order panel method. The output from the WAMIT is plotted using the MATLAB and Simulink software. The following chart shows how the three programs, namely MATLAB, MultiSurf, and WAMIT are used together in the added mass analysis and their respective input and output files with extensions: MS2, PAT, GDP, POT, FRC, OUT.

In the high-order method, different panels' sizes could be used to represent different shapes or patches, hence allowing a different number of panels to represent a surface individually. However, the WAMIT states two precautions when using a higher-order method. First, the iterative method for the solution of the linear system may fail to converge in many cases. Hence, block-iterative solution options are recommended. Second, the result may be less accurate when processing a geometry that has sharp corners.

To maintain the accuracy and consistency of the results, the body of interests (that is, the ROV) is divided into parts and solved incrementally. As the ROV is made up of simple geometrical shapes such as a sphere and cylinder (see Figures 4.12 and 4.13), the analytical results of the added mass on these simple geometrical bodies are used. Studies have been conducted to verify the results between WAMIT and the theoretical results of these simple geometrical shapes.

For example, the theoretical added mass of a sphere is $2\pi\rho r^3/3$ for surge, sway and heave. The added mass output from WAMIT is normalized against the density. Thus, the added mass of a sphere after being normalized against density is $2\pi\rho r^3/3$.

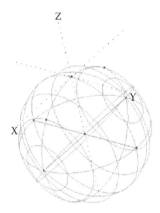

FIGURE 4.12 Sphere drawn in MULITSURF. Origin = (0, 0, 0), radius $r = 1$ m, density $\rho = 1$ [61-62].

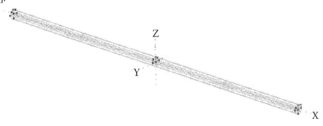

FIGURE 4.13 Cylinder drawn in MULTISURF. Origin = (0, 0, 0), radius = 1 m, length = 80 m [61-62].

TABLE 4.3

Low-Order (Top) and High-Order (Bottom) Method for Sphere

Panel Number	Low-Order Method					
	Theoretical			Numerical		
	Surge	Sway	Heave	Surge	Sway	Heave
256	2.0944	2.0944	2.0944	2.0171	2.0892	2.0892
512				2.0183	2.0972	2.0929
1024				2.0749	2.0929	2.0929
2304				2.0861	2.0939	2.0940
			Total	−0.4%	−0.01%	~0%

Panel Number	High-Order Method					
	Theoretical			Numerical		
	Surge	Sway	Heave	Surge	Sway	Heave
256	2.0944	2.0944	2.0944	2.0952	2.0919	2.0924
512				2.0944	2.0944	2.0945
1024				2.0944	2.0944	2.0945
2304				2.0944	2.0944	2.0945
			Total	0%	0%	0%

As shown in Table 4.3, the results obtained from WAMIT (wave analyst MIT) are within 0.5% of the theoretical results in the lower-order method. The results attained 100% accuracy for the higher-order method. For the case of a cylinder (see Table 4.4), the results from WAMIT are within 1.4% of the theoretical results. The results using a high-order method are accurate and converged faster than the lower-order method. As the body of interests is divided into small pieces and solved individually, the results should converge to a certain value as more numbers of elements (panels) are used. These can be observed in Tables 4.3 and 4.4 where the calculated added mass converges to the theoretical value as the number of panels increased for the low-order method.

In WAMIT, the depth of the submerged body can be specified. The same sphere was used to study the effects of the depth on the added mass. As the theoretical added mass of the sphere in X direction is 2.9044, the results converge at around 10 m as seen in Table 4.5. This allows the subsequent added mass analysis on the ROV to be performed at 10 m water depth.

Another concern for the CFD using WAMIT is the result may be less accurate when processing a geometry that has sharp corners and is more complex. To circumvent this, only half of the ROV is modeled as it has symmetry in the XZ plane. As shown in Figure 4.14, the main components of the ROV are drawn to reduce the complexity of the computation.

As seen in Figure 4.15, other components such as thrusters are not included in the model. The results in Table 4.6 show that the diagonal components of the added mass for the case of two (T2) and four thrusters (T4) are small, as compared to

TABLE 4.4

Low-Order (Top) and High-Order (Bottom) Method for Cylinder [61-62]

	Low-Order Method			
	Theoretical		Numerical	
Panel Number	Surge	Heave	Sway	Heave
768	251.3274	251.3274	249.6437	249.8821
			(−0.7%)	(−0.6%)
3072			248.0583	248.2957
			(−1.3%)	(−1.2%)

	High-Order Method			
	Theoretical		Numerical	
Panel Number (Panel number)	Surge	Heave	Sway	Heave
5 (75)	251.3274	251.3274	247.6803	247.7656
2 (150)			247.4787	247.5613
1 (368)			247.4198	247.5017
		Total	−1.4%	−1.4%

TABLE 4.5

Added Mass of Sphere at Various Depths [61-62]

Depth (m)	Added Mass (Kg)
0	2.5910
1	2.1419
2	2.1073
5	2.0947
10	2.0931
100	2.0929

the ROV without thrusters. This can be verified, as the contribution to the thrusters (see Figure 4.16) alone is quite small ($\approx 10^{-3}$). The added mass contributed by the thruster itself can be seen in the following matrix. As a result, the thrusters' contribution to the added mass matrix is ignored.

$$-\begin{bmatrix} 0.00019 & 0 & 0 & 0 & 0 & 0 \\ 0 & 0.0009 & 0 & 0 & 0 & 0.00004 \\ 0 & 0 & 0.0009 & 0 & 0.00004 & 0 \\ 0 & 0 & 0 & 0 & 0 & 0 \\ 0 & 0 & 0.00004 & 0 & 0.000004 & 0 \\ 0 & 0.00004 & 0 & 0 & 0 & 0.000004 \end{bmatrix}$$

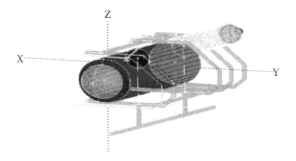

FIGURE 4.14   ROV model drawn in MULTISURF (without thrusters) [61-62].

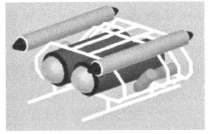

FIGURE 4.15   ROV model drawn in MULTISURF (with two and four thrusters) [61-62].

TABLE 4.6
Magnitude of Error on Diagonal Components of Added Mass Matrix
(T2: Two Thrusters and T4: Four Thrusters) [61-62]

M_A	Col 1		Col 2		Col 3		Col 4		Col 5		Col 6	
	T2	T4	T2	T4	T2	T4	T2	T4	T2	T4	T2	T4
Row 1	−0.03	**0.03**	0	**0**	10	**72**	0	**0**	−4	**−6**	0	**0**
Row 2	0	**0**	0.003	**0**	0	**0**	2	**2**	0	**0**	−28	**−8**
Row 3	1	**−7**	0	**0**	0.02	**0.06**	0	**0**	−32	**−34**	0	**0**
Row 4	0	**0**	2	**2**	0	**0**	0.07	**0.1**	0	**0**	32	**33**
Row 5	−4	**−2**	0	**0**	−34	**−38**	0	**0**	−1.7	**−2**	0	**0**
Row 6	0	**0**	−28	**−28**	0	**0**	27	**29**	0	**0**	−1	**−1**

To obtain the CAD model of the ROV, various locations and orientation of the reference frame could be defined. Table 4.7 shows the effects of changing the orientation and location of the reference plane in the CAD model. It was found that the changes were not significant compared to the diagonal components in the added mass matrix.

In the subsequent section, a scaled model of the ROV was used to obtain the experiment results of the added mass coefficients. To facilitate the study of the scaled ROV, the results of the scaled ROV as compared to the actual size were studied. It was found that the scaled model is scaled accordingly by a factor R as shown

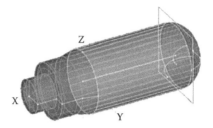

FIGURE 4.16 MULTSURF model for a thruster. Panel Size = 0.02500, Volumes (VOLX, VOLY, VOLZ) 0.115451×10^{-2} 0.115450×10^{-2} 0.115812×10^{-2}, Center of Buoyancy (Xb,Yb,Zb), −0.050563 0.000000 −0.015410 [61-62].

below. If the model is scaled up by the factor of R, and then each element in the added mass matrix is scaled by a ratio as seen in the last column of Table 4.8, the results coincide with the work reported in reference [66].

The effects of multiple bodies such as two spheres and a cylinder are also analyzed. The multiple bodies in MultiSurf could be drawn simultaneously for analysis in WAMIT. The results from MultiSurf and WAMIT are compared with other similar methods such as the Java Amass applet that was constructed for the usage of Marine Hydrodynamics students at MIT. The JAVA A mass applet is often used to approximate the added mass of various objects composed by spheres and cylinders. As shown in Figure 4.17, a comparison between the added masses calculated by the two methods can be seen. The location of zero elements in both added masses is exactly the same. Although the value for nonzero elements could not match exactly, the maximum deviation is less than 20%. Thus, the methodology in using WAMIT and MultiSurf are considered to be quite accurate.

To verify the solution of the linear system converges, the convergences of the solutions were plotted. As WAMIT solves the ROV over finite panels using the higher-order panel method, the convergences of the solution can be seen in Figure 4.18. As observed, the added mass converges to the desired values at around 500–1000 unknowns or panels in the linear system. The computed added mass parameters give a positive definite matrix, in agreement with the earlier assumption made. All the eigenvalues (that is equal to 21.1403, 51.7012, 92.4510, 3.6191, 2.6427, 2.3033) of the added mass matrix are greater than zero. Besides, the data indicates that the added mass is smallest in the surge DOF and largest in the heave DOF. This is consistent with the fact that the vehicle's cross-section area is smallest in the surge DOF and largest in heave DOF.

The final added mass of the RRC ROV becomes:

$$
\mathbf{M_A} = -\begin{bmatrix}
21.1403 & 0 & 0.0619 & 0 & -0.5748 & 0 \\
0 & 51.7012 & 0 & -2.0928 & 0 & -0.3767 \\
0.0917 & 0 & 92.4510 & 0 & 0.5871 & 0 \\
0 & -2.0090 & 0 & 3.6191 & 0 & 0.0235 \\
-0.5237 & 0 & 0.5594 & 0 & 2.6427 & 0 \\
0 & -0.3783 & 0 & 0.0275 & 0 & 2.3033
\end{bmatrix} \quad (4.34)
$$

TABLE 4.7
Effects of the Changes in Added Mass Matrix [61-62]

Items	Case Studies	Results	Plots
1	Effect of changing origin of the reference frame	No changes in the added mass coefficients in translation directions. Some changes in the rotational directions and the off-diagonal terms of the added mass matrix.	Plots

Distance = 2.000 m
[dX,dY,dZ] = (2.000 m, 0.000 m, −0.000 m)

(Continued)

TABLE 4.7 (Continued)
Effects of the Changes in Added Mass Matrix [61-62]

Items	Case Studies	Results	Plots
2	Effect of changing orientation of the reference frame	No changes in the added mass matrix.	

Positive - x   Negative - x

TABLE 4.8
Scaling Applied to Full Scale Model [61-62]

[1] Scale	[2] Added mass matrix	[3] Matrix (to apply on scaled model)
[4] Full-scale (original)	[5]	[6]
[7] Half-scale (R = 2)	[8]	[9]
[10] Quarter-scale (R = 4)	[11]	[12]

[5]

$$\begin{bmatrix}
21.1403 & 0 & 0.0619 & 0 & -0.5748 & -0.3767 \\
0 & 51.7012 & 0 & -2.0928 & 0 & 0 \\
0.0917 & 0 & 92.4510 & 0 & 0.5871 & 0.0235 \\
0 & -2.0090 & 0 & 3.6191 & 0 & 0 \\
-0.5237 & 0 & 0.5594 & 0 & 2.6427 & 0 \\
0 & -0.3783 & 0 & 0.0275 & 0 & 2.3033
\end{bmatrix}$$

[6]

$$\begin{bmatrix}
R^3 & 0 & R^3 & 0 & R^4 & 0 \\
0 & R^3 & 0 & R^4 & 0 & R^4 \\
R^3 & 0 & R^3 & 0 & R^4 & 0 \\
0 & R^4 & 0 & R^5 & 0 & R^5 \\
R^4 & 0 & R^4 & 0 & R^5 & 0 \\
0 & R^4 & 0 & R^5 & 0 & R^5
\end{bmatrix}$$

[8]

$$\begin{bmatrix}
2.6425 & 0 & 0 & 0 & -0.0359 & -0.0235 \\
0 & 6.4626 & 0 & -0.1308 & 0 & 0 \\
0.0115 & 0 & 11.5562 & 0 & 0.0367 & 0.0007 \\
0 & -0.1256 & 0 & 0.1131 & 0 & 0 \\
-0.0327 & 0 & 0.0350 & 0 & 0.0826 & 0 \\
0 & -0.0236 & 0 & 0.0009 & 0 & 0.0720
\end{bmatrix}$$

[9]

$$\begin{bmatrix}
R^3 & 0 & R^3 & 0 & R^4 & 0 \\
0 & R^3 & 0 & R^4 & 0 & R^4 \\
R^3 & 0 & R^3 & 0 & R^4 & 0 \\
0 & R^4 & 0 & R^5 & 0 & R^5 \\
R^4 & 0 & R^4 & 0 & R^5 & 0 \\
0 & R^4 & 0 & R^5 & 0 & R^5
\end{bmatrix}$$

[11]

$$\begin{bmatrix}
0.3303 & 0 & 0.0010 & 0 & -0.0022 & -0.0015 \\
0 & 0.8078 & 0 & -0.0082 & 0 & 0 \\
0.0014 & 0 & 1.4445 & 0 & 0.0023 & 0 \\
0 & -0.0078 & 0 & 0.0035 & 0 & 0 \\
-0.0020 & 0 & 0.0022 & 0 & 0.0026 & 0 \\
0 & -0.0015 & 0 & 0 & 0 & 0.0022
\end{bmatrix}$$

[12]

$$\begin{bmatrix}
R^3 & 0 & R^3 & 0 & R^4 & 0 \\
0 & R^3 & 0 & R^4 & 0 & R^4 \\
R^3 & 0 & R^3 & 0 & R^4 & 0 \\
0 & R^4 & 0 & R^5 & 0 & R^5 \\
R^4 & 0 & R^4 & 0 & R^5 & 0 \\
0 & R^4 & 0 & R^5 & 0 & R^5
\end{bmatrix}$$

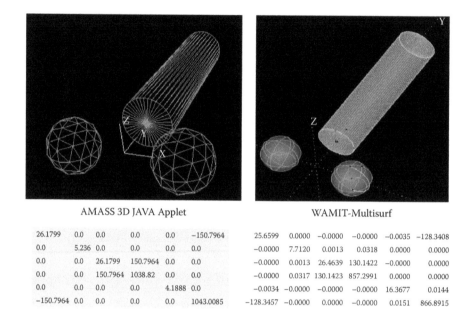

AMASS 3D JAVA Applet						WAMIT-Multisurf					
26.1799	0.0	0.0	0.0	0.0	-150.7964	25.6599	0.0000	-0.0000	-0.0000	-0.0035	-128.3408
0.0	5.236	0.0	0.0	0.0	0.0	-0.0000	7.7120	0.0013	0.0318	0.0000	0.0000
0.0	0.0	26.1799	150.7964	0.0	0.0	-0.0000	0.0013	26.4639	130.1422	-0.0000	0.0000
0.0	0.0	150.7964	1038.82	0.0	0.0	-0.0000	0.0317	130.1423	857.2991	0.0000	0.0000
0.0	0.0	0.0	0.0	4.1888	0.0	-0.0034	-0.0000	-0.0000	-0.0000	16.3677	0.0144
-150.7964	0.0	0.0	0.0	0.0	1043.0085	-128.3457	-0.0000	0.0000	-0.0000	0.0151	866.8915

FIGURE 4.17 Comparisons of methods to obtain added mass coefficients [61-62].

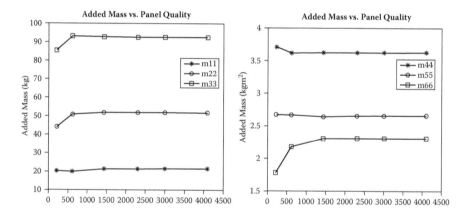

FIGURE 4.18 Convergence test for added mass of RRC ROV [61-62].

and $\mathbf{C_A}(\mathbf{v})$ becomes:

$$\mathbf{C_A}(\mathbf{v}) = \begin{bmatrix} 0 & 0 & 0 & 0 & -a_3 & a_2 \\ 0 & 0 & 0 & a_3 & 0 & -a_1 \\ 0 & 0 & 0 & -a_2 & a_1 & 0 \\ 0 & -a_3 & a_2 & 0 & -b_3 & b_2 \\ a_3 & 0 & -a_1 & b_3 & 0 & -b_1 \\ -a_2 & a_1 & 0 & -b_2 & b_1 & 0 \end{bmatrix} \tag{4.35}$$

FIGURE 4.19 Hydrodynamic added mass matrix in Simulink.

where

$a_1 = 21.1403u + 0.0619w - 0.05748q$
$a_2 = 51.7012v - 2.0928p - 0.3767r$
$a_3 = 0.0619u + 92.4510w + 0.5871q$
$b_1 = -2.0928v + 3.6191p + 0.0235r$
$b_2 = -0.5748u + 0.5871w + 2.6427q$
$b_3 = -0.3767v + 0.0235p + 2.3033r$

The negative signs are because the pressure forces on the ROV would tend to retard the vehicle motion. The real mass (or the rigid body mass) and the virtual added mass are originally on opposite sides of the equation; one is a rigid body property, while the other is related to the (pressure) force experienced by the vehicle when the virtual mass is *subtracted* from the real mass, the net effect (as seen in Figure 4.19) has greater apparent mass in most DOF, hence the virtual mass is *added mass*. Note that all the motions (both linear and angular motion) are defined according to the CG. For example, the added mass matrix term $N_{\dot{r}}$ means that the hydrodynamic moment acts on the ROV as it rotates about the Z-axis.

As observed, the off-diagonal terms in $\mathbf{M_A}$ are small as compared to its diagonal components. Hence, with the application of Assumption 4.1, $\mathbf{M_A}$ can be approximated as:

$$\mathbf{M_A} = - \begin{bmatrix} 21.1403 & 0 & 0 & 0 & 0 & 0 \\ 0 & 51.7012 & 0 & 0 & 0 & 0 \\ 0 & 0 & 92.4510 & 0 & 0 & 0 \\ 0 & 0 & 0 & 3.6191 & 0 & 0 \\ 0 & 0 & 0 & 0 & 2.6427 & 0 \\ 0 & 0 & 0 & 0 & 0 & 2.3033 \end{bmatrix} \quad (4.36)$$

where the Coriolis/centripetal added mass matrix becomes:

$$\mathbf{C_A(v)} = \begin{bmatrix} 0 & 0 & 0 & 0 & -92.4510w & 51.7012v \\ 0 & 0 & 0 & 92.4510w & 0 & -21.1403u \\ 0 & 0 & 0 & -51.7012v & 21.1403u & 0 \\ 0 & -92.4510w & 51.7012v & 0 & -2.3033r & 2.6427q \\ 92.4510w & 0 & -21.1403u & 2.3033r & 0 & -3.6191p \\ -51.7012v & 21.1403u & 0 & -2.6427q & 3.6191p & 0 \end{bmatrix}$$

$$(4.37)$$

In summary, the added mass matrix for the ROV has been determined using WAMIT, a CFD software based on the potential flow theory and panel method.

The added mass for surge, sway, and heave motion is around 21 kg, 51 kg, and 93 kg, respectively. After considering the added mass, the effective inertia for ROV in heave motion is almost double, increasing from 115 kg to 208 kg. As such, the added mass forces are quite significant and thus cannot be neglected.

A series of tests were conducted in Section 4.6.3 to verify the result from WAMIT. The calculated added mass from WAMIT is accurate when compared with the theoretical results for simple body shapes. In the subsequent section, the added mass terms obtained from WAMIT were shown to be in good agreement with the experimental data. Thus, the added mass matrix for ROV obtained from WAMIT is considered to be reliable.

4.6.3 Hydrodynamic Damping Forces

Matrix **D** is the hydrodynamic damping matrix [61-62] consisting of both linear and quadratic terms. This hydrodynamic damping is caused by the potential damping due to linear skin friction (linear damping) drag and vortex shedding (quadratic or nonlinear damping). The sum of these individual components gives the overall hydrodynamic damping effect on the ROV.

Since the ROV is designed to be self-stabilizing in the roll and pitch angle, the magnitude of these angles, especially the pitch angle, is very small. Hence, for such small pitch angle (or angle of attack), there is no cross-flow separation and no boundary layer separation, and the flow remains attached. Therefore, the non-linear part of the forces and moments can be considered as forces and moments due to viscous effect of the flow, which becomes less important as the pitch angle is small. These potential flow forces have a linear relation with the pitch angle. Thus, it may be stated that the linear coefficients are sufficient to represent the force and moments arising from the inviscid part of the flow, particularly, for slow speed (less than 2 m/s) maneuvering of the ROV and with small pitching during maneuvering. For comparison purposes, both linear and nonlinear damping forces were computed.

As observed in the ROV, it is a complex block structure. It may suggest that a more complete approach is required to quantify the hydrodynamic coefficients. An alternative semipredictive approach that uses the open-tank and the CFD method using ANSYS-CFX is used. This method is becoming increasingly tractable due to the development of CFD packages and the advancement of computer technology such as the use of MATLAB to facilitate the determination of the parameters obtained from the tank test and later use the ANSYS-CFX to verify the hydrodynamic coefficients. However, in any method used, it will always be subject to experimental and numerical errors (around 30%). Thus, due to these error or uncertainties, the model derived by either method has to depend on some form of robust control scheme (as seen in Sections 5.3 and 5.4) to control the ROV.

The hydrodynamic damping coefficients of the scaled ROV [61–62] were obtained using ANSYS-CFX software. The accuracy of the numerical approach using ANSYS-CFX was established through comparison of the current study with empirical and other computational results of the sphere. The damping coefficients were verified using the free-decaying experiment on the scaled ROV. By applying laws of Similitude, the hydrodynamics parameters of the scaled model can be scaled up to predict the corresponding values for the true model (or the actual ROV model).

With the assumptions commonly stated in [60], the hydrodynamic damping matrix can be simplified by using the following Assumption 4.2.

ASSUMPTIONS 4.2 HYDRODYNAMIC DAMPING MATRIX

1. As the ROV is operating around a linear speed of 0.5 m/s (or angular speed of 0.5 rad/s), it is well within the linear damping region of maximum 2 m/s (see circle region in Figure 4.20). Hence, the linear hydrodynamic damping was considered. But for completeness, the quadratic hydrodynamic damping force was determined.
2. The off-diagonal elements in **D** on an underwater vehicle [60] are small compared to the diagonal elements, hence only the hydrodynamic damping in the diagonal form is used:

$$\mathbf{D} = -\mathrm{diag}\{X_u, Y_v, Z_w, K_p, M_q, N_r\} \tag{4.38}$$

RESULT 4.2 HYDRODYNAMIC DAMPING MATRIX

Like the added mass terms, the hydrodynamic damping force and moment coefficients for the ROV are difficult to obtain experimentally without a proper full-scale instrumented tow tank facility. However, it could be expensive and difficult to justify building the facility for this usage only. The problem is circumvented using

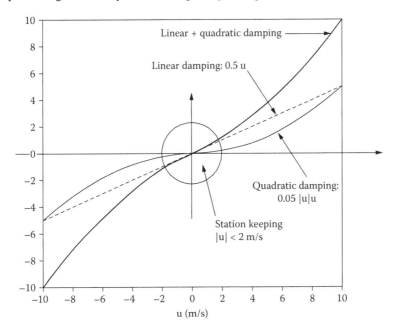

FIGURE 4.20   Hydrodynamic damping at low and high velocity of vehicle. (Cited from T. I. Fossen, *Guidance and Control of Ocean Vehicles*, New York: John Wiley & Sons Ltd, 1994.)

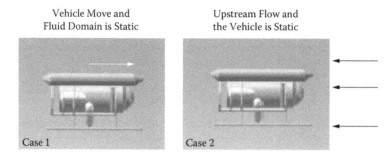

FIGURE 4.21   Physical model and computer simulation model use in CFD [61-62].

ANSYS-CFX and experimental study of a scaled model in a small water tank that is later translated to the full-scale model by using laws of similitude.

In CFD modeling, the physical model for computer simulation has to be determined. Figure 4.21 shows the ROV moving forward at a constant speed with the fluid domain remaining static with no current flow. Instead of the ROV moving, the flow can be made to move at constant speed in the opposite direction with the ROV, which remains static in position. Since the drag force depends only on the relative motion between the ROV and the fluid, the result obtained from the two situations is the same. The advantage of keeping the ROV static is the boundary conditions can set up more easily for computer simulation. On the other hand, moving the ROV creates more meshes and thus is more complex to handle.

In most ROV operations, considering its scale and speed, the typical Reynolds number will be greater than 1.0×10^6, indicating turbulence in flow field as shown in Case 2 (see Figure 4.21). In turbulence flow, the fluid motion is characterized by a highly random, unsteady three-dimensional flow. The turbulent length and timescales are much smaller than the smallest finite volume meshes that are used in most numerical analysis. To generate such a small mesh is beyond the computing power currently available in the laboratory. The problem is solved by employing a turbulence model.

The boundary condition and model environment setup are established as follows. The flow in the domain is expected to be turbulent and approximately isothermal. The SST turbulence model with automatic wall function treatment will be used because of its highly accurate predictions of flow separation. To take advantage of the SST (shear stress transport) model, the boundary layer should be resolved with at least 10 mesh nodes. This is done by inspecting the y+ value on the surface of the ROV that must be around 1.

Water at 20°C is used as the fluid and the reference pressure is set at 1 atmosphere. The main difficulty is to determine the fluid domain dimension. In order to study damping force acting on the ROV in an unbound fluid domain, an infinite large fluid domain is needed. However, this is not practical both in CFD and even in experiment. As shown in Figure 4.22, the fluid domain dimensions of around 20 times the length, width, and height of the ROV were used after the sensitivity analysis was conducted. It is shown that the deviations in the results are negligible after these dimensions.

A flow speed is to be specified at the inlet boundary as shown in Figure 4.23. Physically, the flow upstream is uniform. It is set to be normal to the plane and

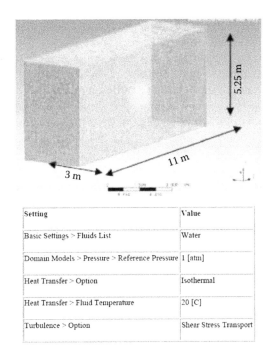

Setting	Value
Basic Settings > Fluids List	Water
Domain Models > Pressure > Reference Pressure	1 [atm]
Heat Transfer > Option	Isothermal
Heat Transfer > Fluid Temperature	20 [C]
Turbulence > Option	Shear Stress Transport

FIGURE 4.22 Meshing for flow domain [61-62].

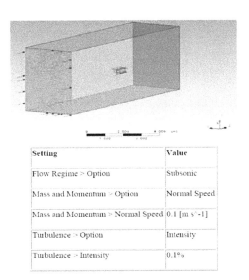

Setting	Value
Flow Regime > Option	Subsonic
Mass and Momentum > Option	Normal Speed
Mass and Momentum > Normal Speed	0.1 [m s^-1]
Turbulence > Option	Intensity
Turbulence > Intensity	0.1%

FIGURE 4.23 Input settings [61-62].

has a speed of 0.1 m/s. Turbulence properties at the inlet flow are hard to specify because the turbulent kinetic energy and dissipation data are unknown. Therefore, their values have to be specified using sensible engineering assumptions and the effect of the choices made should be examined by sensitivity tests. In this study, the turbulent properties are specified using turbulent intensity.

To do this accurately it is good to have some form of measurements or previous experience. As shown below, common estimations of the incoming turbulence intensity are as follows.

- High-turbulence case: High-speed flows inside complex geometries like heat exchangers and flow inside rotating machinery (turbines and compressors) typically have turbulence intensity between 5% and 20%.
- Medium-turbulence case: Flows in not-so-complex devices like large pipes, ventilation flows, and so forth, or low speed flows typically have turbulence intensity between 1% and 5%.
- Low-turbulence case: External flow across cars, submarines and aircraft in otherwise stationary fluid domain and flow generated by very high-quality wind tunnels typically have a turbulence intensity well below 1%.

The external flow over an underwater vehicle has a turbulence intensity well below 1%. Therefore, the turbulent intensity of 0.1% is selected for the ROV.

The outlet boundary condition is selected to have a weak influence on the upstream flow as shown in Figure 4.24. Specifying static pressure at the outlet plane is mentioned by W. S. Atkins Consultants [67]. At a high Reynolds number, the wall stress cannot be calculated correctly using the gradients on the coarse grids

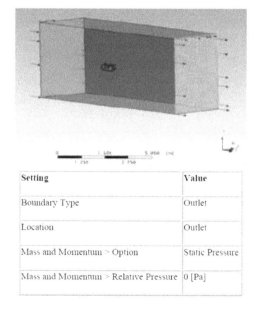

Setting	Value
Boundary Type	Outlet
Location	Outlet
Mass and Momentum > Option	Static Pressure
Mass and Momentum > Relative Pressure	0 [Pa]

FIGURE 4.24 Output settings [61-62].

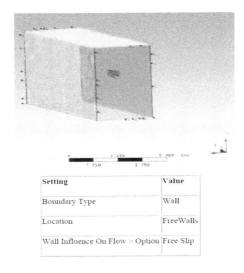

Setting	Value
Boundary Type	Wall
Location	FreeWalls
Wall Influence On Flow - Option	Free Slip

FIGURE 4.25 Free-Slip Boundary Condition on fluid domain wall [61-62].

or meshes. The boundary layer is often too thin to be represented by the meshes. Hence, there is a need to develop a model to approximate the wall-boundary condition. The wall-function used in ANSYS-CFX is an extension of the method in Launder and Spalding's work [68]. In their log-law method, the near-wall tangential velocity is logarithmically related to the wall shear stress. This relationship helps to predict the velocity profile in the boundary layer and hence, fine meshes are not necessary.

The top, bottom, and side surfaces of the rectangular fluid domain can be modeled using free-slip wall-boundary conditions as shown in Figure 4.25. On free-slip walls, the shear stress is set to zero so that the fluid is not retarded. The velocity normal to the wall is also set to zero. However, it is not an ideal boundary condition, as the flow around the body will be affected by its close proximity to the walls.

And due to fluid viscosity, the flow speed on the vehicle's surfaces is zero. Hence, a nonslip wall boundary condition is applied on the vehicle's body as shown in Figure 4.26. As observed, the ROV has symmetry on the XZ plane. By using the symmetry on the XZ plane, it greatly reduces the computational time and resources. On the symmetry plane, the gradients perpendicular to the plane are zero.

Initial values for all variables have to be set before the computation begins. For a steady-state calculation, the initial variable values serve to give the ANSYS-CFX Solver a flow field to start its calculations. Convergence is more rapidly achieved if sensible initial values are provided. However, converged results should not be affected by the initialization. Unless there is information about the turbulent eddy dissipation, it is recommended to use the automatic initial guess as shown in Table 4.9.

In the ANSYS-CFX's Solver control, the selection of an appropriate time step size is essential to obtain good convergence rates. Physical time step is used to provide sufficient relaxation of the nonlinearity so that a converged steady-state solution can be obtained. A physical time step that is too large is characterized by oscillatory convergence or

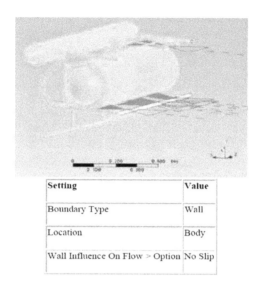

Setting	Value
Boundary Type	Wall
Location	Body
Wall Influence On Flow > Option	No Slip

FIGURE 4.26   Nonslip Boundary Condition on ROV surface [61-62].

TABLE 4.9
Initial Condition of the Flow Domain [61-62]

Setting	Value
Initial Conditions > Cartesian Velocity Components > Option	Automatic with Value
Initial Conditions > Cartesian Velocity Components > U	$0.1~\mathrm{ms}^{-1}$
Initial Conditions > Turbulence Eddy Dissipation	(Selected)

TABLE 4.10
CFX's Solver Control [61-62]

Setting	Value
Convergence Control > Max Iterations	60
Convergence Control > Fluid Timescale Control > Timescale Control	Physical Timescale
Convergence Control > Fluid Timescale Control > Physical Timescale	5 s
Convergence Criteria > Residual Target	1e-05

---

results that do not converge. A small time step is characterized by very slow, steady convergence. For advection-dominated flows, the physical time step size is some fraction of a length scale divided by a velocity scale. A good approximation is the dynamical time for the flow is the time taken for the flow to make its way through the fluid domain. A reasonable estimate is around one-third of the length of the fluid domain L and the mean velocity U. In this study, the time step is taken as: $\Delta t = L/3U = 10\mathrm{m}/3 \times 0.5~\mathrm{ms}^{-1} = 6.67\mathrm{s}$. For purposes of convenience, the time step is set to 5 s as shown in Table 4.10.

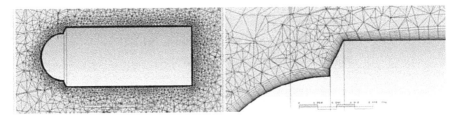

FIGURE 4.27 Cut Surface Mesh Plot at the Middle Plane of the Cylinder Hull [61-62].

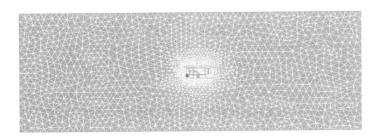

FIGURE 4.28 Meshing Domain [61-62].

Before running the CFD simulation, face spacing that determines the surface area of a mesh in contact with the ROV's surface has to be defined. It must be small enough to capture the geometry of the ROV. The inflation layer is important in the capture of the boundary layer flow and prediction of the flow separation. The number of inflation layers increases on the surface of the ROV.

Figure 4.27 illustrates the effect of the face spacing expansion factor. One could observe that the mesh sizes increase exponentially away from the surface of the cylinder. It is important to have a smaller number of meshes far away from the body and keeping the meshes fine near to the body. The right-hand side of Figure 4.27 shows an enlarged image of the inflation layers. The boundary layer is contained within a thin layer near to the body. The inflation layer is thus important to capture the flow near the boundary layer by providing a finer mesh in the region.

The mesh domain around the ROV has around 2,219,437 elements. They consist of 1,067,375 tetrahedras, 1,140,046 wedges, and 12,016 pyramids as shown in Figure 4.28.

The accuracy of the numerical approach on the ROV was established through comparison of the current study with empirical and other computational results of the sphere. Flow at three different Reynolds numbers was selected for comparison. They are $Re = 100$, in the steady and axis-symmetric regime; $Re = 300$, in the laminar periodic vortex-shedding regime; $Re = 1.1 \times 10^6$ is highly unsteady and the boundary layer may be turbulent. The mesh statistics and mesh settings used for the study are shown in Table 4.11.

Table 4.12 presents the streamline plots of the sphere at different Reynolds numbers. The drag coefficients obtained in this study closely matched those published results for Reynolds numbers 100, 300, and 1.1×10^6. These give the assurance that

TABLE 4.11

Mesh Statistics and Mesh Settings [61-62]

No. of Mesh Elements	1255570
Face Spacing	0.001 m^2
Face Spacing Expansion Factor	1.2
No. of Inflation Layer	35
Thickness Multiplier	1
Inflation Expansion Factor	1.2

the procedures in setting up the simulation are done correctly. For the study of a sphere's drag, it is observed that the y+ value on the sphere surface must be close to unity.

By applying the settings on the scale model, the similar phenomena caused by the flow of the fluid around the body is formed as seen in Figure 4.29. This is known as a wake that is the region of disturbed flow (usually turbulent) downstream of the body. As shown in the pressure distribution in Table 4.12, the wake creates a low-pressure region at the rear end and thus high pressure in the front resists the vehicle motion. The results exhibit similar phenomena compared to the streamline plots for the sphere in Table 4.12.

Figure 4.30 shows the normal force and the tangential force acting on the scaled model. As observed, the forces converge to a constant value. It is an important indication that the flow has reach steady state and thus in accordance with the assumption of steady flow.

Besides, the computed drag coefficient converges to a constant value as shown in Figure 4.30. The results imply that the drag coefficient is insensitive to the change of the Reynolds number at the ROV operating range (i.e., $Re = 1 \times 10^6$) and can be treated as a constant drag coefficient (Figure 4.31).

It can be seen in Figure 4.32 that the front of the ROV facing the flow experiences the largest pressure force. The total pressure difference between front and back causes the pressure drag, which is the dominant drag acting on the bluff body like the ROV's body.

Applying the similar steps to the ROV, the formation of wake at the rear of the ROV can be seen in Figure 4.33. This turbulence region produces the nonlinear damping effect on the ROV. However, as the pitch angle is small, the nonlinear damping forces acting on the ROV can be quite small.

The drag force in surge, sway and heave direction are plotted against the velocity in Figure 4.34. The drag value in heave is plotted between 0–0.2 m/s because the ROV operates around this speed in actual operation. On the other hand, the vehicle operates between 0–0.5 m/s in the surge and sway direction. As expected, the vehicle has largest drag in heave followed by sway and surge. Heave has the largest frontal area normal to the flow direction of about 0.9 m^2 compare to 0.5 m^2 for sway and 0.45 m^2 for surge. The drag in sway is only slightly larger than the drag in surge. This may be explained by the relative size of the frontal areas. From the second-order polynomial fit of the graph, the linear damping and quadratic damping for respective directions are obtained. The results are tabulated in Table 4.13 following.

TABLE 4.12
Streamline Plots for Various Reynolds Number (for Sphere) [61-62]

Reynolds Number	Source	Damping Coefficient (Cd)	Streamline Plots from CFD
100	Current study (using ANSYS-CFX)	1.26	
	[83] Empirical	1.09 1.18	
300	Current study (using ANSYS-CFX)	0.814	
	[84] Empirical	0.626 0.826	

(Continued)

TABLE 4.12 (*Continued*)
Streamline Plots for Various Reynolds Number (for Sphere) [61-62]

Reynolds Number	Source	Damping Coefficient (Cd)	Streamline Plots from CFD
1.1×10^6	Current study (using ANSYS-CFX)	0.117	
	[85,86]	0.12–0.14 0.084	

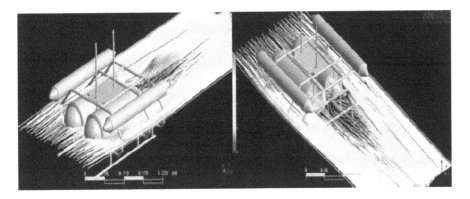

FIGURE 4.29 Streamline plot of the scaled model [61-62].

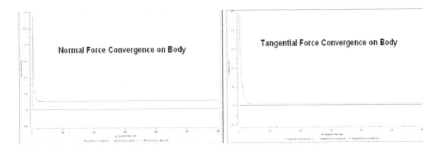

FIGURE 4.30 Iteration history of the normal force and tangential force acting on the scaled model [61-62].

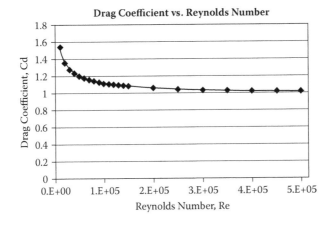

FIGURE 4.31 Drag Coefficient versus Reynolds Number [61-62].

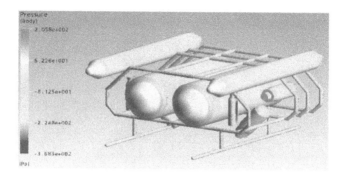

FIGURE 4.32   Pressure distributions on the ROV Body [61-62].

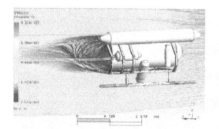 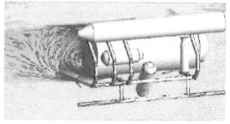

FIGURE 4.33   Streamline and vector plot at the center plane of the cylinder hull at flow speed (0.5 m/s) [61-62].

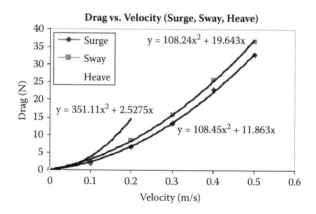

FIGURE 4.34   Drag Force as the function of Velocity in Surge, Sway, and Heave [61-62].

As seen in Figure 4.35, the results suggest that the linear damping in yaw is ignored and the quadratic damping alone is sufficient to describe the drag correctly for the ROV. As shown in Table 4.13, the linear damping coefficients are smaller than the quadratic damping terms. This is due to the effect of the skin friction, which was not included in the CFD simulation. However, the contributions of the quadratic

TABLE 4.13

Drag Coefficient of Actual or Real ROV (by ANSYS-CFX)

Direction Range	Surge (0–0.5 m/s)		Sway (0–0.5 m/s)		Heave (0–0.2 m/s)		Yaw (0–0.5 rad/s)	
Parameter	K_L	K_Q	K_L	K_Q	K_L	K_Q	K_L	K_Q
ANSYS-CFX	11.863	108.45	19.64	108.24	2.3756	351.98	0	10.39

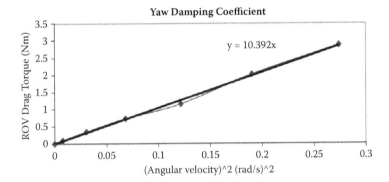

FIGURE 4.35 Drag Torque s a Function of Square Angular Velocity [61-62].

damping forces are lower than the linear damping forces due to the *square* of the velocity terms. This can be seen in the computations of the forces in Table 4.17.

In the previous sections, the parameters associated with the hydrodynamic forces acting on the ROV had been estimated using CFD simulations. However, as stated in CFD guidelines [67], these parameter values should be corroborated by other means. Therefore, in this chapter, verification of the added mass parameter and the drag force coefficient were performed by comparing the value predicted by CFD and that to be obtained experimentally using the scaled down model of the ROV (which hereafter will be referred to as the *scaled model*). By applying laws of Similitude, the hydrodynamics parameters of the scaled model can be scaled up to predict the corresponding values for the true model (actual ROV model).

Experimental data on the ROV's dynamics is often gathered using different test equipment. Most of the dynamic testing is conducted with the model undergoing forced lateral or vertical plane motions to determine added masses, damping, and some other derivatives; for surface models these are invariably frequency-dependent. Routine dynamic testing was introduced with the Planar Motion Mechanism (PMM) [69] in the late 1950s. The PMM imparts harmonic oscillations to the model. Vertical PMMs for subsurface vehicles are, with few exceptions, low amplitude devices for obtaining linear hydrodynamic coefficients; horizontal PMMs are more often large amplitude. A new generation of test apparatus is represented by the Marine Dynamic Test Facility [70]. This device imparts large amplitude and high-rate arbitrary motions to a subsurface or surface model in up to six degrees of freedom. To take maximum advantage of its capabilities requires new approaches to experimental design, and extensive use of SI techniques. The model contains a 6-DOF (degrees of freedom) balance to measure total forces and moments.

However, for initial design and prototype testing, a smaller scale testing is often desirable and economic to run during the developmental stage. The hydrodynamics parameters can be extracted from an experiment as shown in Figure 4.36, in which the scaled model performs a free decay motion in water. The least-square approach was then used to determine the hydrodynamic parameters. As the number of sample size is greater than the number of estimated parameters, the least-square approach is still appropriate for this application. As observed later, the system is still properly excited (does not go to zero) within the time zone of interest. The free-decaying test can be used for a class of medium and small class of underwater vehicles; it could be a viable alternative to estimate some pertinent hydrodynamic parameters without extensive and expensive facilities and instrumentation at a very early stage of the ROV development.

As seen in Figure 4.37, a set of experiments was carried out in a $0.3 \times 0.2 \times 0.3$ m open tank giving a characteristics length ratio of about 0.3:1 for the ROV to investigate the motion characteristics of the scaled ROV in surge, sway, heave, and yaw in the positive direction. The scaled ROV is attached to one end of a pendulum in

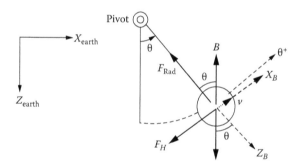

FIGURE 4.36 Free body diagram of the setup [61-62].

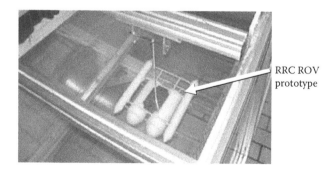

FIGURE 4.37 RRC ROV prototype in water tank (orientated in surge direction) [61-62].

the water and fully submerged in the water by an aluminum frame mounting on the water tank. The pendulum will be set in motion in an arc at a certain radius. The motion is captured by a digital camera. In the earth-fixed frame, the scaled model is constrained to rotate in one plane about the pivot point. However, in the body-fixed frame, the flow is with respect to the body in one direction.

Each experiment was repeated a number of times at each velocity and the average damping force readings (the longitudinal force, the transverse force, altitudinal force and moment are the surge force, sway force, heave force, and yaw moment, respectively) were tabulated in Table 4.15. As a result, the curves of the thrust versus the motion variables u, v, w, and r were obtained in Figures 4.41 through 4.44. With the vehicle surge, sway, or heave set to a constant velocity, the thrust can be considered to be equal to the hydrodynamic force, and the moments can be considered to be equal to the hydrodynamic moments when it yaws at a constant angular velocity. Therefore, the curves in Figures 4.41 through 4.44 of the thrust versus the motion variables are assumed to be the same as the curves of hydrodynamic loads versus these motion variables. As the hydrodynamics forces resist the motion, the amplitude of the swing will decay slowly over time. With the motion data obtained, the hydrodynamics forces could be deduced using the least-square method.

In the experiment, the amplitude of the wave created by the ROV is smaller than the amplitude created by the pendulum itself. The changes in the velocity (i.e., the speed of the free-decaying test) do not significantly affect the hydrodynamic parameters. In this respect, the test regime is still within the *steady state* with little wave contributed by the scaled ROV at a low speed of 0.55 m/s (maximum).

For the surge direction, it has only a velocity component in the surge direction that can be determined. To determine the parameters in the sway direction, the model is orientated perpendicular to the direction of the arc motion. For the yaw motion, the model is allowed to perform a decaying twisting motion. The symbols (m_a, K_L, K_Q) are used to represent the added mass, linear damping and quadratic damping, respectively, in each of these equations in each of the separate tests. The three experiments were conducted for each of the surge, heave, and yaw directions. Since the free-stream velocity is referred to the earth-fixed frame, Euler's coordinate transformation is used to obtain the ROV body-fixed velocity. The rotation speed was about 0.5 rad/s and maximum speed of the surge and heave was at 0.55 m/s.

With a small black mark on the rod, the motion of the pendulum can be captured by a video camera. The recorded trajectory of the black mark can be digitized using an open-source program VirtualDubMod [71]. For each frame, the X and Y coordinates of the black mark were acquired for image processing using MATLAB Image Processing Toolbox. The time history of θ is determined from the X and Y coordinates. The least-square algorithm was then used to calculate the respective (m_a, K_L, K_Q) terms. The nomenclatures used in the equations are given in Table 4.14.

The added mass m_a and damping coefficients (linear, K_L and quadratic, K_Q) due to the hydrodynamic force F_H are defined in a body-fixed frame as:

$$F_H = m_a \ddot{x} + K_L \dot{x} + K_Q |\dot{x}| x \tag{4.39}$$

TABLE 4.14

Nomenclature of Terms Used in the Free-Decay Experiment

Symbols	Descriptions
m	Mass of the Scaled Model
m_a	Added Mass in Single DOF
g	Gravity Term
B	Buoyancy
θ	Angle of Rotation of the Pendulum
r	Length of the Pendulum (radius)
k_L	Linear Damping Coefficient
k_Q	Quadratic Damping Coefficient
F_H	Hydrodynamics Force
x	Tangential Velocity

The equation of motion using Newton's second law of motion is given:

$$-mg\sin\theta + B\sin\theta - m_a\ddot{x} - K_L\dot{x} - K_Q|\dot{x}|x = m\ddot{x} \tag{4.40}$$

Rearranging Equation (4.40) gives the translational motion:

$$(B-mg)\sin\theta - K_L\dot{x} - K_Q|\dot{x}|x = (m+m_a)\ddot{x}$$

$$\ddot{x} = \frac{(B-mg)}{(m+m_a)}\sin\theta - \frac{K_L}{(m+m_a)}\dot{x} - \frac{K_Q}{(m+m_a)}|\dot{x}|x \tag{4.41}$$

For rotational motion, $\dot{x} = r\dot{\theta}$ and $\ddot{x} = r\ddot{\theta}$:

$$\ddot{\theta} = \frac{(B-mg)}{(m+m_a)r}\sin\theta - \frac{K_L}{(m+m_a)}\dot{\theta} - \frac{K_Q}{(m+m_a)}r|\dot{\theta}|\dot{\theta} \tag{4.42}$$

Let $\alpha = \dfrac{(B-mg)}{(m+m_a)r}$, $\beta = \dfrac{K_L}{(m+m_a)}$, $\gamma = \dfrac{K_Q r}{(m+m_a)}$, it becomes

$$\ddot{\theta} = \alpha\sin\theta - \beta\dot{\theta} - \gamma|\dot{\theta}|\dot{\theta} \tag{4.43}$$

The least-square method is used to obtain the estimated values α, β, and γ:

$$\begin{bmatrix} \ddot{\theta}_1 \\ \ddot{\theta}_2 \\ \vdots \end{bmatrix} = \underbrace{\begin{bmatrix} \sin\theta_1 & \dot{\theta}_1 & |\dot{\theta}_1|\dot{\theta}_1 \\ \sin\theta_2 & \dot{\theta}_2 & |\dot{\theta}_2|\dot{\theta}_2 \\ \vdots & \vdots & \vdots \end{bmatrix}}_{\mathbf{H}} \cdot \underbrace{\begin{bmatrix} \alpha \\ \beta \\ \lambda \end{bmatrix}}_{\theta} + error \tag{4.44}$$

Where subscript i = 1, 2, 3... represents the number of samples collected from the experiment result,

$$\hat{\theta}_{LS} = (\mathbf{H}^T\mathbf{H})^{-1}\mathbf{H}^T \qquad (4.45)$$

The buoyancy of the scaled model is first determined by measuring the weight of the fully submerged model with a force sensor. The camera is used to record the pendulum motion and the video is split into multiple frames up to 30 frames per second. Figure 4.38 shows some image frames of the free-decay motion of the pendulum in the surge direction.

Despite the simple experiment setup, the result obtained is highly repeatable and consistent. Figure 4.39 shows the measured values of the pendulum trajectory in water and the scale model using the estimated parameters.

A similar test in the heave direction, with the scaled model rotated 90° facing the direction of the motion, was done to estimate the parameters in the yaw direction. In order to identify the parameters in the yaw motion, the pendulum's rod was replaced by a torsion spring. The image sequence of the scaled model in the yaw direction can be seen in Figure 4.40. The scaled model exhibits pure rotational motion in the water. Similar to the above approach, the dynamics equation becomes:

$$\ddot{\theta} = \frac{K}{I+I_a}\theta - \frac{K_L}{I+I_a}\dot{\theta} - \frac{K_Q}{I+I_a}|\dot{\theta}|\dot{\theta} \qquad (4.46)$$

where K is the torsion spring constant (using a period of oscillation that is linearly proportional to $2\pi\sqrt{I/K}$, the moment of inertia for the scaled model I is 0.0026 kgm^2 and oscillation period is 10.45s), K_L is the linear rotational drag coefficient, K_Q is the quadratic rotational drag coefficient, I_a is the added moment of inertia, and θ is the angle of rotation.

FIGURE 4.38 Image sequence of the scaled model under pendulum motion—in surge direction [61-62].

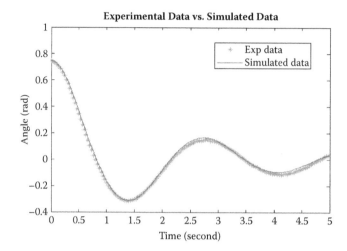

FIGURE 4.39 Experiment data versus simulated data in surge direction [61-62].

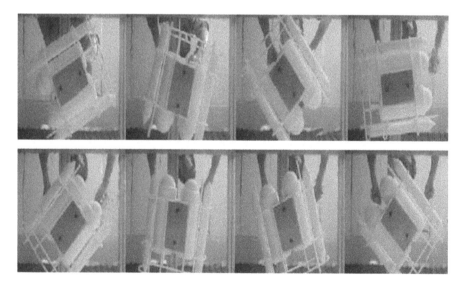

FIGURE 4.40 Image sequence of the scaled model in the yaw direction [61-62].

The results for the scaled model in surge, sway, heave, and yaw at different swing velocity were tabulated in Table 4.15. The root mean square (RMS) error is computed to determine the difference between the angles measured from the simulated and experimental test. It can be seen that the variations of the hydrodynamic parameters or the RMS errors are quite small and consistent. It indicates that the amplitude of the wave created is smaller than the amplitude created by the pendulum motion. Thus, the changes in the speed during the free-decaying test do not affect the hydrodynamic parameters or in other words, the Keulegan–Carpenter number

TABLE 4.15a

Hydrodynamic Parameters at Different Velocity (Surge, Sway and Heave Direction) [61-62]

Direction	Max Speed	Added Mass	K_L	K_Q	RMS Error
Surge	0.55 m/s*	0.5581	1.2736	9.9392	0.0568
		0.6054	1.7768	9.1907	0.0500
		0.5765	1.6040	9.5018	0.0552
		(Avg 0.580)	(Avg 1.550)	(Avg 9.540)	
	0.50 m/s	0.5134	1.2011	9.667	0.0572
		0.5011	1.5800	9.008	0.0685
		0.5125	1.4123	8.995	0.0717
	0.35 m/s	0.3541	1.0192	7.0113	0.0632
		0.4078	0.9988	7.0393	0.0541
		0.4109	1.0078	7.4931	0.0797
Sway	0.55 m/s	1.5491	3.3456	8.0991	0.0266
		1.5578	3.4577	7.9981	0.0372
		1.5713	3.5678	8.0113	0.0213
	0.50 m/s*	1.4347	3.1388	7.8691	0.0277
		1.5465	3.6694	7.4974	0.0234
		1.4878	3.4671	7.3418	0.0269
		(Avg 1.489)	(Avg 3.420)	(Avg 7.560)	
	0.35 m/s	1.2735	3.0069	6.1170	0.0294
		1.1198	2.9977	7.0008	0.0333
		1.1899	2.8997	7.0912	0.0459
Heave	0.55 m/s	5.4789	8.0711	10.0125	0.0240
		4.9989	8.1113	10.0034	0.0331
		5.0811	7.9871	9.9874	0.0423
	0.50 m/s	4.9976	8.0212	9.0982	0.0341
		4.8126	7.8971	9.0124	0.0556
		4.7894	7.8132	9.1492	0.0421
	0.35 m/s*	3.0760	6.5011	8.9937	0.0239
		3.1407	6.8327	8.1637	0.0219
		2.9909	6.1905	11.035	0.0277
		(Avg 3.069)	(Avg 6.520)	(Avg 9.390)	

is considerably low in comparison to those of the offshore structure subjected to wave disturbances. Based on the root mean squared errors (RMS) computations, the values that correspond to the lowest and consistent errors are chosen; as the results, the hydrodynamic parameters at velocity 0.55 m/s (for surge), 0.50 m/s (for sway), 0.35 m/s (for heave), and 0.55 rad/s (for yaw) are used.

As a sanity check on the method, the ANSYS-CFX study of the scaled model for similar flow condition was included. In the CFD test using ANSYS-CFX, the flow in the domain surrounding the vehicle was simulated for various speeds as shown in Table 4.16 and the hydrodynamic loads were obtained by integrating the pressure on its surface, which was obtained via the simulation.

TABLE 4.15b
Hydrodynamic Parameters at Different Velocity (for Yaw Direction) [61-62]

Direction	Max Speed	Added Mass	K_L	K_Q	RMS Error
Yaw	0.35 rad/s	0.0037	0.00036	0.012	0.0311
		0.0023	0.0012	0.009	0.0425
		0.0029	0.0027	0.0074	0.0511
	0.50 rad/s	0.0037	0.007	0.0011	0.0427
		0.0043	0.0024	0.0092	0.0588
		0.0057	0.0017	0.0084	0.0403
	0.55 rad/s*	**0.0065**	**0.00073**	**0.0220**	**0.0320**
		0.0090	**0.0031**	**0.0180**	**0.0300**
		0.0110	**0.0054**	**0.0150**	**0.0380**
		(Avg 0.008)	**(Avg 0.003)**	**(Avg 0.019)**	

TABLE 4.16
Damping (and Added Mass) Coefficients Obtained from Free-Decay Experiment and ANSYS-CFX [61-62]

Methods	Surge (0–0.5 m/s)		Sway (0–0.5 m/s)		Heave (0–0.2 m/s)		Yaw (0–0.5 rad/s)	
	Damping Coefficients							
	K_L	K_Q	K_L	K_Q	K_L	K_Q	K_L	K_Q
ANSYS-CFX	11.863	108.45	19.640	108.24	2.3756	351.98	0	10.390
Experiment (Scaled-up)	17.240	106.03	38.060	84.100	72.530	104.41	1.180	7.5100
Experiment (Scaled)	1.55	9.54	3.42	7.56	6.52	9.39	0.003	0.019
	Added Mass Coefficients							
WAMIT	21.140		51.700		92.450		2.3030	
Experiment (Scaled-up)	21.480		55.170		113.60		0.2960	
Experiment (Scaled)	0.5800		1.4890		3.0690		0.0080	

    In summary, the experimentally or theoretically identified quadratic damping (and even the added mass) parameters are shown in Table 4.16. The average readings at the respective velocity in Table 4.15 are used. Both the theoretical and experimental results indicate that the damping is the smallest in the surge DOF and largest in the heave DOF. This is reasonable given that the vehicle's cross-section area is smallest in the surge DOF and largest in heave DOF. However, the vehicle's linear damping behaves in the opposite manner with a smaller value. This is due to the skin friction being more dominant in the surge DOF than in the heave DOF.

It is difficult to replicate the same surface roughness of the actual ROV model on the scaled model. This can be seen in the values of the linear damping coefficients, which are generally less than the quadratic damping coefficients. However the quadratic damping forces are lower than the linear damping forces due to the *square* of the velocity terms in its computations.

As shown in Table 4.16, the forces show good agreement when the vehicle surges and sways; however, the calculated heave and moments are greater than the experimental results. There are two main reasons for the differences. The first is that the thruster is not considered in the CFD model, which may lead to an error in the simulated hydrodynamic center of rotation and heave motion. The hydrodynamic interaction between the thrusters and the rest of the body is strong, so the effects of the thrusters while the thrusters are operating must be considered. Detailed work on the effects of the operating thrusters in CFD will need to be carried out in future work. The other is that some small changes were made to adjust the center of gravity and buoyancy in the experiments due to the fixture that held the ROV during the test, which implies that they may be different from their design points. Both of these factors have some effect on the dynamic of the vehicle, whereas their effect on the translational dynamics is much smaller than the rotational and vertical dynamics. In addition, it must be noted that this vehicle was operated in the pendulum rig setup with some pitching effect during the experiment, and therefore the pitching may increase the experimental values. This can be seen in the slight increase over the experiment results.

The determination of the added mass terms is more consistent between the free-decay method and CFD values if all the features of the dominating structures are modeled. On the other hand, due to the differences in surface textures, determination of the damping coefficients can be difficult. Nevertheless, in spite of these discrepancies and constraints, the simulated responses compare quite well with the experimentally measured responses both in magnitude and trend, within the limits of experimental errors.

Thus, the parameters determined for the scaled model will be useful only if they can be related to that of the actual or real ROV. As observed in Table 4.16, the scaled-up results for the actual ROV are tabulated. The following scaling method is used to obtain the scaled-up results. For the drag forces, this can be done using the laws of similitude. The magnitude of the drag force acting on a submerged object moving in a rectilinear direction is given by:

$$D = \frac{1}{2}\rho C_d A U^2 \qquad (4.47)$$

where ρ is the fluid density, A is the frontal area, U is the forward velocity and C_d is the drag coefficient, which is a function of Reynolds number, Re [72] for a submerged vehicle.

The scaled model was built with a characteristics length of about $1/0.3$ of the ROV to give geometric similarity. It was tested at a speed of about 0.30–0.5 m/s to give the same Re of about 1.6×10^5 as the ROV operated at a speed of about 0.3 m/s in seawater. This maintains the dynamic similitude between the models. As shown in

Figure 4.34, the drag coefficient is nearly constant for a range of Reynolds numbers from 2.0×10^5 to about 5.0×10^5. Using (4.47), the ratio of drag force experienced by the real or actual model of the ROV (subscript a) to the damping force experienced by the scaled model (subscript s) can be obtained.

$$\frac{D_a}{D_s} = \frac{\rho_a C_d A_a U_a^2}{\rho_s C_d A_s U_s^2} \tag{4.48}$$

The drag force in the rotational direction [73] can be scaled up as shown.

$$T_d = \sum_i r_i f_{di}$$

$$= \frac{1}{2} C_d \rho A_p L^3 \dot{\theta}^2$$

$$\frac{T_a}{T_s} = \frac{\rho_a C_d A_a L_a^3 \dot{\theta}_a^2}{\rho_s C_d A_s L_s^3 \dot{\theta}_s^2} \tag{4.49}$$

The added mass term, scaling factors from the scaled model to ROV can be obtained from [73–74]; the scaling factor was determined heuristically using CFD computations and these can be independently verified [74]. As shown in Table 4.8, the added mass coefficient can be scaled up using the following factors:

$$\begin{bmatrix} R^3 & R^3 & R^3 & R^4 & R^4 & R^4 \\ R^3 & R^3 & R^3 & R^4 & R^4 & R^4 \\ R^3 & R^3 & R^3 & R^4 & R^4 & R^4 \\ R^4 & R^4 & R^4 & R^5 & R^5 & R^5 \\ R^4 & R^4 & R^4 & R^5 & R^5 & R^5 \\ R^4 & R^4 & R^4 & R^5 & R^5 & R^5 \end{bmatrix} \tag{4.50}$$

where the dimension of a real model to scaled model has ratio $R = 1/0.3 = 3.33$.

The graphical plots of the drag force can be seen in Figures 4.41 through 4.44. Comparison should not be done on the basic of K_L versus K_Q. Indeed, only the combination effect of both should be compared by plotting the total drag force within the velocity range of operation. Figures 4.41 and 4.42 show that the drag forces for surge and sway match closely with values from CFX although they are consistently larger. The main reason is that a smooth surface is assumed in ANSYS-CFX computation and this probably underestimates the drag value due to skin friction.

For the heave direction, the experimental values are consistently larger than the ANSYS-CFX values by a constant for velocity range from 0.1 m/s to 0.2 m/s. This is expected because there are two rectangular metal plates attached to the scaled model, which increases the drag in the heave direction (as one can see from Figure 4.43). In ANSYS CFX study, the two iron plates do not exist in a full-scale ROV.

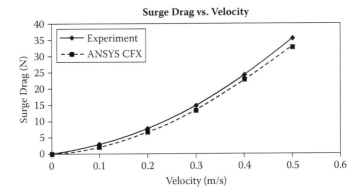

FIGURE 4.41   Comparison of damping force along surge direction [61-62].

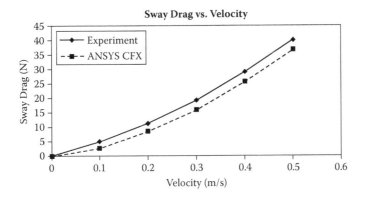

FIGURE 4.42   Comparison of damping force along sway direction [61-62].

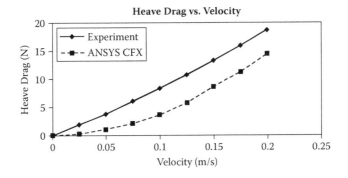

FIGURE 4.43   Comparison of damping force along heave direction [61-62].

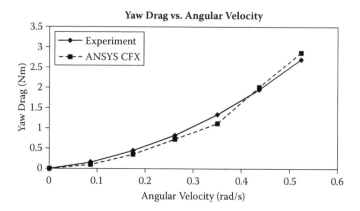

FIGURE 4.44   Comparison of damping force in yaw direction [61-62].

The results for the yaw direction from both experiment and ANSYS CFX are closely matched as shown in Figure 4.44. This is an unexpected result as calculating the drag moment is considered difficult by the CFD community. The experiment data for ROV was obtained by scaling up the drag moment from scaled model tests that involve a very big scaling factor. Therefore, any error in scaled model tests will inevitably scale up. This is probably a limitation in the use of the free-decay test for a scale model. Nevertheless, the magnitude of the moment drag is reasonable and could be a good estimation of the true moment drag that is experienced by the ROV.

The damping coefficients for the roll and pitch are obtained using the ANSYS-CFX software. No experimental results are available for these coefficients due to experiment limitations. Recall that the linear damping forces are assumed for the ROV application at low speed. As shown in Table 4.17, the contribution of the linear damping force is greater than the quadratic force on the ROV.

Hence, the linear hydrodynamic damping matrix is considered in the subsequent control and can be written as:

$$
\mathbf{D} = -
\begin{bmatrix}
17.20 & 0 & 0 & 0 & 0 & 0 \\
0 & 38.06 & 0 & 0 & 0 & 0 \\
0 & 0 & 72.50 & 0 & 0 & 0 \\
0 & 0 & 0 & 1.665 & 0 & 0 \\
0 & 0 & 0 & 0 & 1.456 & 0 \\
0 & 0 & 0 & 0 & 0 & 1.180
\end{bmatrix}
\tag{4.51}
$$

The linear hydrodynamic damping matrix can be modeled in Simulink as shown in Figure 4.45.

4.6.4   Buoyancy and Gravitational Forces

The $\mathbf{G_r(\eta)}$ term is used to describe the gravitational and buoyancy [60] vector exerted on the ROV in water. As seen below, the gravitational and buoyancy forces are functions

TABLE 4.17

Contribution of Linear Damping Forces as Compared to Quadratic Damping Force

Percent of Linear to Quadratic Damping Forces

Surge	Sway	Heave	Yaw
0	0	0	0
31%	11%	6%	356%
62%	22%	12%	713%
92%	33%	17%	1069%
123%	44%	23%	1425%
154%	55%	29%	1782%
185%	66%	35%	2138%
216%	77%	41%	2494%
247%	88%	46%	2851%
277%	99%	52%	3207%
308%	110%	58%	3564%
339%	122%	64%	3920%

FIGURE 4.45 Linear hydrodynamic damping matrix in Simulink.

of orientation and are independent of vehicle motion. Transforming the restoring forces to the body-fixed coordinate system, the weight and buoyancy forces can be written as:

$$\mathbf{f_G}(\eta) = \mathbf{J_1^T}(\eta)\begin{bmatrix} 0 \\ 0 \\ W \end{bmatrix} \tag{4.52a}$$

$$\mathbf{f_B}(\eta) = -\mathbf{J_1^T}(\eta)\begin{bmatrix} 0 \\ 0 \\ B \end{bmatrix} \tag{4.52b}$$

where the weight of the ROV is given by $W = mg$ and g is the gravitational acceleration (positive downward) term. When fully submerged, the ROV's buoyancy is equal to the weight of water displaced, that is, $B = \rho g \nabla$ where ρ is the fluid density and ∇ is the volume displaced by the submerged ROV. Here, $\mathbf{J_1}(\eta)$ is the Euler angle coordinate transformation matrix defined in (4.17). In the body-fixed coordinate system, the restoring force vector becomes:

$$\mathbf{G_f}(\eta) = -\begin{bmatrix} \mathbf{f_G}(\eta) + \mathbf{f_B}(\eta) \\ \mathbf{r_G} \times \mathbf{f_G}(\eta) + \mathbf{r_B} \times \mathbf{f_B}(\eta) \end{bmatrix} \qquad (4.53)$$

Expanding (4.53) gives:

$$\mathbf{G_f}(\eta) = \begin{bmatrix} (W - B)\sin\theta \\ -(W - B)\cos\theta\sin\phi \\ -(W - B)\cos\theta\cos\phi \\ -(y_G W - y_B B)\cos\theta\cos\phi + (z_G W - z_B B)\cos\theta\sin\phi \\ (z_G W - z_B B)\sin\theta + (x_G W - x_B B)\cos\theta\cos\phi \\ -(x_G W - x_B B)\cos\theta\sin\phi - (y_G W - y_B B)\sin\theta \end{bmatrix} \qquad (4.54)$$

ASSUMPTIONS 4.3 BUOYANCY AND GRAVITATIONAL FORCE MATRIX

The buoyancy and gravitational force hydrodynamic damping matrix, $\mathbf{G_f}(\eta)$ was computed using the following design rules.

1. In designing ROVs, it is desirable to make the ROV neutrally buoyant or slightly positive buoyant by adding additional float or balancing mass. With that, the RRC ROV becomes neutrally buoyant, $W = B$.
2. By placing additional mass on the ROV to make the X-Y coordinates of the CB coincide with the X-Y coordinate of the CG, that is, $x_G = x_B = 0$, $y_G = y_B = 0$. Equation (4.54) becomes:

$$\mathbf{G_f}(\eta) = \begin{bmatrix} 0 \\ 0 \\ 0 \\ (z_G - z_B)W\cos\theta\sin\phi \\ (z_G - z_B)W\sin\theta \\ 0 \end{bmatrix} \qquad (4.55)$$

The buoyancy and gravitational force matrix can be modeled in Simulink as shown in Figure 4.46.

Result 4.3 Buoyancy and Gravitational Force Matrix The buoyancy force and CB was found via PRO/ENGINEER. A CAD assembly, similar to the one used to generate the rigid-body parameters was created. The difference in this new assembly is that all the empty space in each of its parts is filled up. In order to generate the

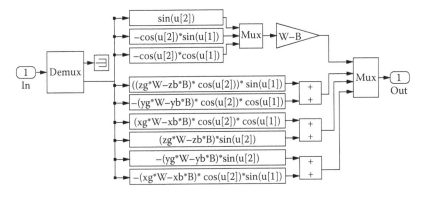

FIGURE 4.46 Buoyancy and gravitational force hydrodynamic damping matrix in Simulink.

buoyancy parameters, the density of each component is equated to the density of water. Therefore, the model now represents the mass of water displaced by the ROV.

Step 1: Before adding additional float (or balancing mass), the buoyancy force and CB with respect to the CG (see Figure 4.47) is found to be:

$$B = 95.25 \times 9.81 = 934 \text{N}$$

$$W = 115 \times 9.81 = 1128 \text{N}$$

$$x_G - x_B = 217 - 231 = -14 \text{ mm} \qquad (4.56)$$

$$y_G - y_B = 0 \text{ mm}$$

$$z_G - z_B = 253 - 283 = -30 \text{ mm}$$

where x_B, y_B, z_B are measured with respect to the CB. Notice that the gravitational force is greater than the buoyancy force $(W > B)$; the ROV is negatively buoyant.

Step 2: After adding additional float (or balancing mass), the ROV becomes neutrally buoyant. The gravitational force due to ROV weight equals the buoyancy force $(W = B)$ due to water displaced by the submerged ROV. Additional mass is placed on the ROV to make the X-Y coordinate of the CB coincides with the X-Y coordinate of the CG, that is, $x_G = x_B = 0$, $y_G = y_B = 0$. Hence, the new buoyancy force and CB with respect to the CG (see Figure 4.48) is found to be:

$$W = B = 115 \times 9.81 = 1128 \text{N}$$

$$x_G - x_B = 0 \text{ mm} \qquad (4.57)$$

$$y_G - y_B = 0 \text{ mm}$$

$$z_G - z_B = -48 \text{ mm}$$

where x_B, y_B, z_B are measured with respect to the center of gravity. The RRC ROV has been made neutrally buoyant $(W = B)$ with CG and CB coinciding in the X and Y

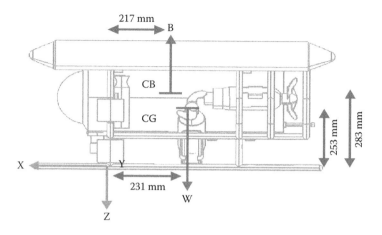

FIGURE 4.47 RRC ROV side view (before addition of balancing mass).

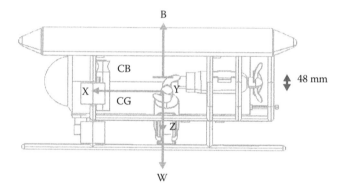

FIGURE 4.48 RRC ROV side view (after addition of balancing mass).

axes. This simply means the ROV would not sink and exhibit self-restoring forces in roll and pitch angular motion when deployed into the water.

4.6.5 THRUSTER'S CONFIGURATION MODEL

The position of the thruster on the ROV (as seen in Figure 4.49) is defined by the thruster's configuration matrix. This is followed by establishing the thrust versus the input voltage relationship model. The details of the steady-state experiment performed on the thruster can be found in Chin et al. [54] and Chin and Lau [75].

Recall that in the nonlinear ROV dynamic equation, the left-hand side of the equation refers to the input forces and moments to the ROV. These input force and moments are determined based on summation of the force and moment equations in the six DOFs:

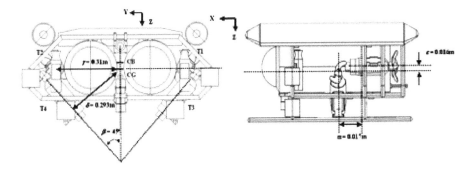

FIGURE 4.49 Thruster's position on the RRC ROV platform.

$$u_1 + u_2 = \tau_x$$

$$u_3 \sin\beta - u_4 \sin\beta = \tau_y$$

$$u_3 \cos\beta + u_4 \cos\beta = \tau_z$$

$$u_3 \delta - u_4 \delta = \tau \tag{4.58}$$

$$-u_1\varepsilon - u_2\varepsilon + u_3\alpha\cos\beta - u_4\alpha\cos\beta = \tau_\theta$$

$$u_1\gamma - u_2\gamma + u_3\alpha\sin\beta - u_4\alpha\sin\beta = \tau_\psi$$

where the $\alpha = 0.017$ m, $\beta = 45°$, $\gamma = 0.31$ m, $\delta = 0.293$ m, $\varepsilon = 0.016$ m are the geometrical parameters based on the thrusters' location on the ROV platform, $u_{min} \le u_i \le u_{max}$, $i = 1, 2, 3, 4$ are the minimum and maximum thrust output by each thruster and $\tau_A = [\tau_x \tau_y \tau_z \tau_\phi \tau_\theta \tau_\psi]^T$ is the force and moment vector generated in the six DOFs.

As observed in (4.58), the roll and pitch motions (that is the fourth and fifth equations) are not fully actuated as the thrusters are positioned in the ROV such that the contributions to these motions are not dominant. The values of α and ε are small compared to the rest of the parameters. In this context, the ROV is underactuated since there are insufficient thrusters (only four thrusters) to fully maneuver the ROV in the six DOFs. Fortunately, the two unactuated DOFs of roll and pitch velocity are asymptotically stable and the roll and pitch angles are bounded. To simplify Equation (4.58) into matrix form:

$$
\begin{bmatrix} \tau_x \\ \tau_y \\ \tau_z \\ \tau \\ \tau_\theta \\ \tau_\psi \end{bmatrix}
=
\overset{T}{\begin{bmatrix}
1 & 1 & 0 & 0 \\
0 & 0 & \sin\beta & -\sin\beta \\
0 & 0 & \cos\beta & \cos\beta \\
0 & 0 & \delta & -\delta \\
-\varepsilon & -\varepsilon & \alpha\cos\beta & -\alpha\cos\beta \\
\gamma & -\gamma & \alpha\sin\beta & -\alpha\sin\beta
\end{bmatrix}}
\overset{u}{\begin{bmatrix} u_1 \\ u_2 \\ u_3 \\ u_4 \end{bmatrix}}
\tag{4.59}
$$

$$\tau_A = Tu$$

$\mathbf{u} = \mathbf{F_T}\bar{\mathbf{u}} \in \Re^4$ with $\mathbf{F_T} = f_T \mathbf{I}_{4\times4}$, $\bar{\mathbf{u}} \in \Re^4$ the commanded voltage input to the thrusters and $\mathbf{I}_{4\times4}$ is an identity matrix of dimensions four. In summary, the relationship between these parameters can be shown in the block diagrams in Figures 4.50 and 4.51.

As shown later, the thruster has a relatively shorter response time compared with the ROV and thus can be modeled as a gain term. This is similar to the approach found in Smallwood and Whitcomb [55–56] where a steady-state thruster model was considered due to its faster response time as compared to the ROV dynamic. An experiment was done to establish the thruster's steady-state model (f_T) between the thrust (u) and the commanded voltage input ($\bar{\mathbf{u}}$).

In the thruster's experiment setup, a PWM servo amplifier was used to provide controlled voltage to drive the thruster. A force sensor (ATI Industrial Automation Force/Torque Sensor model Gamma) with a sensitivity of 0.1 N/count and a range of –130 to +130 N was utilized to measure the thrust exerted by the thruster. The thruster (see Figure 4.52) was saddled on one end of a lever while the force sensor was mounted at the opposite end. The lever transmits the force exerted by the thruster to the sensor at a ratio of 1:1. In order to measure the rotational speed of the propeller, a water-resistance type, optical-fiber speed sensor (Fuji Electric PH21A Series) was used together with a frequency counter.

From the steady-state experimental results [54,75], the lumped hydroelectromechanical dynamic model for both the DC motor shaft speed (assuming small electrical time constant as compared to the mechanical time constant) and propeller's dynamic can be written as:

$$J_m \dot{\Omega} + \frac{K_m K_e}{R_m} \Omega = \frac{K_m}{R_m} \bar{u}_i - Q, \quad i = 1, 2, 3 \text{ or } 4 \tag{4.60a}$$

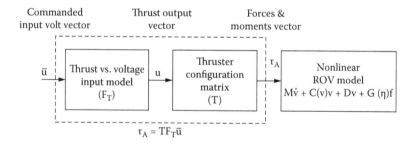

FIGURE 4.50 Block diagram of the thruster's parameters.

FIGURE 4.51 Thruster's configuration matrix in Simulink.

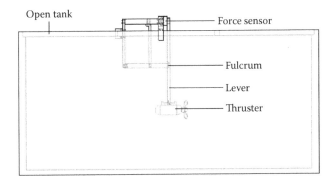

FIGURE 4.52 Thruster test rig.

$$\bar{u}_i = R_m I_t + L_m \frac{dI_t}{dt} + K_e \Omega, \quad i = 1, 2, 3 \text{ or } 4 \quad (4.60b)$$

The steady-state propeller's thrust and torque model can be written as:

$$u_i = K_{Td}\Omega^2, \quad i = 1, 2, 3 \text{ or } 4 \quad (4.60c)$$

$$Q = K_{Qd}\Omega^2 \quad (4.60d)$$

where $\bar{u}_i$ is the voltage input to thruster in V, R_m is the armature resistance in Ohms, I_t is the current to the armature in ampere, L_m is the inductance in henry, K_m is the motor torque constant N·m/A, K_e is the motor back emf in V·s/rad, J_m is the rotor moment of inertia in Nm·s²/rad, Ω is the rotational speed of the propeller in rad/s, Q is the propeller's torque in N·m, u_i is the thrust from each propeller i in Newton, K_{Td} is the thrust constant in N·s/rad and K_{Qd} is the torque constant in N·m·s/rad. Figure 4.18 summarizes the thruster constants used.

By examining (4.60a) through (4.60d), an approximated linear relationship between the thrust (u_i) and the commanded voltage input ($\bar{u}_i$) for the thruster can be plotted as shown in the Figure 4.56. The constants used in (4.60a) through (4.60d) can be tabulated in Table 4.18.

$$u_i = f_T \bar{u}_i, \quad i = 1, 2, 3 \text{ or } 4 \quad (4.61)$$

where $f_T = 0.92$N/V (forward thrust) and $f_T = 0.61$N/V (reverse thrust). The $\mathbf{F_T}$ in Figure 4.56 for both forward and reverse thrust can be determined as follows:

$$\mathbf{F_T} = \begin{bmatrix} 0.92 & 0 & 0 & 0 \\ 0 & 0.92 & 0 & 0 \\ 0 & 0 & 0.92 & 0 \\ 0 & 0 & 0 & 0.92 \end{bmatrix}, \quad (4.62a)$$

TABLE 4.18
Summary of Thruster Constants [75]

Constant	Forward Thrust	Reverse Thrust
K_m	12.562 N·m/A	8.739 N·m/A
R_m	2.4 Ω	2.4 Ω
K_e	0.245 V·s/rad	0.245 V·s/rad
K_{Td}	0.0022 N·s/rad	−0.0013 N·s/rad
K_{Qd}	6.0×10^{-5} N·m·s/rad	-2.0×10^{-5} N·m·s/rad
L_m	0.491 H	0.491 H

$$\mathbf{F_T} = - \begin{bmatrix} 0.61 & 0 & 0 & 0 \\ 0 & 0.61 & 0 & 0 \\ 0 & 0 & 0.61 & 0 \\ 0 & 0 & 0 & 0.61 \end{bmatrix} \tag{4.62b}$$

As shown in Figure 4.53, there is a reasonable correlation between the two variables $\bar{u}_i$ and u_i as the typical value of R falls in the range 0.7 to 1. In Figure 4.53, the forward and linear thrusts were approximated by different linear functions. As observed in the time response of the thrust—theoretical and experimentation in Figure 4.54, there is a good match in the steady-state response though the errors due to this approximation in the region less than 20 volts are quite high. The thruster's time constant is about 0.5 second and is relatively shorter than the ROV response time (due to large inertia). Hence, the thruster dynamic (especially input up to 20 volts) can be ignored, as it is not as dominant as compared to the ROV dynamic. The dead band does affect the position errors, while in practice, it may not necessarily be zero.

4.7   PERTURBED RRC ROV MODEL

The nonlinear equation with perturbation on (4.24a) can be expressed as:

$$\dot{\mathbf{x}} = \bar{\mathbf{f}}(\mathbf{x},t) + \bar{\mathbf{g}}(\bar{\mathbf{u}},t) + \ \mathbf{f}(\mathbf{x},t) + \ \mathbf{g}(\bar{\mathbf{u}},t), \quad \mathbf{x}(0) = \mathbf{x}_0 \tag{4.63}$$

where the full-order states $\mathbf{x} = [\boldsymbol{\eta} \quad \mathbf{v}]^T$ and functions in (4.63) can be defined as:

$$\bar{\mathbf{f}}(\mathbf{x},t) = \begin{bmatrix} \mathbf{0}_{6\times 6} & \mathbf{J}(\boldsymbol{\eta}) \\ \mathbf{0}_{6\times 6} & -\mathbf{M}_{RB}^{-1}[\mathbf{C}_{RB}(\mathbf{v}) + \mathbf{D}] \end{bmatrix} \begin{bmatrix} \boldsymbol{\eta} \\ \mathbf{v} \end{bmatrix} + \begin{bmatrix} \mathbf{0}_{6\times 1} \\ -\mathbf{M}_{RB}^{-1}\mathbf{G}_f(\boldsymbol{\eta}) \end{bmatrix}$$

$$\bar{\mathbf{g}}(\bar{\mathbf{u}},t) = \begin{bmatrix} \mathbf{0}_{6\times 1} \\ -\mathbf{M}_{RB}^{-1}\mathbf{TF_T}\bar{\mathbf{u}} \end{bmatrix}, \quad \bar{\mathbf{u}} \in \Re^4 \tag{4.64}$$

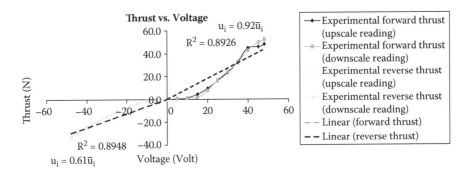

FIGURE 4.53   Thrust versus voltage input to a thruster (R^2 is the correlation factor).

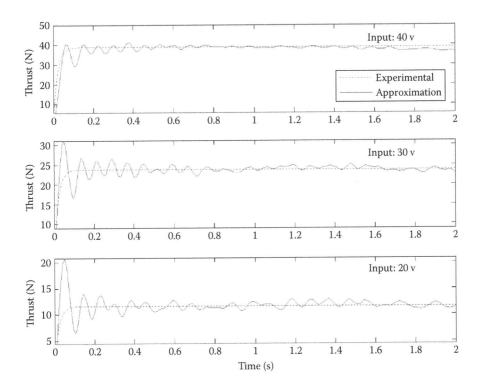

FIGURE 4.54   Time response of the thrust—theoretical versus experimental results.

The perturbation terms in the $\mathbf{f}(\mathbf{x},t)$, $\mathbf{g}(\mathbf{x},\bar{\mathbf{u}},t)$ due to the hydrodynamic added mass and damping forces are defined as:

$$\mathbf{f}(\mathbf{x},t) = \begin{bmatrix} \mathbf{0}_{6\times 1} \\ \mathbf{M}_A^{-1}[\mathbf{C}_A(\mathbf{v}) + \mathbf{D}]\mathbf{v} \end{bmatrix}, \quad \mathbf{g}(\bar{\mathbf{u}},t) = \begin{bmatrix} \mathbf{0}_{6\times 1} \\ \mathbf{M}_A^{-1}\mathbf{T}\mathbf{F}_T\bar{\mathbf{u}} \end{bmatrix}, \quad (4.65)$$

The upper bound for the $\Delta\mathbf{M}_A$, $\Delta\mathbf{C}_A$, and $\Delta\mathbf{D}$ used in (4.65) can be determined by the following sections. As seen in Chapter 5, the perturbed terms are canceled during some of the controllers design stage. And hence, knowledge of the parameters' changes is important for control system design.

4.7.1 PERTURBATION BOUND ON M AND C MATRIX

The uncertainties in the added mass computation can be treated as a perturbation on the nominal model. The sensitivity of the added mass matrix can be seen in various conditions as shown below.

In the first case, the ROV was modeled *without the extra components (such as a small cylinder)* and with no change in the reference frame. The added mass is computed as:

$$\mathbf{M}_A = - \begin{bmatrix} 21.0981 & 0 & 0.0884 & 0 & -2.6127 & 0 \\ 0 & 51.0026 & 0 & 2.9200 & 0 & -1.0534 \\ 0.0692 & 0 & 91.837 & 0 & 1.8883 & 0 \\ 0 & 2.9199 & 0 & 3.5760 & 0 & -0.5795 \\ -2.5988 & 0 & 1.8884 & 0 & 6.7568 & 0 \\ 0 & -1.0537 & 0 & -0.5793 & 0 & 4.4147 \end{bmatrix}$$

$$(4.66)$$

By including the small cylinder and using the same reference frame, the added mass is:

$$\mathbf{M}_A = - \begin{bmatrix} 21.724 & 0 & -0.5683 & 0 & -2.8284 & 0 \\ 0 & 51.5244 & 0 & 3.3080 & 0 & -1.0988 \\ -0.0147 & 0 & 94.300 & 0 & 1.9858 & 0 \\ 0 & 3.3320 & 0 & 3.8803 & 0 & -0.7420 \\ -2.6862 & 0 & 1.9905 & 0 & 7.1148 & 0 \\ 0 & -1.1078 & 0 & -0.7329 & 0 & 4.7673 \end{bmatrix}$$

$$(4.67)$$

In the second case, with the *change in the reference frame to CB*, the added mass becomes:

$$\mathbf{M}_A = - \begin{bmatrix} 21.1398 & 0 & 0.0621 & 0 & 1.0083 & 0 \\ 0 & 51.7013 & 0 & -5.9705 & 0 & 1.588 \\ 0.0918 & 0 & 92.4511 & 0 & -2.9191 & 0 \\ 0 & -5.8866 & 0 & 4.2176 & 0 & -0.1719 \\ 1.0583 & 0 & -2.9489 & 0 & 2.7687 & 0 \\ 0 & 1.5863 & 0 & -0.171 & 0 & 2.3493 \end{bmatrix}$$

$$(4.68)$$

With the change in the reference frame to CG, the final $\mathbf{M}_A$ becomes:

$$\mathbf{M}_A = -\begin{bmatrix} 21.1403 & 0 & 0.0619 & 0 & -0.5748 & 0 \\ 0 & 51.7012 & 0 & -2.0928 & 0 & -0.3767 \\ 0.0917 & 0 & 92.4510 & 0 & 0.5871 & 0 \\ 0 & -2.0090 & 0 & 3.6191 & 0 & 0.0235 \\ -0.5237 & 0 & 0.5594 & 0 & 2.6427 & 0 \\ 0 & -0.3783 & 0 & 0.0275 & 0 & 2.3033 \end{bmatrix}$$

(4.69)

As can be observed from (4.66) through (4.69), changes in the off-diagonal components are quite small compared with the diagonal counterparts. Thus, the perturbation $\Delta\mathbf{M}_A$ comprises only the diagonal terms, with all the maximum values as the upper bound.

$$\overline{\mathbf{M}}_A = -\begin{bmatrix} 21.7240 & 0 & 0 & 0 & 0 & 0 \\ 0 & 51.7013 & 0 & 0 & 0 & 0 \\ 0 & 0 & 94.3000 & 0 & 0 & 0 \\ 0 & 0 & 0 & 4.2176 & 0 & 0 \\ 0 & 0 & 0 & 0 & 7.1148 & 0 \\ 0 & 0 & 0 & 0 & 0 & 4.7673 \end{bmatrix}$$

(4.70)

In addition, as shown in Table 4.19, at different temperatures, the seawater density with 3.5% salinity (as compared to water at 1000 kg/m³) varies. To account for the differences in density at various temperatures, a factor is included in the perturbation $\Delta\mathbf{M}_A$ as follows:

$$\mathbf{M}_A \leq 1.028 \ \overline{\mathbf{M}}_A$$

(4.71)

Similarly, the upper bound on the uncertainties in the Coriolis and centripetal added mass matrix, $\mathbf{C}_A(\mathbf{v})$ can be expressed as:

$$\mathbf{C}_A(\mathbf{v}) \leq 1.028 \ \overline{\mathbf{C}}_A(\mathbf{v})$$

(4.72)

where

$$\overline{\mathbf{C}}_A(\mathbf{v}) = \begin{bmatrix} 0 & 0 & 0 & 0 & -a_3^u & a_2^u \\ 0 & 0 & 0 & a_3^u & 0 & -a_1^u \\ 0 & 0 & 0 & -a_2^u & a_1^u & 0 \\ 0 & -a_3^u & a_2^u & 0 & -b_3^u & b_2^u \\ a_3^u & 0 & -a_1^u & b_3^u & 0 & -b_1^u \\ -a_2^u & a_1^u & 0 & -b_2^u & b_1^u & 0 \end{bmatrix}$$

(4.73)

TABLE 4.19

Seawater Density at Various Temperatures

Medium	0°C	10°C	20°C	30°C
Seawater (3.5% salinity)	1028.48 kg/m³	1026.91 kg/m³	1024.85 kg/m³	1021.81 kg/m³

$$a_1^l = 21.0981u, \quad a_1^u = 21.7240u$$

$$a_2^l = 51.0002v, \quad a_2^u = 51.7013v$$

$$a_3^l = 91.8370w, \quad a_3^u = 94.3000w$$

$$b_1^l = 3.5760p, \quad b_1^u = 4.2176p \tag{4.74}$$

$$b_2^l = 2.7687q, \quad b_2^u = 7.1148q$$

$$b_3^l = 2.4393r, \quad b_3^u = 4.7673r$$

4.7.2 PERTURBATION BOUND ON D MATRIX

The uncertainties may occur due to skin friction and the numerical results obtained from CFD software. To account for this, the perturbation on the linear damping ΔD can be approximated as:

$$D = - \begin{bmatrix} 0.370 & 0 & 0 & 0 & 0 & 0 \\ 0 & 0.291 & 0 & 0 & 0 & 0 \\ 0 & 0 & 0.237 & 0 & 0 & 0 \\ 0 & 0 & 0 & 0.166 & 0 & 0 \\ 0 & 0 & 0 & 0 & 0.145 & 0 \\ 0 & 0 & 0 & 0 & 0 & 0.011 \end{bmatrix} \tag{4.75}$$

The perturbation ΔD is taken to be about 10% of the nominal matrix. The Simulink block diagram for $M, C(v)$ and D perturbation modeling are shown in Figure 4.55. As shown, the perturbation is added onto the nominal model as an additive perturbation.

In summary, the parameters used in the nonlinear RRC ROV can be seen in Table 4.20. These parameters are used in the subsequent ROV control system design and analysis.

4.8 VERIFICATION OF ROV MODEL

With the parameters obtained from the water tank experiment, [61-62] ANSYS-CFX and WAMIT, it is important to verify these parameters in an actual pool test [61-62]. The basic concept is to use the simulation of the real model with parameters obtained from the water tank experiment and software, and subsequently compare them with the actual responses of the ROV in pool tests for the same thrusters' inputs. If the simulation

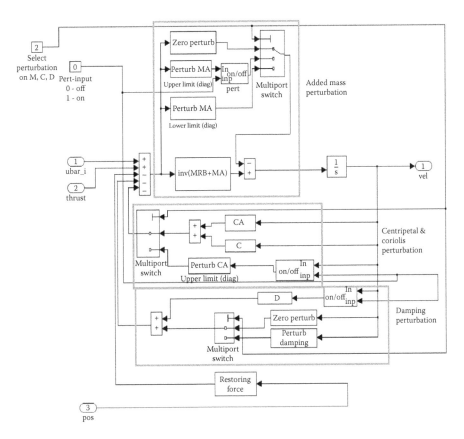

FIGURE 4.55 Perturbation added onto the nominal model as an additive perturbation.

model responses match well with the measured responses in the pool tests, the model and the parameters are acceptable for subsequent control purpose.

The simulation of the dynamic model of the ROV and the thruster's model are implemented with Simulink (Figure 4.56). Due to the limitation of the sensors, the ROV was tested on the surge, heave and yaw motion. The mass and buoyancy force (= 12N) and the three scaled hydrodynamic parameters of added mass, linear drag, quadratic drag are given in Table 4.19 (or in Table 4.17). In the subsequent discussion, *simulated responses* mean results of the model using the estimated parameters. As the DVL (Doppler velocity log) sensor was not working during the time of the experiment, it was not used. The velocity plots for the experiments were obtained through differentiation of the position data using an explicit Runge–Kutta formula. It is a one-step solver; that is, in computing $y(t)$, it needs only the solution at the immediately preceding time point, $y(t-1)$. For this reason, the Runge–Kutta is the default solver used for the ROV model with continuous states.

In the pool tests, the ROV (Figure 4.57) is supplied with a 230 VAC electrical power via an umbilical cable, which also carries the communication and video signal lines. The control system embedded inside the ROV consists of an industrial controller

TABLE 4.20
Parameters Used in Nonlinear RRC ROV Dynamic

[13] Parameters	[14] Values

[15] Mass inertia matrix (rigid body)

$$[16] \quad M_{zz} = \begin{bmatrix} 115.0000 & 0 & 0 & 0 & 0 & 0 \\ 0 & 115.0000 & 0 & 0 & 0 & 0 \\ 0 & 0 & 115.0000 & 0 & 0 & 0 \\ 0 & 0 & 0 & 6.1000 & -0.00016 & -0.1850 \\ 0 & 0 & 0 & -0.00016 & 5.9800 & 0.0006 \\ 0 & 0 & 0 & -0.1850 & 0.0006 & 5.5170 \end{bmatrix}$$

[17] Weight of ROV/Center of gravity

$$[18] \quad W = 115 \text{ kg, } r_G = [x_G, y_G, z_G]^T = [0,0,0]^T$$

[19] Mass inertia matrix (added mass)

$$[20] \quad M_A = - \begin{bmatrix} 21.1403 & 0 & 0 & 0 & 0 & 0 \\ 0 & 51.7012 & 0 & 0 & 0 & 0 \\ 0 & 0 & 924510 & 0 & 0 & 0 \\ 0 & 0 & 0 & 3.6191 & 0 & 0 \\ 0 & 0 & 0 & 0 & 2.6427 & 0 \\ 0 & 0 & 0 & 0 & 0 & 2.3033 \end{bmatrix}$$

[21] Coriolis and centrifugal matrix (rigid body)

$$[22] \quad C_{12}(v) = \begin{bmatrix} 0 & 115.0000w & -115.0000v \\ -115.00004w & 0 & 115.0000u \\ 115.00004v & -115.0000u & 0 \end{bmatrix}$$

$$[23] \quad C_{22}(v) = \begin{bmatrix} 0 & -0.0006g + 0.185\sigma + 5.5170\sigma & 0.0006v - 0.0002\sigma - 5.9800g \\ 0.0006g - 0.185\sigma - 5.5170\sigma & 0 & 0.1850\sigma + 0.0002g + 6.100\sigma \\ -0.0006\sigma + 0.0002g + 5.9800g & -0.1850\sigma - 0.0002g - 6.100\sigma & 0 \end{bmatrix}$$

[24] Coriolis and centrifugal matrix (added mass)

$$[25] \quad C_A(v) = \begin{bmatrix}
0 & 0 & 0 & 0 & -924510v & 51.7013v \\
0 & 0 & 0 & 924510v & 0 & -211400v \\
0 & 0 & 0 & -51.7013v & 211403v & 0 \\
0 & -924510v & 51.7013v & 0 & -2.3033 & 26427\sigma \\
924510v & 0 & -21.1403v & 2.3033 & 0 & -3.6191v \\
-51.7013v & 211403v & 0 & -26427\sigma & 3.6191v & 0
\end{bmatrix}$$

[26] Hydrodynamic damping matrix

$$[27] \quad D = -\begin{bmatrix}
17.20 & 0 & 0 & 0 & 0 & 0 \\
0 & 38.06 & 0 & 0 & 0 & 0 \\
0 & 0 & 72.50 & 0 & 0 & 0 \\
0 & 0 & 0 & 1.665 & 0 & 0 \\
0 & 0 & 0 & 0 & 1.456 & 0 \\
0 & 0 & 0 & 0 & 0 & 1.180
\end{bmatrix}$$

[28] Center of buoyancy (CB)

$$[29] \quad T = \begin{bmatrix}
1 & 1 & 0 & 0 \\
0 & 0 & 0.707 & -0.707 \\
0 & 0 & 0.707 & 0.707 \\
0 & 0 & -0.293 & 0.293 \\
-0.016 & -0.016 & 0.012 & -0.012 \\
0.31 & -0.31 & 0.012 & -0.012
\end{bmatrix}$$

[30] Thruster configuration

$$[31] \quad [x_G - x_F, y_G - y_F, z_G - z_F]^T = [0, 0, -0.048]^T$$

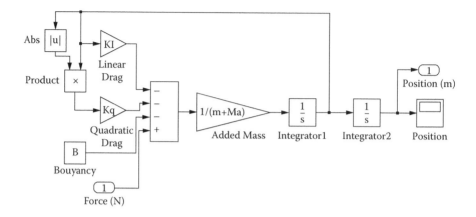

FIGURE 4.56   Simulation block diagram of the real model of the ROV.

FIGURE 4.57   ROV setup for pool test.

(an AMD 5×86-133 processor with 6 MB RAM) running a QNX real-time operating system (QNX-RT OS), thrusters' drivers, and a navigation and controller module. All control algorithms were coded in C-language and that include the thrusters' driver for the thrusters' speed and direction control and the navigation module such as the velocity doppler, magnetic compass and pressure sensors. The graphic user interface (GUI) with a joystick provides the real-time command input to move the ROV, and to direct the orientation of the pan-tilt mechanism of the camera via RS232 or LAN. The outputs from the navigation sensors were saved in the main PC for offline analysis.

The ROV was designed to be tested in the desired directions, with minimum effects due to the tether. The tether was arranged in such a manner that it would not pose an external disturbance to the ROV during the motion. For example, when conducting experiments for surge, the ROV needs to be maintained at a certain depth and only the horizontal thrusters are activated to provide only the forward motion. Input voltages to the thrusters were recorded and the vehicle responses were obtained using the sensors suite given in Table 4.21.

TABLE 4.21

Scaled Up Hydrodynamic Parameters for Surge, Heave, and Yaw of the ROV [61-62]

Axis	Mass/Moment of Inertia	Added Mass	Linear Drag	Quadratic Drag	Buoyancy
Surge	115 kg	21.4 kg	17.2 N/ms^{-1}	106.032 N/(ms^{-1})^2	0
Heave	115 kg	51.7 kg	72.5 N/ms^{-1}	104.4 N/(ms^{-1})^2	12 N
Yaw	9.6 kg-m^2	2.3 kg-m^2	1.2 Nm/rads^{-1}	7.5 Nm/(rads^{-1})^2	0

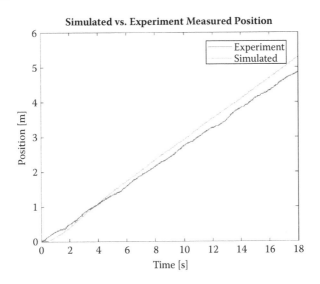

FIGURE 4.58   Experimental versus simulated position in a positive surge movement [61-62].

In the surge DOF test, the ROV was commanded to move 5 m forward (see Figure 4.58) with speed of 0.3 m/s as shown in Figure 4.59. At time equal to zero, the vehicle is static, so a relatively large force input as seen in Figure 4.60 is required to overcome the vehicle inertia. After around 4 s, when the vehicle is moving at 0.3 m/s, the force of 20 N is required to overcome the inertia of the vehicle. The simulated results match the experiment results when the ROV travels around 5 m forward. The maximum error is less than 0.5 m or 10% of the travel distance. However, the match between the simulated responses and experimental results is still quite reasonable.

For the heave DOF test, the vehicle is fully submerged to a depth of about 1.3 m initially. The vehicle is made to hover at depth of about 1.3 m for 40 seconds before moving deeper to 2.6 m. During the hovering period, a thrust of 12 N as shown in Figure 4.61 is needed. As the ROV is designed to be neutral buoyant in water, the buoyancy force is identified to be 12 N. As shown in Figures 4.62 and 4.63, the simulated position and velocity responses follow the measured response closely.

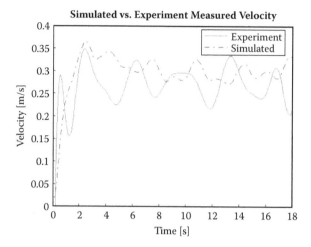

FIGURE 4.59 Experimental versus simulated speed in a positive surge movement [61-62].

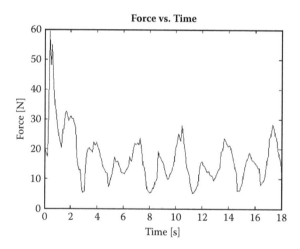

FIGURE 4.60 Thrust for surge verification [61-62].

For the yaw DOF test, the ROV was commanded to rotate 4.8 radians or 275 degrees within 5 seconds as shown in Figure 4.64. The ROV maintained a fixed heading for about 10 seconds. The position and velocity simulated responses in Figures 4.65 and 4.66 match reasonably well with the experimentally measured responses. Compared with the surge and heave, the results of the simulation and the experiment measurements of yaw are slightly better as the hydrodynamic force is lower in this direction.

In conclusion, the free-decay test has been proposed to identify the hydrodynamic parameters of the ROV obtained from the WAMIT and ANSYS-CFX. The test uses the pendulum swing motion to identify the added mass and the drag coefficients

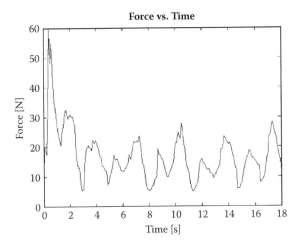

FIGURE 4.61   Thrust for heave verification [61-62].

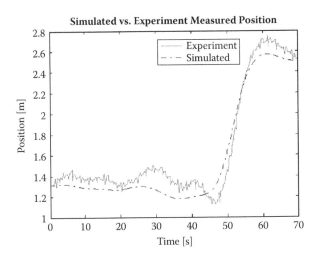

FIGURE 4.62   Experimental versus simulated depth in a heave movement [61-62].

in surge, heave, and yaw directions. The responses of the real model simulation compared relatively well with the experimentally measured responses of the ROV in the pool test. The surge direction results appear the worst, although the estimated parameters compared favorably with the CFD results. A possible reason is the large hydrodynamic forces in this direction and this could imply the possibility of uncertainty in the parameters obtained. Besides, the effect of the tether can be dominant as the distance increased from its initial starting position. According to Whitcomb and Yoerger [76], an estimate of the cable tension is about 10 N. With the maximum force generated by thrusters of about 50 N, the tether force may contribute up to 20% of the total forces and moments applied to the ROV. However, the force exerted by

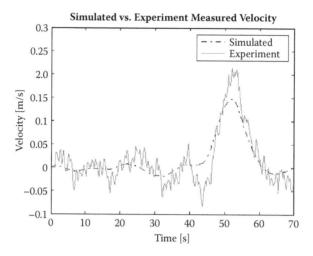

FIGURE 4.63 Experimental versus simulated depth velocity in a heave movement [61-62].

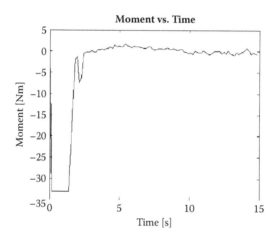

FIGURE 4.64 Torque for yaw verification [61-62].

the tether is not easy to predict as the cable may form different curvature during the
ROV motion and the dynamics are time varying.

As the simulated responses are obtained using the parameters estimated from
the proposed free-decay methods and subsequently scaled up to the real ROV, the
free-decay test in the water tank has shown here could be one of viable alternative
to estimate some pertinent hydrodynamic parameters at the early phase of ROV development. Although the experiment setup is considered to be simple, it estimates the
hydrodynamic parameters in an intuitive manner without exhaustive instrumentation,
manpower, and facilities. In particular, it is useful for the researcher if the test facilities are not easily available. Besides, the CFD method is quite useful and economical
to estimate the hydrodynamic coefficients, especially during the design stages.

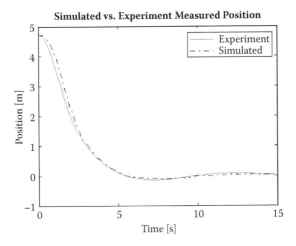

FIGURE 4.65　Experimental versus simulated yaw movement [61-62].

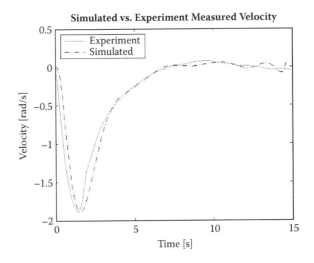

FIGURE 4.66　Experimental versus simulated yaw movement [61-62].

However, the nature and setup of this method imposes a constraint on the maximum size and speed of the ROV in which the parameters can be determined. Experiment errors could result if the motion is inappropriately constrained in the plane of motion and the test conditions are not properly set up. For example, fluid surrounding the scaled ROV needs to be unbounded. It indirectly implies that the water tank experiment is not acceptable since it is not large enough to be considered as unbounded. However, the data obtained can be used as a first estimation of the ROV model for subsequent control systems design. Nevertheless, for control design purposes, there is always a need to first consider a simplified nominal model and later an additive perturbation bounds on the model to represent the associate modeling errors. However, the parameters have to be tested and compared with existing method in the literature.

Lastly, the accuracy of the hydrodynamic coefficients estimated in this study can be improved by using the GPS antenna, ballast releaser, obstacle detection sonar, and transponder (SSBL). In addition, it would be preferable to increase the size of the simulation domain if more computational power were available. Also, the hydrodynamic interaction between the thrusters and the rest of the body is strong, so the effects of the thrusters while the thrusters are operating must be considered. Detailed work on the effects of the operating thrusters in CFD will be carried out in future work. Sensitivity study of the ROV to variations in the hydrodynamic parameters needs to be analyzed.

5 Control of a Remotely Operated Vehicle

5.1 NONLINEAR ROV SUBSYSTEM MODEL

Remotely operated vehicles (ROVs) have become an important tool in diverless subsea operations. This is mainly because of the increased operating range and depth. In addition, operation endurance and less risk to human life are also important factors. Especially in the offshore industry, ROVs have become indispensable. Most oil companies are using such vehicles for various underwater problems in all parts of the oil/gas production line such as: deep water drill rig support, pipeline survey (see Figure 5.1), shallow water rig support, and deep water platform inspection, maintenance and repair (IMR).

Many of the tasks require that the ROVs are equipped with manipulator arms to perform object handling, assembling/operation of underwater apparatus, underwater welding, and so forth. ROVs are usually connected to a surface ship by a tether, through which all communication is wired. It often receives power through the tether. Drag from the tether will influence the vehicle's motion, and may represent some disturbances and energy loss. Hence, besides the model uncertainties and nonlinearities, external disturbances such as the tether drag force could influence the maneuverability of the vehicles. Thus, there are numerous controllers designed to control the ROVs under these adverse conditions during the ROV's pipeline tracking; just to name a few controllers: proportional-integral-derivative (PID) control, adaptive control, sliding-mode control, hybrid switching control, neural network control, fuzzy control, model predictive control, and many other control schemes.

These controllers are often used due to different reasons. For example, it is difficult to fine-tune the control gains during underwater cruising. Therefore, it is sometimes desirable to have an adaptive ROV control system that has a self-adaptive ability when the control performance degrades during operation due to changes in the dynamics of the ROV and its environment. Some engineers prefer to use a fixed model-based type to control the ROV and set a perturbation bound on its nominal model. This creates some robustness in the control system if the model does not deviate too much from the preset bound. However, there are many of the ROV's tasks, which cannot be handled by preprogrammed artificial intelligence. Human skill, judgment, and experience are still required when unpredictable tasks are to be performed. For ROVs, which are controlled in six degrees of freedom (DOFs), local autonomy is required up to some extent. Simultaneously, six DOF station keeping or tracking of pipelines is difficult to perform for human operators. Hence, the need for local intelligence is increased. Supervisory control has become an important aid for high-level teleoperation of ROVs. Besides, the ROV motions are highly coupled.

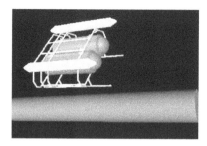

FIGURE 5.1   ROV performing underwater pipelines tracking.

In order to reduce the motion coupling, a decoupled maneuver control scheme is often used. The ROV could move either in a horizontal plane or in a vertical plane. Thus, considering only a few degrees of freedom at a time decouples the motions. This reduces the number of controllable degrees of freedom involved in control system design.

With that, a station-keeping model is used to describe the ROV dynamic during point stabilization on the pipeline while the horizontal and vertical plane models are used for pipeline tracking. Recall, the roll and pitch motions are not fully actuated as the thrusters are positioned in the ROV such that the contributions to these motions are not dominant. In this context, the RRC ROV is underactuated since there are insufficient sets of thrusters (only four thrusters) to fully maneuver the ROV in the six DOFs. However, the unactuated roll and pitch rates are stable and they can be left uncontrolled in the subsequent control system design since they will not destabilize the system stability and nominal performance. This is mainly due to the metacentric height being sufficiently large to provide adequate static stability to enable the gravitational and buoyancy force to act as self-restoring moments in these unactuated motions. The same phenomena can be observed in some working ROVs where the roll and pitch are also self-stabilizable. Intuitively, this reduces the number of DOFs for control as the additional thrusters needed for these unactuated states are not required. Hence, it significantly reduces the computational complexity for subsequent control system analysis and design without affecting the generic dynamic of the ROV.

5.1.1   STATION-KEEPING MODEL

A station-keeping model is used to describe the ROV dynamic during the hovering or (commonly known as) *stabilization about equilibrium*. As the roll and pitch motions are unspecified in the equilibrium and they are asymptotically stable and bounded, they are neglected in the station-keeping mode. It mainly involves four DOFs such as: surge, sway, heave, and yaw. In this section, the nominal and perturbed models for the station-keeping dynamics are formulated.

The nominal nonlinear dynamic equation of the station-keeping model can be written as:

$$(m - X_{\dot{u}})\dot{u} + X_u u + (Y_{\dot{v}} - m)vr = \tau_x \qquad (5.1a)$$

$$(m - Y_{\dot{v}})\dot{v} + Y_v v + (m - X_{\dot{u}})ur = \tau_y \tag{5.1b}$$

$$(m - Z_{\dot{w}})\dot{w} + Z_w w = \tau_z \tag{5.1c}$$

$$(I_z - N_{\dot{r}})\dot{r} + N_r r + (-Y_{\dot{v}}u + X_{\dot{u}}u)v = \tau_\psi \tag{5.1d}$$

The kinematics equation can be written as:

$$\dot{x} = u\cos\psi - v\sin\psi \tag{5.2a}$$

$$\dot{y} = u\sin\psi + v\cos\psi \tag{5.2b}$$

$$\dot{z} = w \tag{5.2c}$$

$$\dot{\psi} = r \tag{5.2d}$$

In compact form, the nonlinear dynamic equation in (5.1a) through (5.1d) can be written as (variables labeled with subscript s):

$$\dot{\mathbf{v}}_s = -\mathbf{M}_s^{-1}[\mathbf{C}_s(\mathbf{v}_s) + \mathbf{D}_s]\mathbf{v}_s + \mathbf{M}_s^{-1}\tau_s \tag{5.3}$$

where

$$\mathbf{v}_s = \begin{bmatrix} u & v & w & r \end{bmatrix}^T, \quad \tau_s = \mathbf{T}_s\mathbf{F}_T\bar{\mathbf{u}} = [\tau_x, \tau_y, \tau_z, \tau_\psi]^T$$

in which

$$\bar{\mathbf{u}} \in \mathfrak{R}^4, \quad \mathbf{F}_T \in \mathfrak{R}^{4\times4},$$

$$\mathbf{T}_s = \begin{bmatrix} 1 & 1 & 0 & 0 \\ 0 & 0 & \sin\beta & -\sin\beta \\ 0 & 0 & \cos\beta & \cos\beta \\ \gamma & -\gamma & \alpha\sin\beta & -\alpha\sin\beta \end{bmatrix};$$

$$\mathbf{M}_s = \begin{bmatrix} m - X_{\dot{u}} & 0 & 0 & 0 \\ 0 & m - Y_{\dot{v}} & 0 & 0 \\ 0 & 0 & m - Z_{\dot{w}} & 0 \\ 0 & 0 & 0 & I_r - N_{\dot{r}} \end{bmatrix}$$

$$\mathbf{D}_s = \begin{bmatrix} X_u & 0 & 0 & 0 \\ 0 & Y_v & 0 & 0 \\ 0 & 0 & Z_w & 0 \\ 0 & 0 & 0 & N_r \end{bmatrix};$$

$$\mathbf{C}_s(\mathbf{v}_s) = \begin{bmatrix} 0 & (Y_{\dot{v}} - m)r & 0 & 0 \\ 0 & 0 & 0 & (m - X_{\dot{u}})u \\ 0 & 0 & 0 & 0 \\ 0 & -Y_{\dot{v}}u + X_{\dot{u}}u & 0 & 0 \end{bmatrix}$$

The kinematics equation in (5.2a) through (5.2d) can be written as compact form:

$$\dot{\boldsymbol{\eta}}_s = \mathbf{J}_s(\boldsymbol{\eta}_s)\mathbf{v}_s \tag{5.4}$$

where

$$\boldsymbol{\eta}_s = \begin{bmatrix} x & y & z & \psi \end{bmatrix}^T \quad \text{and} \quad \mathbf{J}_s(\boldsymbol{\eta}_s) = \begin{bmatrix} \cos\psi & -\sin\psi & 0 & 0 \\ \sin\psi & \cos\psi & 0 & 0 \\ 0 & 0 & 1 & 0 \\ 0 & 0 & 0 & 1 \end{bmatrix}.$$

Equations (5.3) and (5.4) can be rewritten in state-space form:

$$\dot{\mathbf{x}}_s = \overline{\mathbf{f}}_s(\mathbf{x}_s, t) + \overline{\mathbf{g}}_s(\overline{\mathbf{u}}, t) \tag{5.5}$$

where
$\mathbf{x}_s = [\boldsymbol{\eta}_s \mathbf{v}_s]^T$ and

$$\overline{\mathbf{f}}_s(\mathbf{x}_s, t) = \begin{bmatrix} \mathbf{J}_s(\boldsymbol{\eta}_s)\mathbf{v}_s \\ -\mathbf{M}_s^{-1}[\mathbf{C}_s(\mathbf{v}_s) + \mathbf{D}_s]\mathbf{v}_s \end{bmatrix} \tag{5.6}$$

and

$$\overline{\mathbf{g}}_s(\overline{\mathbf{u}}, t) = \begin{bmatrix} \mathbf{0}_{3\times 1} \\ \begin{bmatrix} \overline{\mathbf{g}}_{s1} & \overline{\mathbf{g}}_{s2} & \overline{\mathbf{g}}_{s3} & \overline{\mathbf{g}}_{s4} \end{bmatrix} \mathbf{T}_s \mathbf{F}_T \overline{\mathbf{u}} \end{bmatrix} \tag{5.7}$$

where

$$\overline{\mathbf{g}}_{s1} = \begin{bmatrix} \dfrac{1}{m - X_{\dot{u}}} & 0 & 0 & 0 \end{bmatrix}^T; \quad \overline{\mathbf{g}}_{s2} = \begin{bmatrix} 0 & \dfrac{1}{m - Y_{\dot{v}}} & 0 & 0 \end{bmatrix}^T; \tag{5.8}$$

$$\bar{\mathbf{g}}_{s3} = \begin{bmatrix} 0 & 0 & \dfrac{1}{m - Z_{\dot{w}}} & 0 \end{bmatrix}^{\mathrm{T}} ; \quad \bar{\mathbf{g}}_{s4} = \begin{bmatrix} 0 & 0 & 0 & \dfrac{1}{I_r - N_{\dot{r}}} \end{bmatrix}^{\mathrm{T}} \qquad (5.9)$$

The corresponding perturbed model due to hydrodynamic model uncertainties can be written in the form:

$$\dot{\mathbf{x}}_s = \bar{\mathbf{f}}_s(\mathbf{x}_s,t) + \bar{\mathbf{g}}_s(\bar{\mathbf{u}},t) + \Delta\mathbf{d}_s(\mathbf{x}_s,\bar{\mathbf{u}},t) \qquad (5.10)$$

where

$$\Delta\mathbf{d}_s(\mathbf{x}_s,\bar{\mathbf{u}},t) = \Delta\mathbf{f}_s(\mathbf{x}_s,t) + \Delta\mathbf{g}_s(\bar{\mathbf{u}},t)$$

$$= \begin{bmatrix} \mathbf{0}_{4\times1} \\ \Delta\mathbf{M}_A^{-1}[\Delta\mathbf{D} + \Delta\mathbf{C}_A(\mathbf{v}_s)]\mathbf{v}_s \end{bmatrix} + \begin{bmatrix} \mathbf{0}_{4\times1} \\ \Delta\mathbf{M}_A^{-1}\mathbf{T}_s\mathbf{F}_T\bar{\mathbf{u}} \end{bmatrix} \qquad (5.11)$$

are the system uncertainties in the hydrodynamic added mass and damping forces terms. The upper bound of $\Delta\mathbf{M}_A$, $\Delta\mathbf{C}_A(\mathbf{v}_s)$ and $\Delta\mathbf{D}$ can be obtained in Section 4.7.

5.1.2 HORIZONTAL AND VERTICAL PLANE SUBSYSTEM MODELS

From the ROV mass inertia, centripetal, and Coriolis matrices, it can be seen that the nonlinear ROV motions are highly coupled, although the roll and pitch motions are designed to be self-stabilizable. To further reduce the motion coupling, adopting a decoupled (see Figure 5.2) maneuver scheme is suggested: moving with a horizontal plane motion whereby the vehicle is first steered at constant forward speed along a straight line that passes through the desired target points at a certain vehicle orientation. At the end of this horizontal plane motion, the vehicle is switched to a vertical plane motion: it maintains a constant orientation (heading) and follows a straight line with a change in altitude.

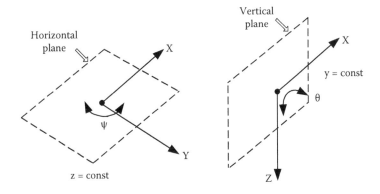

FIGURE 5.2 Horizontal (left) and vertical plane (right) motion.

Thus, considering only a few degrees of freedom at a time decouples the motions. This reduces the number of controllable degrees of freedom involved in subsequent control system analysis and design. With that, the subsystems become actuated.

In summary, the ROV motions can be decoupled into two simple subsystems such as the:

- Horizontal plane subsystem: Velocity states (u, v, r) and position states (x, y, ψ).
- Vertical plane subsystem: Velocity states (u, w, q) and position states (x, z, θ).

These represent the ROV at different plane maneuvering conditions. To begin with, the total mass inertial matrix can be written as:

$$M = M_{RB} + M_A$$

$$= \begin{bmatrix} 93.8597 & 0 & 0 & 0 & 0 & 0 \\ 0 & 63.3000 & 0 & 0 & 0 & 0 \\ 0 & 0 & 22.5490 & 0 & 0 & 0 \\ 0 & 0 & 0 & 2.4809 & -0.0002 & -0.1850 \\ 0 & 0 & 0 & -0.0002 & 3.3373 & 0.0006 \\ 0 & 0 & 0 & -0.1850 & 0.0006 & 7.2867 \end{bmatrix} \qquad (5.12)$$

Extracting the rows and columns of the two subsystem's DOFs in (5.12) gives the total mass inertial matrix of the two subsystems as:

$$M_h = \begin{bmatrix} 93.8597 & 0 & 0 \\ 0 & 63.3000 & 0 \\ 0 & 0 & 7.2867 \end{bmatrix} \qquad (5.13)$$

$$M_v = \begin{bmatrix} 93.8597 & 0 & 0 \\ 0 & 22.5490 & 0 \\ 0 & 0 & 3.3373 \end{bmatrix} \qquad (5.14)$$

The horizontal plane model describes the ROV dynamic during longitudinal motion that encompasses three DOFs such as: surge, sway, and yaw. The roll, pitch, and heave velocities are bounded as the ROV is following along the X-axis with z a constant (that is $w = \dot{w} = 0$) and at a constant surge velocity that defines the horizontal plane dynamic. With the origin of the ROV coinciding with the center of gravity, the nominal horizontal plane dynamic can be written as:

$$(m - X_{\dot{u}})\dot{u} - X_u u + (Y_{\dot{v}} - m)vr = \tau_x \qquad (5.15a)$$

$$(m - Y_{\dot{v}})\dot{v} - Y_v v + (m - X_{\dot{u}})ur = \tau_y \qquad (5.15b)$$

$$(I_z - N_{\dot{r}})\dot{r} - N_r r + (-Y_{\dot{v}}u + X_{\dot{u}}u)v = \tau_\psi \qquad (5.15c)$$

The dynamics in u and v are coupled with r, respectively. The kinematics equation becomes:

$$\dot{x} = u \cos\psi - v \sin\psi \qquad (5.16a)$$

$$\dot{y} = u \sin\psi + v \cos\psi \qquad (5.16b)$$

$$\dot{\psi} = r \qquad (5.16c)$$

The dynamics of the horizontal plane (denoted by subscript h) subsystem as seen in (5.16a) through (5.16c) can be written in compact form:

$$\mathbf{M_h}\dot{\mathbf{v}}_\mathbf{h} + \mathbf{C_h}(\mathbf{v_h})\mathbf{v_h} + \mathbf{D_h}\mathbf{v_h} = \tau_\mathbf{h} \qquad (5.17)$$

where

$$\eta_\mathbf{h} = \begin{bmatrix} x & y & \psi \end{bmatrix}^\mathrm{T}, \quad \mathbf{v_h} = \begin{bmatrix} u & v & r \end{bmatrix}^\mathrm{T},$$

$$\tau_\mathbf{h} = \mathbf{T_h F_T}\bar{\mathbf{u}} = \begin{bmatrix} \tau_x & \tau_y & \tau_\psi \end{bmatrix}^\mathrm{T} \text{ and}$$

$$\mathbf{M_h} = \begin{bmatrix} m - X_{\dot{u}} & 0 & 0 \\ 0 & m - Y_{\dot{v}} & 0 \\ 0 & 0 & I_r - N_{\dot{r}} \end{bmatrix};$$

$$\mathbf{C_h}(\mathbf{v_h}) = \begin{bmatrix} 0 & (Y_{\dot{v}} - m)r & 0 \\ 0 & 0 & (m - X_{\dot{u}})u \\ 0 & -(Y_{\dot{v}} - X_{\dot{u}})u & 0 \end{bmatrix};$$

$$\mathbf{D_h} = -\begin{bmatrix} X_u & 0 & 0 \\ 0 & Y_v & 0 \\ 0 & 0 & N_r \end{bmatrix}; \quad \mathbf{T_h} = \begin{bmatrix} 1 & 1 & 0 & 0 \\ 0 & 0 & \sin\beta & -\sin\beta \\ \gamma & -\gamma & \alpha\sin\beta & -\alpha\sin\beta \end{bmatrix}$$

The kinematics equations of motion in the vertical plane, (5.16a) through (5.16c), can be written as:

$$\dot{\eta}_\mathbf{h} = \mathbf{J_h}(\eta_\mathbf{h})\mathbf{v_h} \qquad (5.18)$$

where

$$J_h(\eta_h) = \begin{bmatrix} \cos\psi & -\sin\psi & 0 \\ \sin\psi & \cos\psi & 0 \\ 0 & 0 & 1 \end{bmatrix} \tag{5.19}$$

Equations (5.17) and (5.18) can be rewritten in state-space form:

$$\dot{x}_h = \bar{f}_h(x_h, t) + \bar{g}_h(\bar{u}, t) \tag{5.20}$$

where

$$x_h = [\eta_h v_h]^T$$

$$\bar{f}_h(x_h, t) = \begin{bmatrix} J_h(\eta_h) v_h \\ -M_h^{-1}[C_h(v_h) + D_h] v_h \end{bmatrix};$$

$$\bar{g}_h(\bar{u}, t) = \begin{bmatrix} 0_{3\times 1} \\ \begin{bmatrix} \bar{g}_{h1} & \bar{g}_{h2} & \bar{g}_{h3} \end{bmatrix} T_h F_T \bar{u} \end{bmatrix};$$

$$\bar{g}_{h1} = \begin{bmatrix} \dfrac{1}{m - X_{\dot{u}}} & 0 & 0 \end{bmatrix}^T ; \quad \bar{g}_{h2} = \begin{bmatrix} 0 & \dfrac{1}{m - Y_{\dot{v}}} & 0 \end{bmatrix}^T ;$$

$$\bar{g}_{h3} = \begin{bmatrix} 0 & 0 & \dfrac{1}{I_r - N_{\dot{r}}} \end{bmatrix}^T$$

The corresponding perturbed model can be written in the form:

$$\dot{x}_h = \bar{f}_h(x_h, t) + \bar{g}_h(\bar{u}, t) + \Delta d_h(x_h, \bar{u}, t) \tag{5.21}$$

where

$$\Delta d_h(x_h, \bar{u}, t) = \Delta f_h(x_h, t) + \Delta g_h(\bar{u}, t)$$

$$= \begin{bmatrix} 0_{3\times 1} \\ \Delta M_A^{-1}[\Delta D + \Delta C_A(v_h)] v_h \end{bmatrix} + \begin{bmatrix} 0_{3\times 1} \\ \Delta M_A^{-1} T_h F_T \bar{u} \end{bmatrix}$$

are the system uncertainties in the hydrodynamic added mass and damping forces terms. The upper bound of ΔM_A, $\Delta C_A(v_h)$, and ΔD can be obtained from Chapter 4.

On the other hand, the vertical plane model describes the ROV dynamic during altitudinal motion that involves three DOFs such as: surge, heave, and pitch. Now, the roll, pitch, and yaw velocities are bounded as the ROV is moving along the Z-axis with a constant y (i.e., $v = \dot{v} = 0$) defining the vertical plane motion dynamic. The nominal vertical plane dynamic for both the translational and rotational direction can be written as:

$$(m - X_{\dot{u}})\dot{u} - X_u u + (m - Z_{\dot{w}})wq = \tau_x \tag{5.22a}$$

$$(m - Z_{\dot{w}})\dot{w} - Z_w w + (X_{\dot{u}} - m)uq = \tau_z \tag{5.22b}$$

$$(I_y - M_{\dot{q}})\dot{q} - M_q q + (Z_{\dot{w}} - X_{\dot{u}})uw + (z_G - z_B)W\sin\theta = \tau_\theta \tag{5.22c}$$

The kinematics equation becomes:

$$\dot{x} = u\cos\theta + w\sin\theta \tag{5.23a}$$

$$\dot{z} = -u\sin\theta + w\cos\theta \tag{5.23b}$$

$$\dot{\theta} = q \tag{5.23c}$$

The dynamics of the vertical plane (denoted by subscript v) as seen in (5.22a) through (5.22c) can be written in compact form:

$$\mathbf{M}_v \dot{\mathbf{v}}_v + \mathbf{C}_v(\mathbf{v}_v)\mathbf{v}_v + \mathbf{D}_v \mathbf{v}_v + \mathbf{G}_v(\boldsymbol{\eta}_v) = \boldsymbol{\tau}_v \tag{5.24}$$

where

$$\boldsymbol{\eta}_v = \begin{bmatrix} x & z & \theta \end{bmatrix}^T, \quad \mathbf{v}_v = \begin{bmatrix} u & w & q \end{bmatrix}^T,$$

$$\boldsymbol{\tau}_v = \mathbf{T}_v \mathbf{F}_T \bar{\mathbf{u}} = \begin{bmatrix} \tau_x & \tau_z & \tau_\theta \end{bmatrix}^T, \quad \text{and}$$

$$\mathbf{M}_v = \begin{bmatrix} m - X_{\dot{u}} & 0 & 0 \\ 0 & m - Z_{\dot{w}} & 0 \\ 0 & 0 & I_y - M_{\dot{q}} \end{bmatrix},$$

$$\mathbf{C}_v(\mathbf{v}_v) = \begin{bmatrix} 0 & (m - Z_{\dot{w}})q & 0 \\ 0 & 0 & (X_{\dot{u}} - m)u \\ 0 & (Z_{\dot{w}} - X_{\dot{u}})u & 0 \end{bmatrix},$$

$$\mathbf{D_v} = -\begin{bmatrix} X_u & 0 & 0 \\ 0 & Z_w & 0 \\ 0 & 0 & M_q \end{bmatrix} \qquad \mathbf{T_v} = \begin{bmatrix} 1 & 1 & 0 & 0 \\ 0 & 0 & \cos\beta & \cos\beta \\ -\varepsilon & -\varepsilon & \alpha\cos\beta & -\alpha\cos\beta \end{bmatrix}$$

$$\mathbf{G_v} = \begin{bmatrix} 0 \\ 0 \\ (z_G - z_B)W\sin\theta \end{bmatrix}$$

The kinematics equations of motion in the vertical plane, (5.23a) through (5.23c), can be written as:

$$\dot{\mathbf{\eta}}_v = \mathbf{J_v}(\mathbf{\eta_v})\mathbf{v_v} \tag{5.25}$$

where

$$\mathbf{J_v}(\mathbf{\eta_v}) = \begin{bmatrix} \cos\theta & \sin\theta & 0 \\ -\sin\theta & \cos\theta & 0 \\ 0 & 0 & 1 \end{bmatrix} \tag{5.26}$$

Equations (5.24) and (5.25) can be rewritten in state-space form:

$$\dot{\mathbf{x}} = \overline{\mathbf{f}}_v(\mathbf{x_v},t) + \overline{\mathbf{g}}_v(\overline{\mathbf{u}},t) \tag{5.27}$$

where

$$\overline{\mathbf{f}}_v(\mathbf{x_v},t) = \begin{bmatrix} \mathbf{J_v}(\mathbf{\eta_v})\mathbf{v_v} \\ -\mathbf{M_v^{-1}}\left[(\mathbf{C_v}(\mathbf{v_v}) + \mathbf{D_v})\mathbf{v_v} + \mathbf{G_v}(\mathbf{\eta_v})\right] \end{bmatrix};$$

$$\overline{\mathbf{g}}_v(\overline{\mathbf{u}},t) = \begin{bmatrix} \mathbf{0}_{3\times1} \\ \begin{bmatrix} \overline{\mathbf{g}}_{v1} & \overline{\mathbf{g}}_{v2} & \overline{\mathbf{g}}_{v3} \end{bmatrix} \mathbf{T_v}\mathbf{F_T}\overline{\mathbf{u}} \end{bmatrix}$$

$$\overline{\mathbf{g}}_{v1} = \begin{bmatrix} \dfrac{1}{m - X_{\dot{u}}} & 0 & 0 \end{bmatrix}^\mathrm{T}; \quad \overline{\mathbf{g}}_{v2} = \begin{bmatrix} 0 & \dfrac{1}{m - Z_{\dot{w}}} & 0 \end{bmatrix}^\mathrm{T};$$

$$\overline{\mathbf{g}}_{v3} = \begin{bmatrix} 0 & 0 & \dfrac{1}{I_y - M_{\dot{q}}} \end{bmatrix}^\mathrm{T}$$

The corresponding perturbed model can be written in the form:

$$\dot{\mathbf{x}} = \overline{\mathbf{f}}_v(\mathbf{x}_v,t) + \overline{\mathbf{g}}_v(\overline{\mathbf{u}},t) + \Delta\mathbf{d}_v(\mathbf{x}_v,\overline{\mathbf{u}},t) \tag{5.28}$$

where

$$\Delta \mathbf{d}_v(\mathbf{x}_v, \bar{\mathbf{u}}, t) = \Delta \mathbf{f}_v(\mathbf{x}_v, t) + \Delta \mathbf{g}_v(\bar{\mathbf{u}}, t)$$

$$= \begin{bmatrix} \mathbf{0}_{3\times 1} \\ \Delta \mathbf{M}_A^{-1}[\Delta \mathbf{D} + \Delta \mathbf{C}_A(\mathbf{v}_v)]\mathbf{v}_v \end{bmatrix} + \begin{bmatrix} \mathbf{0}_{3\times 1} \\ \Delta \mathbf{M}_A^{-1} \mathbf{T}_v \mathbf{F}_T \bar{\mathbf{u}} \end{bmatrix}$$

are the system uncertainties in the hydrodynamic added mass and damping forces terms. The upper bound of $\Delta \mathbf{M}_A$, $\Delta \mathbf{C}_A(\mathbf{v}_v)$ and $\Delta \mathbf{D}$ can be obtained from Section 4.7.

5.2   LINEAR ROV SUBSYSTEM MODEL

It is useful to comprehend the open-loop dynamic properties of the ROV RRC prior to the closed-loop control system design. Since vast established tools have been present for linear system analysis, for a preliminary analysis, a linearized model of the ROV at different operating conditions such as, station-keeping condition, horizontal and vertical plane are provided.

The linear equations of motion are obtained by linearization of the nonlinear ROV equations about a time-varying reference trajectory or an equilibrium point, for instance:

$$\mathbf{v_0}(t) = [\; u_o(t) \quad v_o(t) \quad w_o(t) \quad p_o(t) \quad q_o(t) \quad r_o(t) \;]^T \tag{5.29a}$$

$$\mathbf{\eta_0}(t) = [\; x_o(t) \quad y_o(t) \quad z_o(t) \quad \varphi_o(t) \quad \theta_o(t) \quad \psi_o(t) \;]^T \tag{5.29b}$$

Let the perturbation from the reference trajectory $\mathbf{v_0}(t)$ and $\mathbf{\eta_0}(t)$ be described by the differentials:

$$\Delta \mathbf{v}(t) = \mathbf{v}(t) - \mathbf{v_0}(t) \tag{5.30a}$$

$$\Delta \mathbf{\eta}(t) = \mathbf{\eta}(t) - \mathbf{\eta_0}(t) \tag{5.30b}$$

$$\Delta \mathbf{\tau}_A(t) = \mathbf{\tau}_A(t) - \mathbf{\tau}_{A0}(t) \tag{5.30c}$$

Introducing the following vector notation:

$$\mathbf{F}_c(t) = \mathbf{C(v)v} \tag{5.31a}$$

$$\mathbf{F}_d(t) = \mathbf{Dv} \tag{5.31b}$$

This implies that the ROV nonlinear equation can be linearized according to,

$$\mathbf{M}\Delta \dot{\mathbf{v}} + \frac{d\mathbf{F}_c}{d\mathbf{v}}\bigg|_{\mathbf{v_0}} \Delta \mathbf{v} + \frac{d\mathbf{F}_d}{d\mathbf{v}}\bigg|_{\mathbf{v_0}} \Delta \mathbf{v} + \frac{d\mathbf{G}(\eta)}{d\eta}\bigg|_{\eta_0} \Delta \eta = \Delta \mathbf{\tau}_A \tag{5.32}$$

Perturbating the kinematic equations yield:

$$\dot{\eta}_o + \Delta\dot{\eta} = J(\eta_o + \Delta\eta)[v_o + \Delta v] \tag{5.33}$$

Substituting $\dot{\eta}_o = J(\eta_o)v_o$ into this expression implies that:

$$\Delta\dot{\eta} = J(\eta_o + \Delta\eta)\Delta v + \underbrace{[J(\eta_o + \Delta\eta) - J(\eta_o)]v_o}_{J^*} \tag{5.34}$$

Linear theory implies that second-order terms can be neglected. Hence,

$$\Delta\dot{\eta} = J(\eta_o)\Delta v + J^*(v_o, \eta_o)\Delta\eta \tag{5.35}$$

Defining $x_1 = \Delta v$ and $x_2 = \Delta\eta$, yields the following linear time-varying model:

$$M\dot{x}_1 + C(t)x_1 + D(t)x_1 + G(t)x_2 = \tau_A \tag{5.36a}$$

$$\dot{x}_2 = J(t)x_1 + J^*(t)x_2 \tag{5.36b}$$

where

$$C(t) = \frac{dF_c}{dv}\bigg|_{v_0}$$

$$D(t) = \frac{dF_d}{dv}\bigg|_{v_0}$$

$$G(t) = \frac{dG(\eta)}{d\eta}\bigg|_{\eta_0}$$

$$J(t) = J(\eta_o(t))$$

$$J^*(t) = J^*(v_o(t), \eta_o(t)) \tag{5.37}$$

Now, the linear state-space model of the ROV can be formulated. Defining $x = \begin{bmatrix} x_1^T & x_2^T \end{bmatrix}$, we obtain the following state-space model:

$$\begin{bmatrix} \dot{x}_1 \\ \dot{x}_2 \end{bmatrix} = \overbrace{\begin{bmatrix} -M^{-1}[C(t) + D(t)] & -M^{-1}G(t) \\ J(t) & J^*(t) \end{bmatrix}}^{A(t)} \begin{bmatrix} x_1 \\ x_2 \end{bmatrix} + \overbrace{\begin{bmatrix} M^{-1} \\ 0 \end{bmatrix}}^{B} \tau_A \tag{5.38}$$

This can be written in abbreviated form as:

$$\dot{x} = A(t)x + B\tau_A \tag{5.39}$$

In many ROV applications, it is reasonable to assume that the neutrally buoyant $(W = B)$ ROV is moving in the vertical plane. In this condition, the steady-state linear and angular components are assumed as: $v_o = p_o = q_o = r_o = 0$ and that the equilibrium point is defined by the zero roll and pitch angles, that is $\varphi_o = \theta_o = 0$. Hence, the time-varying matrices for the vertical plane motion (defined by certain u_o, w_o) with a fixed heading angle ψ_o is simplified to the following constant matrices:

$$\mathbf{M} = \begin{bmatrix} m - X_{\dot{u}} & 0 & 0 & 0 & 0 & 0 \\ 0 & m - Y_{\dot{v}} & 0 & 0 & 0 & 0 \\ 0 & 0 & m - Z_{\dot{w}} & 0 & 0 & 0 \\ 0 & 0 & 0 & I_{xx} - K_{\dot{p}} & -I_{xy} & -I_{xz} \\ 0 & 0 & 0 & -I_{xy} & I_{yy} - M_{\dot{q}} & -I_{yz} \\ 0 & 0 & 0 & -I_{xz} & -I_{yz} & I_{zz} - N_{\dot{r}} \end{bmatrix} \tag{5.40}$$

$$\mathbf{D} = -\begin{bmatrix} X_u & 0 & 0 & 0 & 0 & 0 \\ 0 & Y_v & 0 & 0 & 0 & 0 \\ 0 & 0 & Z_w & 0 & 0 & 0 \\ 0 & 0 & 0 & K_p & 0 & 0 \\ 0 & 0 & 0 & 0 & M_q & 0 \\ 0 & 0 & 0 & 0 & 0 & N_r \end{bmatrix} \tag{5.41}$$

$$\mathbf{C} = \begin{bmatrix} 0 & 0 & 0 & 0 & (m - Z_{\dot{w}})w_o & 0 \\ 0 & 0 & 0 & -(m - Z_{\dot{w}})w_o & 0 & (m - X_{\dot{u}})u_o \\ 0 & 0 & 0 & 0 & -(m - X_{\dot{u}})u_o & 0 \\ 0 & (m - Z_{\dot{w}})w_o & 0 & 0 & 0 & 0 \\ -(m - Z_{\dot{w}})w_o & 0 & (m - X_{\dot{u}})u_o & 0 & 0 & 0 \\ 0 & -(m - X_{\dot{u}})u_o & 0 & 0 & 0 & 0 \end{bmatrix} \tag{5.42}$$

$$\mathbf{G} = \begin{bmatrix} 0 & 0 & 0 & 0 & 0 & 0 \\ 0 & 0 & 0 & 0 & 0 & 0 \\ 0 & 0 & 0 & 0 & 0 & 0 \\ 0 & 0 & 0 & -z_B B & 0 & 0 \\ 0 & 0 & 0 & 0 & -z_B B & 0 \\ 0 & 0 & 0 & 0 & 0 & 0 \end{bmatrix} \tag{5.43}$$

If we assume that ψ_o is constant and $\varphi_o = \theta_o = 0$, the kinematic transformation matrix $\mathbf{J}$ takes the form:

$$\mathbf{J} = \begin{bmatrix} \cos(\psi_o) & -\sin(\psi_o) & 0 & 0 & 0 & 0 \\ \sin(\psi_o) & \cos(\psi_o) & 0 & 0 & 0 & 0 \\ 0 & 0 & 1 & 0 & 0 & 0 \\ 0 & 0 & 0 & 1 & 0 & 0 \\ 0 & 0 & 0 & 0 & 1 & 0 \\ 0 & 0 & 0 & 0 & 0 & 1 \end{bmatrix} \tag{5.44}$$

whereas $\mathbf{J}^* = 0$. Consequently, the linear time-invariant model for both velocity $\mathbf{x}_1$ and position $\mathbf{x}_2$ can be written as:

$$\dot{\mathbf{x}} = \mathbf{A}\mathbf{x} + \mathbf{B}\tau_A \tag{5.45}$$

or

$$\begin{bmatrix} \dot{\mathbf{x}}_1 \\ \dot{\mathbf{x}}_2 \end{bmatrix} = \begin{bmatrix} -\mathbf{M}^{-1}[\mathbf{C}+\mathbf{D}] & -\mathbf{M}^{-1}\mathbf{G} \\ \mathbf{J} & \mathbf{0} \end{bmatrix} \begin{bmatrix} \mathbf{x}_1 \\ \mathbf{x}_2 \end{bmatrix} + \begin{bmatrix} \mathbf{M}^{-1} \\ \mathbf{0} \end{bmatrix} \tau_A \tag{5.46}$$

where $\mathbf{A}$ and $\mathbf{B}$ are constant matrices.

On the other hand, the time-varying matrices for the *horizontal plane motion* can be obtained by using the following constant $\mathbf{C}$ matrix (and the remaining matrices are identical to the case of the vertical plane motion):

$$\mathbf{C} = \begin{bmatrix} 0 & 0 & 0 \\ 0 & 0 & 0 \\ 0 & 0 & 0 \\ 0 & 0 & -(m-Y_{\dot{v}})v_o \\ 0 & 0 & (m-X_{\dot{u}})u_o \\ (m-Y_{\dot{v}})v_o & -(m-X_{\dot{u}})u_o & 0 \end{bmatrix}$$

$$\begin{matrix} 0 & 0 & -(m-Y_{\dot{v}})v_o \\ 0 & -(m-X_{\dot{u}})u_o & (m-X_{\dot{u}})u_o \\ (m-Y_{\dot{v}})v_o & (I_z-N_{\dot{r}})r_o & 0 \\ 0 & 0 & I_{yz}r_o \\ -(I_z-N_{\dot{r}})r_o & 0 & -I_{xz}r_o \\ -I_{yz}r_o & I_{xz}r_o & 0 \end{matrix} \tag{5.47}$$

5.3 NONLINEAR ROV CONTROL SYSTEMS DESIGN

In this section, various control systems designs are used to control the positions and/or velocities of the RRC ROV. We show the basic of the control systems algorithms and how they can be modeled and simulated using MATLAB/Simulink software. The disturbance due to the parametric uncertainties and external disturbances such as underwater currents are not considered. However, in the control systems design like the sliding mode and state-feedback linearization method or even the proportional–integral–derivative (PID) and fuzzy-logic control, the disturbances are assumed to be known and included as nonlinear terms that are eventually canceled or attenuated during the design stage. Throughout the simulation, the ROV was commanded to move diagonally in the X-Y plane, that is $x_d = y_d = 5m$.

5.3.1 MULTIVARIABLE PID CONTROL DESIGN

The Proportional–Integral–Derivative (PID) controller is the most widely used controller structure in industrial applications. Its structural simplicity and sufficient ability for solving many practical control problems have greatly contributed to this wide acceptance. Over the past decades, many PID design techniques [77] have been proposed for industrial use or even on the ROV. Most of these design techniques are based on simple characterizations of process dynamics, such as the characterization by a first-order model with time delay. In spite of this, for plants having higher order such as a ROV, there are very few generally accepted design methods existing. Robust performance design is one of fundamental problems in control. The problem of robust performance design is to synthesize a controller for which the closed-loop system is internally stabilized and the desired performance specifications are satisfied despite plant uncertainty.

Most existing ROV systems use a series of single-input, single-output (SISO) controllers of the PID type where each controller is designed for one DOF. In this case, the RRC ROV has six DOFs and hence it should have six PID controllers. This implies that the controller gains matrices: $\mathbf{K_p}$, $\mathbf{K_d}$, $\mathbf{K_i}$ in the PID-control law, which becomes:

$$\mathbf{u} = \mathbf{T}^+ \left(\mathbf{K_p}e(t) + \mathbf{K_d}e(t) + \mathbf{K_i} \int_0^t e(\tau)d\tau \right) \tag{5.48}$$

where $e(t)$ is the position error vector and $\mathbf{T}^+$ is the Moore–Penrose pseudo-inverse matrix. The pseudo-inverse is used for ROV where the thruster configuration matrix T is nonsquare, that is, there are equal or more control inputs than controllable DOFs. It represents an optimal distribution of control energy for each DOF. The simplified result is computed using $\mathbf{T^T(TT^T)^{-1}}$.

In Simulink, the PID controller can be modeled as shown in Figure 5.3. As the roll and pitch are self-stabilizable, we are not controlling these DOFs. The PID controller block diagram uses the controller as shown in (5.48). In Figure 5.3, it should be noted that the form of the PID control law used by Simulink is not the typical form that

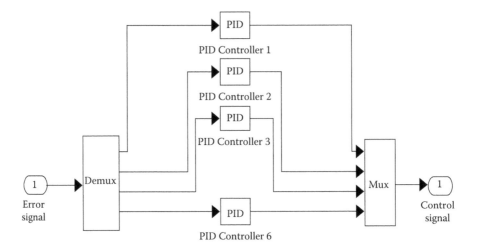

FIGURE 5.3   PID controller for four DOFs.

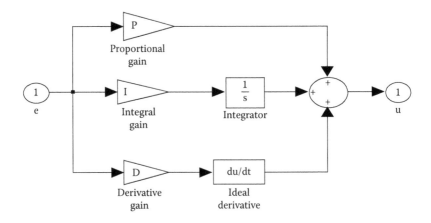

FIGURE 5.4   Details of PID controller under mask.

most control engineers use. The PID controller can be found by right-clicking the icon and Select>Look under Mask to reveal the actual controller representation in Figure 5.4. Here, the P, I and D refer to the respective positive controller gains in the PID-control law.

Below is a short note on the selection of the solver prior to the ROV simulation. The ROV differential equations can be solved by ODE Solvers such as ODE23, ODE45, ODE23s, ODE113, or ODE15s. However, the routine can potentially cause problems because many optimization routines depend on computing the gradient, the change of the function value when it is perturbed. When the function value is determined by the numerical solution of an ODE, the function value may not depend smoothly on the perturbation. If you move the perturbation up a little bit, the function moves one way; if you move it up a little bit more, the function may move the opposite way.

This nonsmoothness is caused by the variable step size used by the ODE solvers. It is possible, however, to eliminate the error checking done by the ODE Solver by forcing the solver to take fixed steps. This eliminates the need for error checking to choose a step size, but may cause accuracy problems. For a *fixed-step* simulation with the ODE Suite, you can use ODESET to turn up both the RelTol and the AbsTol to large values (e.g., 1), and set the MaxStep and the InitialStep to some fixed-step value. For this book, the solver options used are fixed as follows: variable-step type, ODE45 (Dormand–Prince solver and relative tolerance of 0.001). The configuration parameters settings can be seen in Figure 5.5. It shows a simulation time of 60 s, which can be changed depending on the system requirements.

As shown in Figure 5.6, the PID controller is modeled using Simulink. The saturation block diagram is used to limit the thrust outputs from the thrusters. The four inputs on the left-hand side are the desired position inputs. In this simulation, the ROV was commanded to move diagonally in X-Y plane, that is $x_d = y_d = 5m$. The PID controllers receive the error signals and provide the control action to the ROV. The respective PID controller gains used for the four DOFs are as follows:

$$\text{Surge direction: } K_{p,1} = 2, \ K_{d,1} = 0.5, \ K_{i,1} = 0.5;$$

$$\text{Sway direction: } K_{p,1} = 3, \ K_{d,1} = 0.5, \ K_{i,1} = 0.5;$$

$$\text{Heave direction: } K_{p,1} = 3, \ K_{d,1} = 0.5, \ K_{i,1} = 0.5;$$

$$\text{Yaw direction: } K_{p,1} = 2, \ K_{d,1} = 0.5, \ K_{i,1} = 0.5.$$

The position and velocity time responses of the ROV can be seen in Figures 5.7 and 5.8. The ROV is able to position itself to the desired inputs. The velocities do not

FIGURE 5.5   Configuration parameters setting for simulation.

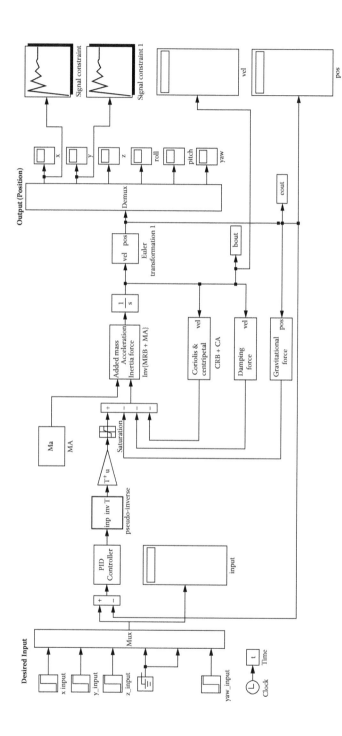

FIGURE 5.6 PID control system for ROV.

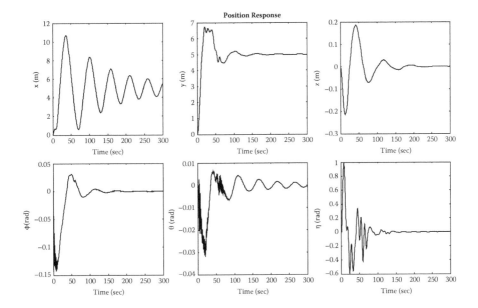

FIGURE 5.7   Position responses of ROV using PID controller ($x_d = y_d = 5m$).

exceed 0.2 m/s during the maneuvering. The velocities of the ROV decrease to zero after it has reached the target positions at around 50 s. As seen in the figures, the roll and pitch DOFs are stable without any control action.

We will demonstrate how to improve designs by estimating and tuning the PID Simulink model parameters using numerical optimization. Simulink Design Optimization toolbox offers a comprehensive interface for setting up and running the optimization problems in Simulink. Using Simulink Design Optimization enables you to reduce the time to calibrate a model and tune a compensator such as a PID controller, and helps you ensure a better system design of the ROV as compared to a trial and error method that is time consuming and inefficient. The initial PID gains are required for the optimization toolbox to run. In this case, the controller gains were set equal to the previous PID value. The desired output response, in particular, the surge and sway position responses were bounded as shown in Figures 5.9 and 5.10, respectively. The optimization began by selecting the `Optimization/Start` command in Figure 5.10. After a few runs, the optimized PID gains are obtained as shown (Figure 5.11).

Surge direction: $K_{p,1} = 2.1107$, $K_{d,1} = 0.5029$, $K_{i,1} = 0.0315$;

Sway direction: $K_{p,1} = 3.000$, $K_{d,1} = 0.5000$, $K_{i,1} = 0.5022$;

Heave direction: $K_{p,1} = 3.000$, $K_{d,1} = 0.5001$, $K_{i,1} = 0.5001$;

Yaw direction: $K_{p,1} = 2.0135$, $K_{d,1} = 0.4886$, $K_{i,1} = 0.5805$.

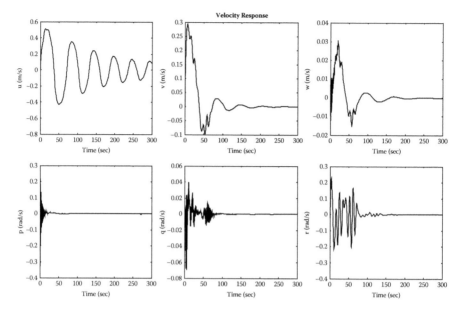

FIGURE 5.8   Velocity responses of ROV using PID controller.

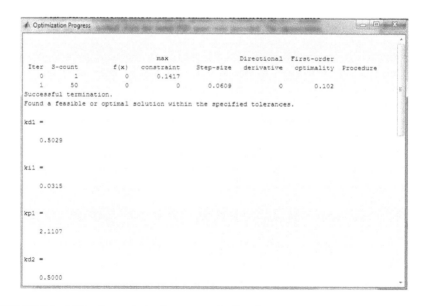

FIGURE 5.9   PID gains optimization process in Simulink Design Optimization Toolbox.

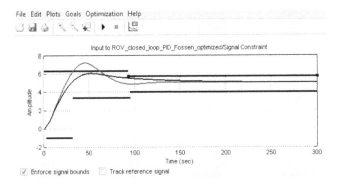

FIGURE 5.10   Desired signal bound input (for surge direction).

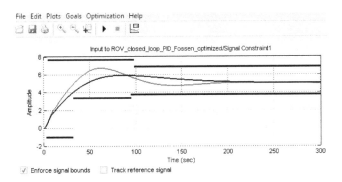

FIGURE 5.11   Desired signal bound input (for sway direction).

By simulating the PID control system using the optimized gains, the position time responses can be seen in Figure 5.12. The position and velocity time responses (see Figure 5.13) have improved with less oscillation. The settling time for the surge position has significantly decreased and it is shown to reach its steady-state values in a shorter time period. As seen in both position and velocity plots, the roll and pitch DOFs are stable without any control action. Further improvement on the time responses can be made by tightening the desired bound in the responses to achieve better time responses.

However, most ROV systems for offshore applications use only simple P, PD, and PI controllers for automatic heading and depth control since it is difficult to measure the velocity vector. A standard PID controller in (5.48) can be improved by using the vehicle kinematics together with gravity compensation. Now, let the control law be chosen as just a simple PD-control law (instead of the PID controller) where the term $\mathbf{G_f(\eta)}$ is included to compensate for gravity and buoyancy, that is:

$$\mathbf{u} = \mathbf{T}^+[\mathbf{J}^T(\eta)\mathbf{K}_p\mathbf{e} - \mathbf{J}^T(\eta)\mathbf{K}_d\mathbf{J}(\eta) + \mathbf{G_f(\eta)}] \tag{5.49}$$

where $\mathbf{e} = \mathbf{\eta_d} - \mathbf{\eta}$ is the tracking error and the PD gains should be positive or diagonal (if the matrices are used).

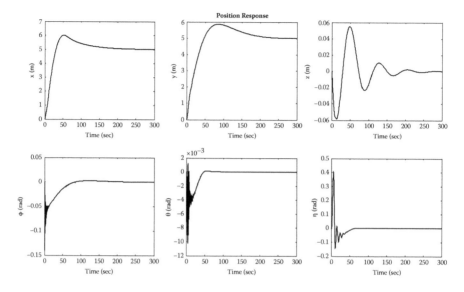

FIGURE 5.12   Simulated position time responses ($x_d = y_d = 5m$).

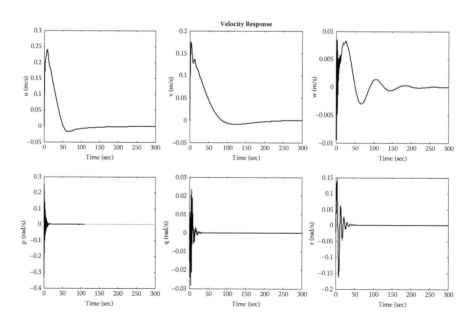

FIGURE 5.13   Simulated velocity time responses.

This control law is motivated from time differentiation of a Lyapunov function candidate:

$$V(\mathbf{v}, \mathbf{e}) = \frac{1}{2}(\mathbf{v}^T \mathbf{M} \mathbf{v} + \mathbf{e}^T \mathbf{K}_p \mathbf{e}) \qquad (5.50)$$

which yields

$$\dot{V} = \mathbf{v}^T [\mathbf{M}\dot{\mathbf{v}} - \mathbf{J}^T(\eta)\mathbf{K}_p \mathbf{e}) \qquad (5.51)$$

Here we have used the fact that $\dot{\mathbf{e}} = -\dot{\eta} = -\mathbf{J}(\eta)\mathbf{v}$. Substituting the vehicle dynamics into this expression for $\dot{V}$, yields:

$$\dot{V} = \mathbf{v}^T [\mathbf{T}\mathbf{u} - \mathbf{D}(\mathbf{v})\mathbf{v} - \mathbf{G}_f(\eta) - \mathbf{J}^T(\eta)\mathbf{K}_p \mathbf{e}] \qquad (5.52)$$

Notice that $\mathbf{v}^T \mathbf{C}(\mathbf{v})\mathbf{v} = 0$ for all $\mathbf{v} \in \mathfrak{R}^6$. From this it is seen that the proposed PD-control law with appropriate choices of $\mathbf{K}_p = \mathbf{K}_p^T > 0$ and $\mathbf{J}^T \mathbf{K}_d \mathbf{J} > 0$ ensures that:

$$\dot{V} = -\mathbf{v}^T [\mathbf{D}(\mathbf{v}) + \mathbf{J}^T(\eta)\mathbf{K}_d \mathbf{J}(\eta)]\mathbf{v} \le 0 \qquad (5.53)$$

This means that the power is dissipated passively by the damping matrix D and actively by virtual damping matrix $\mathbf{J}^T(\eta)\mathbf{K}_d \mathbf{J}(\eta)$. We now only have to check that the system cannot get stuck at V equal to zero, whenever $\mathbf{e} \ne 0$. From the preceding equation, we see that $\dot{V} = 0$ implies that $\mathbf{v} = 0$. Hence, substituting the control law into the vehicle dynamics yields:

$$\dot{\mathbf{v}} = \mathbf{M}^{-1}\mathbf{J}^T(\eta)\mathbf{K}_p \mathbf{e} \qquad (5.54)$$

Consequently $\dot{\mathbf{v}}$ will be nonzero if $\mathbf{e} \ne 0$ and $\dot{V} = 0$ only if $\mathbf{e} = 0$. Therefore, the system cannot get stuck and the system-state vector η will always converge to η_d in view of $V \to 0$.

As shown in the above equations, Euler's transformation matrix requires transpose and inverse operations. Hence, the transpose of Euler's transformation matrix is given by:

$$\mathbf{J}^T(\eta_2) = \begin{bmatrix} \mathbf{J}_1^T(\eta_2) & 0 \\ 0 & \mathbf{J}_2^T(\eta_2) \end{bmatrix} \qquad (5.55)$$

where

$$c(.) = \cos, \ s(.) = \sin, \ t(.) = \tan.$$

$$\mathbf{J}_1^T(\eta_2) = \begin{bmatrix} c(\psi)c(\theta) & s(\psi)c(\theta) & -s(\theta) \\ -s(\psi)c(\phi)+c(\psi)s(\theta)s(\phi) & c(\psi)c(\phi)+s(\phi)s(\theta)s(\psi) & c(\theta)s(\phi) \\ s(\psi)s(\phi)+c(\psi)c(\phi)s(\theta) & -c(\psi)s(\phi)+s(\theta)s(\psi)c(\theta) & c(\theta)c(\phi) \end{bmatrix}$$

$$\mathbf{J}_2^T(\eta_2) = \begin{bmatrix} 1 & 0 & 0 \\ s(\phi)t(\theta) & s(\phi) & \dfrac{s(\phi)}{c(\theta)} \\ c(\phi)t(\theta) & -s(\phi) & \dfrac{c(\phi)}{c(\theta)} \end{bmatrix} \tag{5.56}$$

The inverse of Euler's transformation matrix yields:

$$\mathbf{J}^{-1}(\eta_2) = \begin{bmatrix} \mathbf{J}_1^{-1}(\eta_2) & 0 \\ 0 & \mathbf{J}_2^{-1}(\eta_2) \end{bmatrix}; \quad \mathbf{J}_1^{-1}(\eta_2) = \mathbf{J}_1^T(\eta_2); \tag{5.57}$$

where

$$\mathbf{J}_1^{-1}(\eta_2) = \begin{bmatrix} c(\psi)c(\theta) & s(\psi)c(\theta) & -s(\theta) \\ -s(\psi)c(\phi)+c(\psi)s(\theta)s(\phi) & c(\psi)c(\phi)+s(\phi)s(\theta)s(\psi) & c(\theta)s(\phi) \\ s(\psi)s(\phi)+c(\psi)c(\phi)s(\theta) & -c(\psi)s(\phi)+s(\theta)s(\psi)c(\theta) & c(\theta)c(\phi) \end{bmatrix}$$

$$\mathbf{J}_2^{-1}(\eta_2) = \begin{bmatrix} 1 & 0 & -s(\theta) \\ 0 & c(\phi) & c(\theta)s(\phi) \\ 0 & -s(\phi) & c(\theta)c(\phi) \end{bmatrix} \tag{5.58}$$

The derivative of Euler's transformation matrix yields:

$$\dot{\mathbf{J}}(\eta_2) = \begin{bmatrix} \dot{\mathbf{J}}_1(\eta_2) & 0 \\ 0 & \dot{\mathbf{J}}_2(\eta_2) \end{bmatrix} \tag{5.59}$$

where

$$\dot{\mathbf{J}}_1(\eta_2) = \begin{bmatrix} J_{11} & J_{12} & J_{13} \\ J_{21} & J_{22} & J_{23} \\ J_{31} & J_{32} & J_{33} \end{bmatrix} \quad \dot{\mathbf{J}}_2(\eta_2) = \begin{bmatrix} 0 & J_{45} & J_{46} \\ 0 & J_{55} & J_{56} \\ 0 & J_{65} & J_{66} \end{bmatrix} \tag{5.60}$$

and the following parameters in (5.60) are given by:

$$J_{11} = -s(\psi)\dot{\psi}c(\theta) - c(\psi)s(\theta)\dot{\theta};$$

$$J_{12} = -c(\psi)\dot{\psi}c(\varphi) + s(\psi)s(\varphi)\dot{\varphi} - s(\psi)\dot{\psi}s(\theta)s(\varphi) + c(\psi)c(\theta)\dot{\theta}s(\varphi) + c(\psi)s(\theta)c(\varphi)\dot{\varphi};$$

$$J_{13} = c(\psi)\dot{\psi}s(\varphi) + s(\psi)c(\varphi)\dot{\varphi} - s(\psi)\dot{\psi}c(\varphi)s(\theta) - c(\psi)s(\varphi)\dot{\varphi}s(\theta) + c(\psi)c(\varphi)c(\theta)\dot{\theta};$$

$$J_{21} = c(\psi)\dot{\psi}c(\theta) + s(\psi)s(\theta)\dot{\theta};$$

$$J_{22} = -s(\psi)\dot{\psi}c(\varphi) - c(\psi)s(\varphi)\dot{\varphi} + c(\varphi)\dot{\varphi}s(\theta)s(\psi) + s(\varphi)c(\theta)\dot{\theta}s(\psi) + s(\varphi)s(\theta)c(\psi)\dot{\psi};$$

$$J_{23} = s(\psi)\dot{\psi}s(\varphi) - c(\psi)c(\varphi)\dot{\varphi} + c(\theta)\dot{\theta}s(\psi)c(\varphi) + s(\theta)c(\psi)\dot{\psi}c(\varphi) - s(\theta)s(\psi)s(\varphi)\dot{\varphi};$$

$$J_{31} = -c(\theta)\dot{\theta};$$

$$J_{32} = -s(\theta)\dot{\theta}s(\varphi) + c(\theta)c(\varphi)\dot{\varphi};$$

$$J_{33} = -s(\theta)\dot{\theta}c(\varphi) - c(\theta)s(\varphi)\dot{\varphi};$$

$$J_{45} = c(\varphi)\dot{\varphi}t(\theta) + s(\varphi)[1 + t^2(\theta)]\dot{\theta}; \quad J_{46} = -s(\varphi)\dot{\varphi}t(\theta) + c(\varphi)[1 + t^2(\theta)]\dot{\theta};$$

$$J_{55} = -s(\varphi)\dot{\varphi}; \quad J_{56} = -c(\varphi)\dot{\varphi};$$

$$J_{65} = \frac{c(\varphi)\dot{\varphi}}{c(\theta)} + \frac{s(\varphi)s(\theta)\dot{\theta}}{c^2(\theta)}; \quad J_{66} = \frac{s(\varphi)\dot{\varphi}}{c(\theta)} + \frac{c(\varphi)s(\varphi)\dot{\theta}}{c^2(\theta)}.$$

The block diagram of the PD controller using the vehicle kinematics and the gravity compensation is shown in Figure 5.14. The block diagram is quite similar to the one using the PID except the PD controller with the vehicle kinematics represented by the transpose of Euler's transformation matrix is used. The controller block diagram uses the controller as shown in (5.49). The controller gains for each DOF are given as: $K_p = 1$, $K_d = 0.8$. The details of the transpose of Euler's transformation matrix for the translational motion (i.e., $\mathbf{J}_1$ matrix) can be seen in Figure 5.15a through Figure 5.15e. A similar method can be applied for the $\mathbf{J}_2$ matrix as well.

As Simulink works on column/row vector manipulation, Euler's transformation consists of two submatrices: $\mathbf{J}_1$ and $\mathbf{J}_2$ representing the columns 1 to 3 and 4 to 6, respectively. Each matrix has three rows of parameters as shown in Figure 5.15b. The values of rows 1 to 3 can be shown in Figure 5.15c through Figure 5.15e, respectively. This is not the only method for modeling a matrix in Simulink; however, it is an intuitive method for simulation and trouble shooting purposes.

In this simulation, the ROV was commanded to move diagonally in the X-Y plane, that is, $x_d = y_d = 5m$. By simulating the PD controller that uses the vehicle kinematics and gravity compensation type, the position and velocity time responses can be plotted in Figures 5.16 and 5.17. The position and velocity time responses have improved with less oscillation. But the steady-state error has increased by around 2 m and the heave position has a steady-state error of 4 m. The error may be due to the numerical error resulting from the transpose of Euler's transformation matrix.

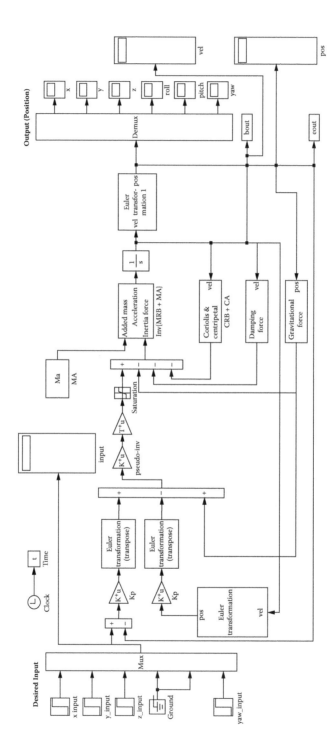

FIGURE 5.14   Vehicle kinematics with gravity compensation PID-type control system for ROV.

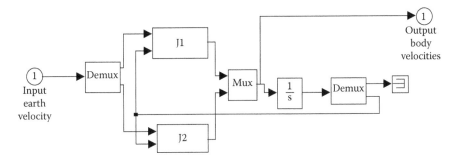

FIGURE 5.15a   Transpose of Euler's transformation.

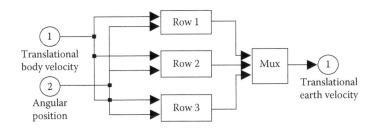

FIGURE 5.15b $J_1(\eta_2)$ matrix.

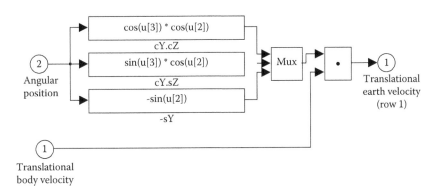

FIGURE 5.15c   Row 1 of $J_1(\eta_2)$ matrix.

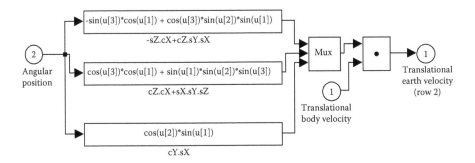

FIGURE 5.15d   Row 2 of $\mathbf{J}_1(\mathbf{\eta}_2)$ matrix.

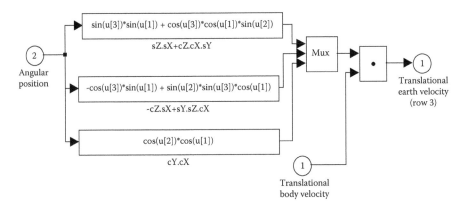

FIGURE 5.15e   Row 3 of $\mathbf{J}_1(\mathbf{\eta}_2)$ matrix.

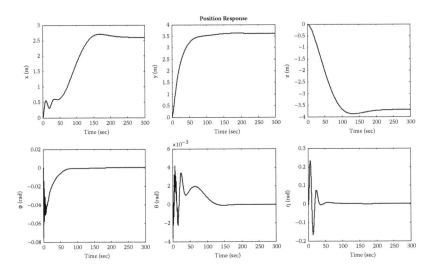

FIGURE 5.16   Position time responses of vehicle kinematics with gravity compensation PD-type control ($x_d = y_d = 5m$).

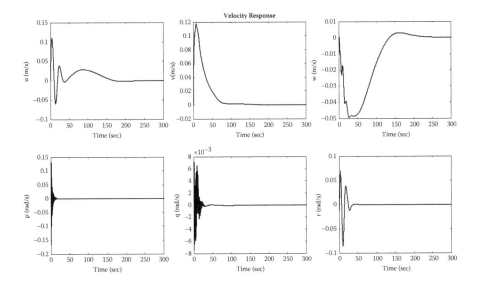

FIGURE 5.17   Velocity time responses of vehicle kinematics with gravity compensation PD-type control.

As seen in the figures, the roll and pitch DOF are stable without any control action. Further improvement can be made by tuning the PD controller gains.

5.3.2   SLIDING-MODE CONTROL

In this section, the sliding-mode controller [78] is used. In this control scheme, we need to define a sliding surface in terms of error. Any states reaching the sliding surface imply that the states have converged to the desired value. Let this error be denoted by $\tilde{\eta} = \eta - \eta_d$. The sliding surfaces are then given by $\mathbf{s} = \dot{\tilde{\eta}} + \lambda\tilde{\eta}$ where bandwidth $\lambda > 0$. The Lyapunov function candidate can be written as such:

$$V(\mathbf{s}(t)) = \frac{1}{2}\mathbf{s}^{\mathrm{T}}\mathbf{M}\mathbf{s} \tag{5.61}$$

The time derivative yields:

$$\dot{V}(\mathbf{s}(t)) = \mathbf{s}^{\mathrm{T}}\mathbf{M}\dot{\mathbf{s}} \tag{5.62}$$

Here $\mathbf{M}\dot{\mathbf{s}} = \mathbf{M}(\ddot{\tilde{\eta}} + \lambda\dot{\tilde{\eta}})$ since $\ddot{\eta}_d = 0, \dot{\eta}_d = 0$ and $\ddot{\eta} = \mathbf{J}(\eta)\dot{\mathbf{v}} + \dot{\mathbf{J}}(\eta)\mathbf{v}$, it can be written as:

$$\dot{V}(\mathbf{s}(t)) = \mathbf{s}^{\mathrm{T}}\mathbf{M}[\mathbf{J}(\eta)\dot{\mathbf{v}} + \dot{\mathbf{J}}(\eta)\mathbf{v} + \lambda\dot{\tilde{\eta}}]$$

$$= \mathbf{s}^{\mathrm{T}}\mathbf{M}\{\mathbf{J}(\eta)\mathbf{M}^{-1}[-\mathbf{C}(\mathbf{v})\mathbf{v} - \mathbf{D}(\mathbf{v})\mathbf{v} - \mathbf{G}_{\mathrm{f}}(\eta) + \mathbf{f}(\mathbf{v},t)] \tag{5.63}$$

$$+ \mathbf{s}^{\mathrm{T}}\mathbf{M}[\mathbf{J}(\eta)\mathbf{M}^{-1}\mathbf{T}\mathbf{F}_{\mathrm{T}}\bar{\mathbf{u}} + \dot{\mathbf{J}}(\eta)\mathbf{v} + \lambda\dot{\tilde{\eta}}]\}$$

where $\mathbf{f}(\mathbf{v}, t)$ is the perturbation on the damping and added mass terms.

The control law can be defined as:

$$\bar{\mathbf{u}} = \mathbf{F}_T^{-1}\mathbf{T}^+[\mathbf{C}(\mathbf{v})\mathbf{v} + \mathbf{D}(\mathbf{v})\mathbf{v} + \mathbf{G_f}(\eta) - \hat{\mathbf{f}}(\mathbf{v},t)\overbrace{-K_s\,\mathrm{sgn}(\mathbf{s}) - K_{sd}\mathbf{s} - K_{sv}\mathbf{v}}^{controller}] \quad (5.64)$$

where k, k_d, k_v are the controller gains and sgn($\mathbf{s}$) is given by:

$$\mathrm{sgn}(\mathbf{s}) = \begin{cases} +1, & for \quad \mathbf{s} \geq 0 \\ -1 & for \quad \mathbf{s} < 0 \end{cases} \quad (5.65)$$

Substituting the control law into $\dot{V}$ yields:

$$\dot{V}(\mathbf{s}(t)) = \mathbf{s}^T\mathbf{M}[-K_s\,\mathrm{sgn}(\mathbf{s}) - K_{sd}\mathbf{s} - K_{sv}\mathbf{v} + \Delta\mathbf{f}(\mathbf{v},t) + \dot{\mathbf{J}}(\eta)\mathbf{v} + \lambda\dot{\eta}] \quad (5.66)$$

where

$\Delta\mathbf{f}(\mathbf{v},t) = \mathbf{f}(\mathbf{v},t) - \hat{\mathbf{f}}(\mathbf{v},t)$. For $\mathbf{f}(\mathbf{v},t) = \hat{\mathbf{f}}(\mathbf{v},t)$, $\dot{V}$ becomes:

$$\dot{V}(\mathbf{s}(t)) = \mathbf{s}^T\mathbf{M}[-K_s\,\mathrm{sgn}(\mathbf{s}) - K_{sd}\mathbf{s} - K_{sv}\mathbf{v} + \dot{\mathbf{J}}(\eta)\mathbf{v} + \lambda\mathbf{J}(\eta)\mathbf{v}] \quad (5.67)$$

Choose K_s, K_{sd}, K_{sv} large enough, $K_{sv}\mathbf{I_6} \geq \dot{\mathbf{J}}(\eta) + \lambda\mathbf{J}(\eta)$ and λ is small such that $\dot{V}(\mathbf{s}(t)) \leq 0$. Notice that $|\dot{\mathbf{J}}(\eta)|$ and $|\mathbf{J}(\eta)|$ do not exceed 1 and $\mathbf{v}(t)$ is small, $\dot{V} \leq 0 \Rightarrow V(t) \leq 0$ and therefore s is bounded. This implies that $\ddot{V}$ must be uniformly continuous. Application of Barbalat's lemma shows that $\mathbf{s} \rightarrow 0$ and thus $\tilde{\eta} \rightarrow 0$ as $t \rightarrow \infty$.

This technique can yield perfect tracking with price "excessive discontinuous switching" across $\mathbf{s}(.) = 0$ or chattering. In order to avoid chattering, a boundary layer around the $\mathbf{s}(.) = 0$ surface with thickness $\Phi \in \mathfrak{R}^6(\Phi(.) > 0)$ is adopted. While the vehicle dynamics remain "inside" this boundary layer, no switching is made but interpolation takes place. If the region outside the layer is reached then switching is carried out according to

$$sat(\mathbf{s}/\Phi) = \begin{cases} \mathrm{sgn}(\mathbf{s}/\Phi), & for \quad |\mathbf{s}/\Phi| \geq 1 \\ \mathbf{s}/\Phi & for \quad |\mathbf{s}/\Phi| < 1 \end{cases} \quad (5.68)$$

In the above expression, we avoid $\mathbf{s}(.)$, $\Phi(.)$ notation, for the sake of simplicity and the absolute function |.| and the division operations are to be evaluated element-wise. The control law becomes:

$$\bar{\mathbf{u}} = \mathbf{F}_T^{-1}\mathbf{T}^+[\mathbf{C}(\mathbf{v})\mathbf{v} + \mathbf{D}(\mathbf{v})\mathbf{v} + \mathbf{G_f}(\eta) - \hat{\mathbf{f}}(\mathbf{v},t) - K_s\,sat(\mathbf{s}/\Phi) - K_{sd}\mathbf{s} - K_{sv}\mathbf{v}] \quad (5.69)$$

The Simulink block diagram for the sliding-mode control can be seen in Figure 5.18. It consists of a subsystem to compute the sliding surface, s (see Figure 5.19) and a subsystem to compensate for the nonlinear function $f(.)$

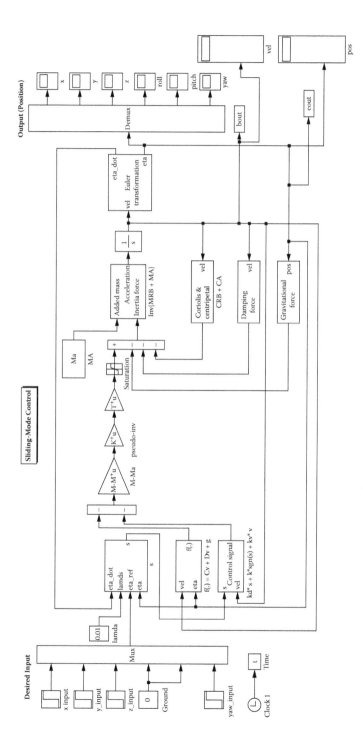

FIGURE 5.18 Sliding-mode controller model using Simulink.

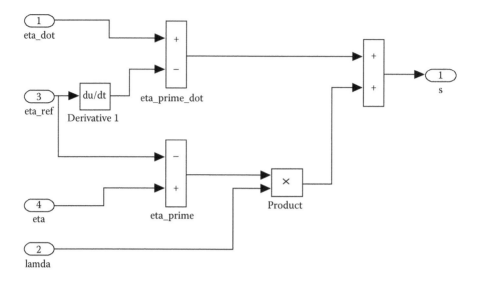

FIGURE 5.19 Sliding surface $s(.)$ for sliding-mode controller.

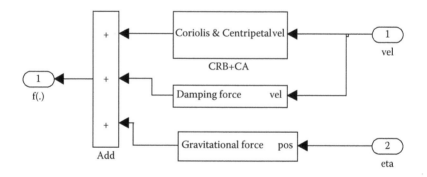

FIGURE 5.20 Nonlinear function $\mathbf{f}(v,t)$ in sliding-mode controller.

in Figure 5.20. The controller in Figure 5.21 is used to control the ROV to the desired position $(x_d = y_d = \approx 5m)$, and it has the following parameters: $K_s = 26$; $K_{sv} = 42$; $K_{sd} = 37$. The sliding-mode control works as follows: the tracking error is negative (positive), indicating that the position is too small, the control input is increased (decreased) in order to increase the output. The "energy" of the system diminishes slowly as it reaches the sliding surface (near to zero tracking error).

In order to avoid chattering, a boundary layer around the sliding surface with thickness Φ is adopted. In this simulation, the ROV was commanded to move diagonally in the X-Y plane, that is, $x_d = y_d = 5m$. By simulating the sliding-mode controller, the position and velocity time responses can be seen in Figures 5.22 and 5.23. As compared to the previous control systems design, the simulation time was extended to 600 s for clarity. As shown in the switching method used, it yields good tracking but with excessive discontinuous switching or chattering as

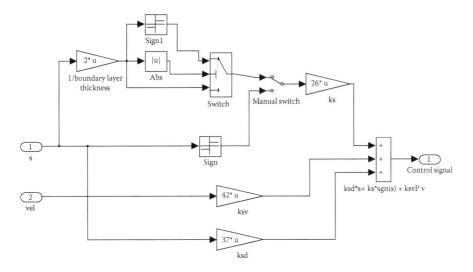

FIGURE 5.21 Compensator term in sliding-mode controller.

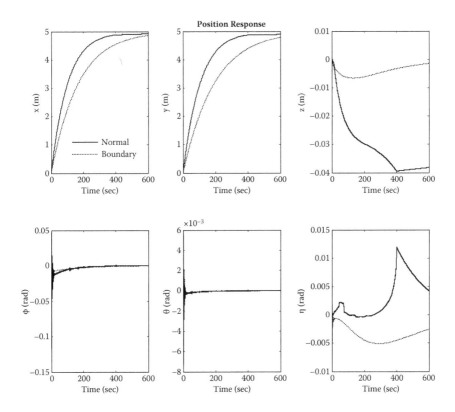

FIGURE 5.22 Position time responses of sliding-mode controller with and without using boundary thickness.

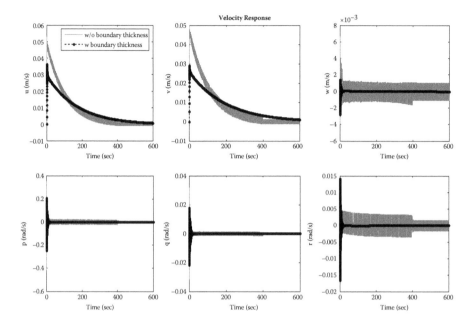

FIGURE 5.23 Velocity time responses of sliding-mode controller with and without using boundary thickness.

seen in the velocity plots. With the use of the boundary layer around the sliding surface, the chattering has disappeared and results in less oscillatory responses in the velocity as observed in Figure 5.23. As seen in Figure 5.22, the responses have improved with less overshoot but the settling times have increased. As seen in the figures, the roll and pitch DOFs are stable without any control action. Further improvement can be made by tuning the sliding-mode controller gains.

5.3.3 Velocity State-Feedback Linearization

The basic idea with feedback linearization [79] is to transform the nonlinear systems dynamics into a linear system. Conventional control techniques like pole placement and linear quadratic optimal control theory can then be applied to a linear system. In robotics, this technique is commonly referred to as *computed torque control*. Adaptive computed torque control has been applied to robot manipulators and to underwater vehicles by Fossen [60]. In this book, the feedback linearization is easily applicable to underwater vehicles. We will discuss application to the body-fixed reference frame only.

The control objective is to transform the vehicle dynamics into a linear system $\dot{\mathbf{v}} = \mathbf{a}_v$ where $\mathbf{a}_v$ can be interpreted as a commanded acceleration vector. The body-fixed vector representation should be used to control the vehicle's linear and angular velocities. Consider the nonlinear ROV dynamics (4.14) that can be compactly expressed as:

$$\mathbf{M}\dot{\mathbf{v}} + \mathbf{N}(\mathbf{v}, \eta) = \tau_A \qquad (5.70a)$$

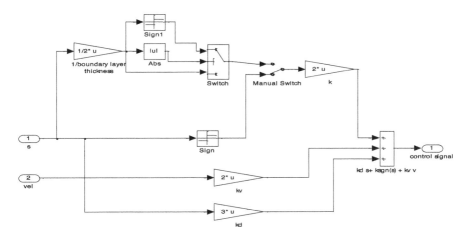

where the nonlinear function:

$$N(v, \eta) = C(v)v + D(v)v + G_f(\eta) \tag{5.70b}$$

The nonlinearities can be canceled out by simply selecting the control law as:

$$\tau_A = Ma_v + N(v, \eta) \tag{5.71}$$

where the commanded acceleration vector a_v can be chosen by pole placement or linear quadratic optimal control theory, v_d is the desired linear and angular velocity vector and $\tilde{v} = v - v_d$ is the velocity tracking error. Then, the commanded acceleration vector:

$$a_v = \dot{v}_d - \lambda\tilde{v} \tag{5.72}$$

obtains the first-order error dynamics:

$$M(\dot{v} - a_v) = M[\dot{v} - \dot{v}_d + \lambda(v - v_d)] = M(\dot{\tilde{v}} + \lambda\tilde{v}) \tag{5.73}$$

The reference model is simply chosen as a first-order model with time constants and τ_v as the commanded input vector. Note that in steady state:

$$\lim_{t\to\infty} v_d(t) = \tau_v \tag{5.74}$$

The Simulink block diagram for the state feedback controller can be seen in Figure 5.24. It consists of a subsystem to compute the commanded acceleration vector (see Figure 5.25) and a subsystem to compensate for the nonlinear function

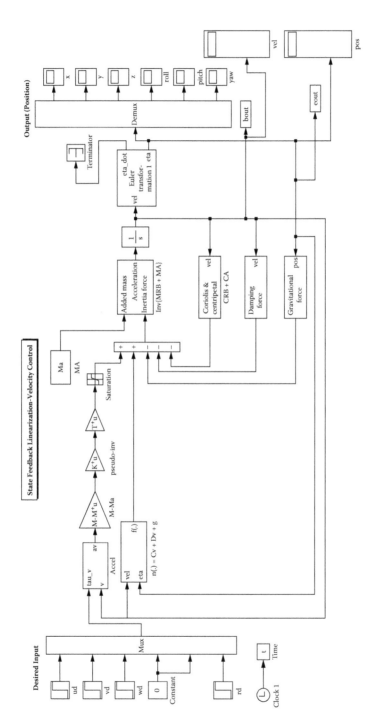

FIGURE 5.24   Velocity state feedback linearization controller using Simulink.

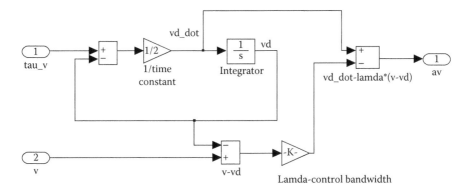

FIGURE 5.25   Calculation of the commanded acceleration vector in velocity state feedback linearization controller.

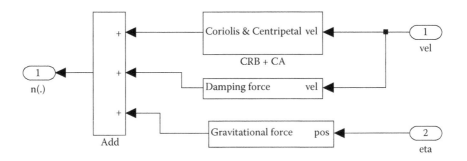

FIGURE 5.26   Nonlinear function in velocity state feedback linearization controller.

in Figure 5.26. The state feedback linearization controls the ROV to the desired velocity ($u_d = v_d = 5$m/s), with the first-order error dynamic (characterized by time constant = 2 s and gain $\lambda = 100$).

By simulating the velocity state-feedback linearization controller, the position and velocity time responses can be seen in Figures 5.27 and 5.28, respectively. As compared to the previous control systems design, the simulation time was reduced to 15 s for clarity. As observed in the velocity plots, the surge and sway velocity reaches 5 m/s in 2 s. It exhibits a good response for a first-order system. On the other hand, the positions show a divergent response as the velocities were maintained at 5 m/s. As observed in the figures, the roll and pitch DOFs are stable without any control action. Further improvement on the time responses can be made by tuning the time constant and gain terms. Besides controlling the velocity, the position of the ROV can be made to follow a desired error dynamic. The position state-feedback linearization controller (not shown here) can be used to control the position as well.

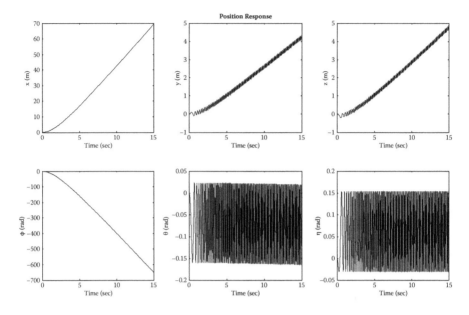

FIGURE 5.27   Position time responses of velocity state feedback linearization ($x_d = y_d = 5m$).

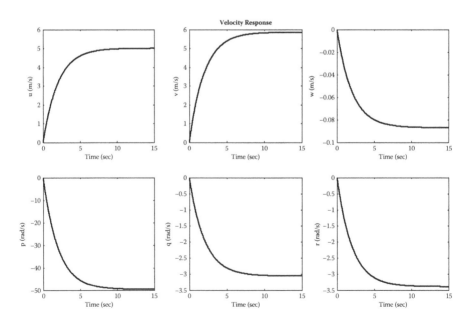

FIGURE 5.28   Velocity time responses of state feedback linearization ($x_d = y_d = 5m$).

5.3.4 FUZZY LOGIC CONTROL

There have been various efforts to develop the controller for the URV, which include both the (conventional) linear and the (modern) intelligent control schemes. Intelligent control methods, which include neural networks (NN), sliding mode (SLC), and fuzzy logic controllers (FLC) are more robust and are able to adopt the hydrodynamic uncertainties. In addition, they exhibit excellent immunity to disturbances. Although intelligent control is very promising for URV application, it requires substantial computational power, due to the complex decision-making processes. For example, an FLC has to deal with fuzzification, rule-based storage, inference mechanism and defuzzification operations. Despite these issues, it is known that an FLC has a simple control structure and offers a higher degree of freedom in tuning its control parameters compared to other nonlinear controllers [80].

The system under control is described in terms of some linguistic variables. The control tasks are performed by using some rules set forth in a rule base. The inference rule used is a fuzzy inference scheme. As seen in Figure 5.29, it takes three steps to design a fuzzy controller: fuzzyfication, inference rules, and defuzzyfication. In the first step, the values obtained through a sensor are transformed into values of the corresponding linguistic variable ranging from 0 to 1. The second step performs the fuzzy inference giving the linguistic values of the control variables. In the third step, these linguistic values are transformed to the numerical value of the control variable in order to perform the required task. After executing the three steps, the controller is fine-tuned in an iterative way.

First, the FLC has two linguistic input variables, sometimes called *controlled inputs*, namely error (e) and the change of error ($\dot{e}$). The error and change of error fuzzy sets each have seven membership functions as shown in Figure 5.30 (top). The triangular and trapezoidal membership functions for inputs have seven linguistic variables defined as: Negative Big (NB), Negative Medium (NM), Negative Small (NS), Zero (Z), Positive Small (PS), Positive Medium (PM), and Positive Big (PB). As observed, all linguistic variables have the same widths. PL and NL are the saturation limits imposed on the FLC.

On the other hand, the linguistic output variable is the voltage applied to the thrusters. The output membership functions have seven singleton functions as shown in Figure 5.30 (bottom). It has the same linguistic variables like NB, NM, NS, Z, PS, PM, and PB. The singleton is represented by an individual point in the output space.

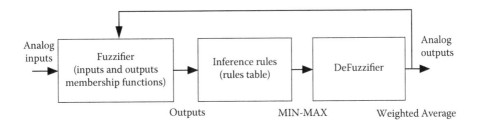

FIGURE 5.29 Fuzzy logic controller design flow.

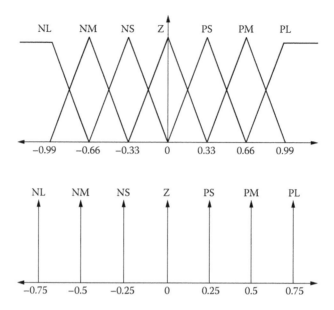

FIGURE 5.30 Input (top) and output (bottom) membership functions for a Sugeno type.

Truncation of singletons results in reduction of the input. For example, for PM equal to 0.5, the singleton value can be 0.5 instead of 1 as shown in Figure 5.30 (bottom).

The fuzzy inference operation or process is implemented by using 49 rules (or If-Then statements). The min-max (or implication-aggregation) compositional rule of inference and the weighted average method have been used in the defuzzifier process based on the Sugeno inference process. For example:

- If $e(t)$ is NL AND $\dot{e}(t)$ is NS Then u is 0.75.
- If $e(t)$ is NL AND $\dot{e}(t)$ is NM Then u is 0.75.
- If $e(t)$ is NL AND $\dot{e}(t)$ is NL Then u is 0.75.
- ....

and go on for all inputs.

Here the AND method truncates the output fuzzy set (that is, the minimum of the linguistic variables). Conversely, if the OR operator is used, it selects the maximum of the two variables. Similarly, the fuzzy inference process can be captured by the fuzzy interference diagram as shown in Figure 5.31.

For reference, the rule table for the Sugeno type is shown in Table 5.1. The 49 rules are created on a two-dimensional space of the phase-plane $(e, \dot{e})$. As observed in the table, the rule table has the same output membership in a diagonal direction, which is known as the Toeplitz structure.

Now, how to model and simulate the fuzzy logic as mentioned above? We will be building it using the graphical user interface (GUI) tools provided by the Fuzzy Logic Toolbox. Although it is possible to use the Fuzzy Logic Toolbox by working strictly from the command line, in general it is much easier to build a system

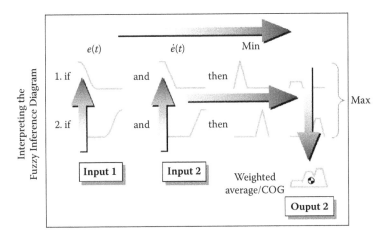

FIGURE 5.31   Fuzzy interference diagram.

TABLE 5.1
Rule Table with Toeplitz Structure for Sugeno

e \ $\dot{e}$	PL	PM	PS	Z	NS	NM	NL
NL	0	0.25	0.5	0.75	0.75	0.75	0.75
NM	0.25	0	0.25	0.5	0.75	0.75	0.75
NS	0.5	0.25	0	0.25	0.5	0.75	0.75
Z	0.25	0.5	0.25	0	0.25	0.5	0.75
PS	0.75	0.25	0.5	0.25	0	0.25	0.5
PM	0.75	0.75	0.25	0.5	0.25	0	0.25
PL	0.75	0.75	0.75	0.25	0.5	0.25	0

graphically. There are five primary GUI tools for building, editing, and observing fuzzy inference systems in the Fuzzy Logic Toolbox: the Fuzzy Inference System (FIS) Editor, the Membership Function Editor, the Rule Editor, the Rule Viewer, and the Surface Viewer. These GUIs are dynamically linked, in that changes you make to the FIS using one of them, can affect what you see on any of the other open GUIs. You can have any or all of them open for any given system.

First, type the following after the MATLAB command prompt.

```
>> fuzzy
```

The FIS Editor will pop up as shown in Figure 5.32. Note that the Sugeno FIS was chosen.

The next step is to create the inputs membership functions for e and $\dot{e}$. This is followed by creating the output membership function for the control voltage to the thruster. The completed membership functions can be seen in the Membership Function Editor in Figure 5.33.

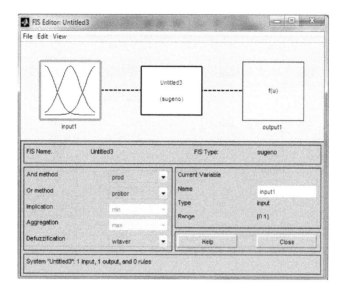

FIGURE 5.32   FIS Editor.

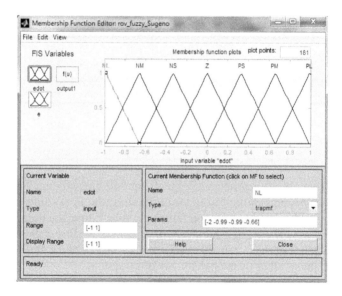

FIGURE 5.33   Membership Function Editor.

The FIS is not completed until the rules in Table 5.1 are implemented in the Rule Editor as seen in Figure 5.34.

As shown in Figure 5.34, there are 49 rules or IF-Then statements. To display the rules graphically, the Rule Viewer in Figure 5.35 is used. It displays a roadmap of the whole fuzzy inference process. It is based on the fuzzy inference diagram described in the previous section. You see a single figure window with 49 small plots

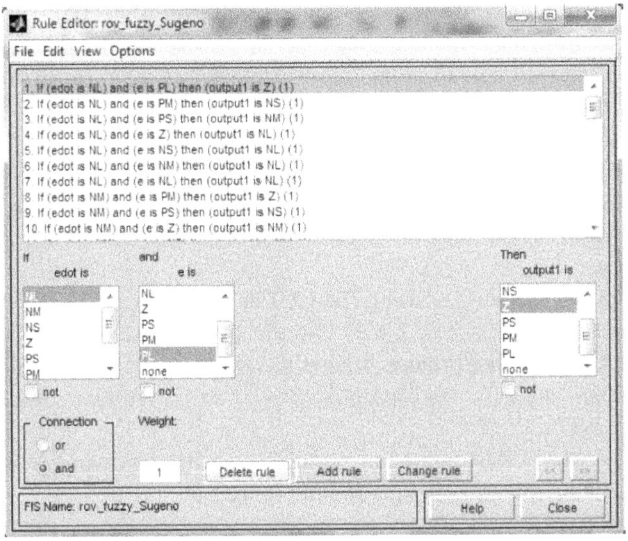

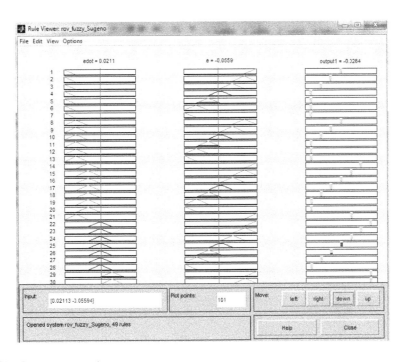

FIGURE 5.34   Rule Editor.

FIGURE 5.35   Rule Viewer.

nested in it. Each rule is a row of plots, and each column is a variable. The first two columns of plots show the membership functions referenced by the antecedent, or the If part of each rule. The third column of plots shows the membership functions referenced by the consequent, or the Then part of each rule.

The control surface in three-dimensional view can be plotted using the Surface Viewer as shown in Figure 5.36. For each input e and $\dot{e}$, the corresponding output can be plotted. The FIS is commonly known as a two-input, one-output fuzzy-logic system. As seen in Figure 5.36, the Sugeno control surface can closely approximate a linear surface.

Besides plotting the surface using the GUI or the Surface Viewer, the user can select to export to workspace or file so that the data can be used later for plotting in the MATLAB command prompt. To do this, select File/Export/To Workspace in Figure 5.36. Then name the workspace as: `rov_fuzzy_Sugeno` and click OK as seen in Figure 5.37.

Now, in the MATLAB command prompt, type the following to generate the surface plot without the gray background in Figure 5.38.

```
>> a=readfis('rov_fuzzy_Sugeno')
>> gensurf(a)
```

And other useful commands such as showing the inputs and output membership functions and displaying the rules in MATLAB command prompts can be seen as follows.

```
>> getfis(a,'output',1)
>> plotmf(a,'input',1)
>> plotmf(a,'input',2)
>> showrule(a)
```

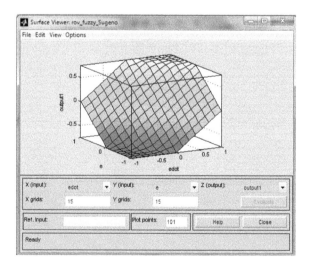

FIGURE 5.36 Surface Viewer.

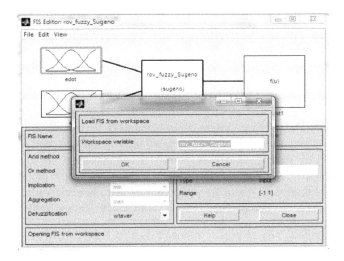

FIGURE 5.37   Exporting data to workspace using the FIS Editor.

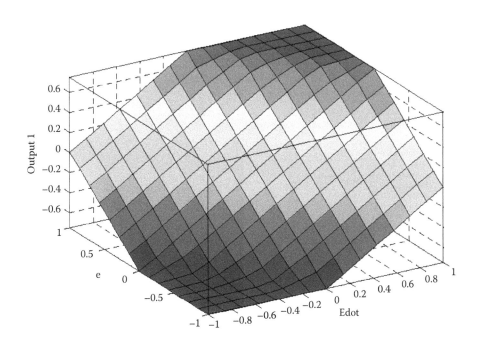

FIGURE 5.38   Surface plot using the MATLAB command prompt.

Besides using the GUI of the Fuzzy Logic System and MATLAB commands, the Fuzzy Logic Toolbox in Simulink can also be used to simulate FLC. In another words, it can be integrated into simulations with Simulink. Prior to that, the FLC designed through the FIS Editor is transferred to the MATLAB Workspace by the same command `Export to Workspace`. Then, the Simulink environment provides a direct access to the FLC through the Simulink Library as shown in Figure 5.39. Figure 5.40 shows the Simulink diagram of the Fuzzy Logic controller (FLC).

In the simulation, the FLC system controls the ROV to the desired velocity ($x_d = y_d = 5$m). It consists of a subsystem to compute the commanded voltage vector (see Figure 5.41) for the thrusters and a subsystem that contains the fuzzy logic controller as shown in Figure 5.42. The remaining block diagrams belong to the ROV models. As noticed in Figure 5.42, there are six ports for each linguistic variable and each has a fuzzy logic controller attached in series with it. The remaining block diagrams in Figure 5.40 for the open-loop ROV are similar.

By simulating the fuzzy-logic control system, the position and velocity time responses can be seen in Figures 5.43 and 5.44, respectively. As observed in the position plots, the surge position reaches 5 m in a shorter time period as compared to the sway position. It exhibits less oscillation during maneuvering. The surge and sway velocity reduce to zero as the positions reach the desired values. As observed in the figures, the roll and pitch DOFs are stable without any control action. Further improvement on the time responses can be made by adjusting the rules table values.

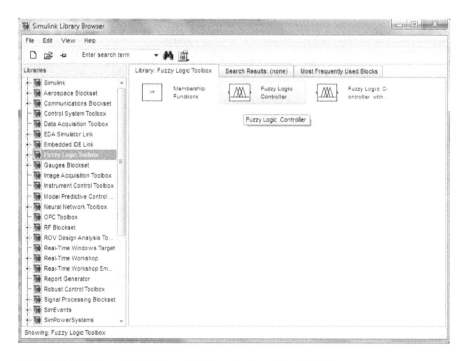

FIGURE 5.39   Fuzzy Logic Controller function from Simulink library.

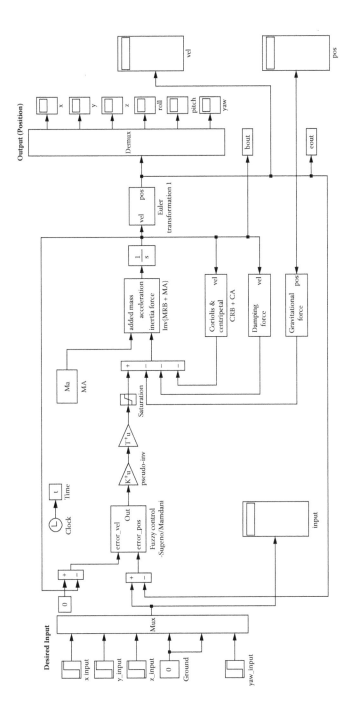

FIGURE 5.40 Simulink block diagram for FLC—Sugeno type on ROV.

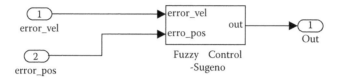

FIGURE 5.41   Simulink block diagram for FLC.

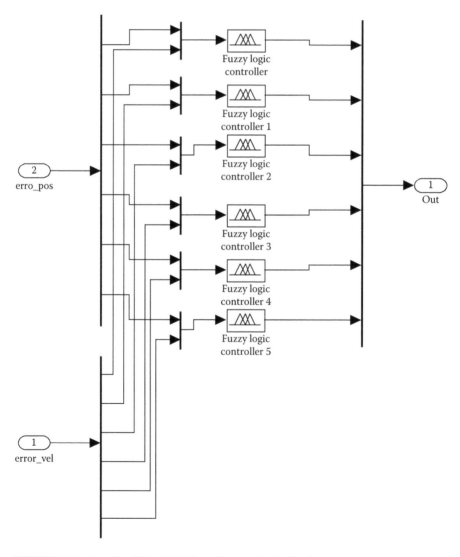

FIGURE 5.42   Details of Simulink block diagram for FLC—Sugeno type.

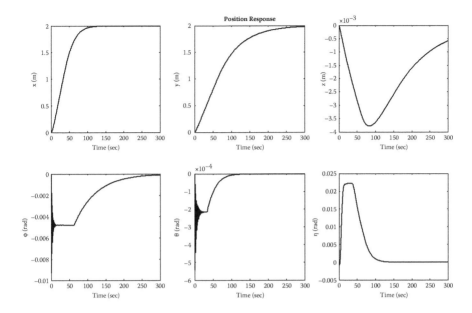

FIGURE 5.43   Position time responses of FLC—Sugeno type.

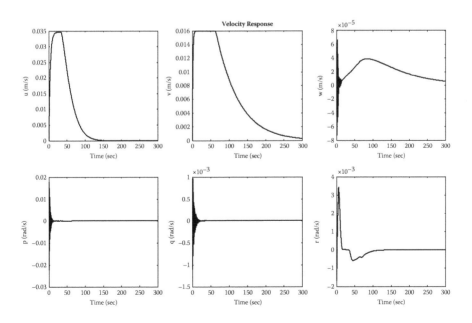

FIGURE 5.44   Velocity time responses of FLC—Sugeno type.

5.3.5 CASCADED SYSTEM CONTROL ON THE REDUCED ROV MODEL

Due to the nature of pipeline inspection that involves position tracking and hovering at points on the pipeline (to locate the actual location of a pipe's leakage or crack), a control system structure that is able to perform both tracking and stabilization at the same time is needed. To address this, the cascaded control system to perform simultaneous stabilization (using the inner loop) and position tracking (using the outer loop) is proposed as seen in Figure 5.45. The control system design for a full six-DOF operation becomes more involved as the ROV dynamics are nonlinear, highly coupled in motion, underactuated, and susceptible to hydrodynamic uncertainties. In particular, performance can suffer significantly when the vehicle executes nonlinear and modes-couplings maneuvers.

To begin, the cascaded structure for the RRC ROV systems needs to be established. In the cascaded structure, there is an inner loop for stabilization and a cascaded loop for position tracking.

The inner loop equation represents the dynamics during the station-keeping operation whereby the ROV's body-fixed velocity is regulated about the desired velocity. The velocity error states, $\mathbf{x}_{2e} = [u_e \ v_e \ w_e \ p_e \ q_e \ r_e]^T$ in the inner loop equation can be defined as:

$$\mathbf{x}_{2e} = \mathbf{x}_{2d} - \mathbf{x}_2 \tag{5.75}$$

where $\mathbf{x}_{2d} = [u_d \ v_d \ w_d \ p_d \ q_d \ r_d]^T$ is the desired velocity states vector and $\mathbf{x}_2 = [u \ v \ w \ p \ q \ r]^T$ is the actual velocity state vector $\mathbf{x}_2$. The inner loop equations for the full-order ROV system can be written as:

$$\dot{\mathbf{x}}_{2e} = \bar{\mathbf{f}}_{2e}(\mathbf{x}_{2e},t) + \bar{\mathbf{g}}_{2e}(\bar{\mathbf{u}}_I,\mathbf{x}_{2e},t) \tag{5.76}$$

where $\bar{\mathbf{u}}_I \in \mathfrak{R}^4$ is the inner-loop control law with roll and pitch velocity uncontrolled.

The outer loop equation, on the other hand, represents the ROV dynamics during its pipeline tracking maneuver that requires the ROV to move in either a horizontal or vertical plane.

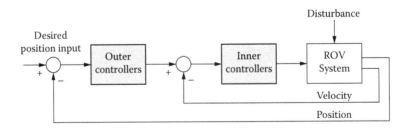

FIGURE 5.45   Cascaded system control.

In ROV tracking control, it is convenient to compute the tracking error in the body-fixed frame as shown in Figure 5.46 as:

$$\mathbf{x}_{1e} = \mathbf{J}^{-1}(\mathbf{x}_1)(\mathbf{x}_{1d} - \mathbf{x}_1) \tag{5.77}$$

where $\mathbf{x}_{1d} = [x_d \; y_d \; z_d \; \phi_d \; \theta_d \; \psi_d]^{\mathrm{T}}$ is the desired position states, $\mathbf{x}_1 = [x \; y \; z \; \phi \; \theta \; \psi]^{\mathrm{T}}$ is the estimate of position state $\mathbf{x}_1$ and $(\mathbf{x}_{1d} - \mathbf{x}_1)$ is the position tracking error in the earth-fixed frame. The $\mathbf{J}^{-1}(\mathbf{x}_1)$ is the inverse of Euler's transformation (and its dependence of $\mathbf{x}_1$ is not shown in (5.77) for clarity). The tracking error dynamic, $\dot{\mathbf{x}}_{1e}$ becomes:

$$\dot{\mathbf{x}}_{1e} = -\mathbf{J}^{-1}[\mathbf{J}^{-1}\dot{\mathbf{J}}(\mathbf{x}_{1d} - \mathbf{x}_1)] + \mathbf{J}^{-1}(\dot{\mathbf{x}}_{1d} - \dot{\mathbf{x}}_1)$$
$$= -\mathbf{J}^{-1}\dot{\mathbf{J}}\mathbf{x}_{1e} + \mathbf{J}^{-1}(\dot{\mathbf{x}}_{1d} - \dot{\mathbf{x}}_1) \tag{5.78}$$

where $\dot{\mathbf{x}}_{1d} = \mathbf{J}(\mathbf{x}_1)\mathbf{x}_{2d}$, and $\dot{\mathbf{x}}_1 = \mathbf{J}(\mathbf{x}_1)\mathbf{x}_2$. Note that the components of the vector $\mathbf{x}_{1e}$ correspond to the error in the ROV longitudinal direction, the cross-track error, the altitude error, roll, pitch, and heading error. On the other hand, the outer loop equations for the full-order system can be written as:

$$\dot{\mathbf{x}}_{1e} = \overline{\mathbf{f}}_{1e}(\mathbf{x}_{1e}, \mathbf{x}_{2e}, \overline{\mathbf{u}}_o, t) \tag{5.79}$$

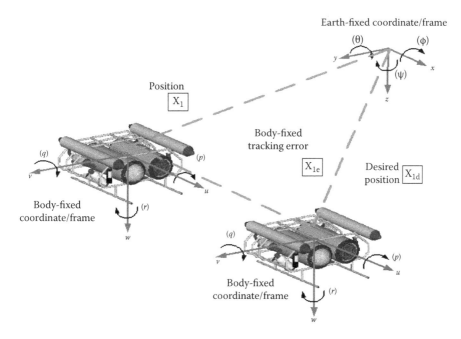

FIGURE 5.46 Tracking error in body-fixed frame.

or alternatively, as $\dot{\mathbf{x}}_{1e} = \overline{\mathbf{f}}_{1e}(\mathbf{x}_{2e},t) + \overline{\mathbf{g}}_{1e}(\mathbf{x}_{1e},\overline{\mathbf{u}}_o,t)$ where the $\overline{\mathbf{u}}_o \in \mathfrak{R}^4$ refers to the control law. Note that the former equation as it involved one less variable term, $\overline{\mathbf{g}}_{1e}(\mathbf{x}_{1e},\overline{\mathbf{u}}_o,t)$ needs to be used. Since the roll and pitch rate is not controlled, its angles are not controlled as well.

We will work on the four DOFs model where the velocity state vector becomes: $\mathbf{x}_{s1} = [x \ y \ z \ \psi]^T$ and position state vector becomes: $\mathbf{x}_{s2} = [u \ v \ w \ r]^T$. The corresponding error state vectors are reduced to four, which are denoted by a subscript s.

The inner control law $\overline{\mathbf{u}}_I$ has the following form as shown below. The inner loop controller makes the state velocity error go to zero and keeps it small until the ROV reaches the desired position in space. In order for the cascaded system to work, the inner loop must be able to react faster than the outer loop. With that in mind, a simple proportional control is used for the inner loop.

$$\overline{\mathbf{u}}_I = \overbrace{\begin{bmatrix} k_{pu} & 0 & 0 & 0 \\ 0 & k_{pv} & 0 & 0 \\ 0 & 0 & k_{pw} & 0 \\ 0 & 0 & 0 & k_{pr} \end{bmatrix}}^{\text{Proportional}} \mathbf{x}_{s2e} \tag{5.80}$$

The outer control law $\overline{\mathbf{u}}_o$ has the following form as shown below.

$$\overline{\mathbf{u}}_o = \overbrace{\begin{bmatrix} k_{px} & 0 & 0 & 0 \\ 0 & k_{py} & 0 & 0 \\ 0 & 0 & k_{pz} & 0 \\ 0 & 0 & 0 & k_{p\psi} \end{bmatrix}}^{\text{Proportional}} \mathbf{x}_{s2e} + \overbrace{\begin{bmatrix} k_{ix} & 0 & 0 & 0 \\ 0 & k_{iy} & 0 & 0 \\ 0 & 0 & k_{iz} & 0 \\ 0 & 0 & 0 & k_{i\psi} \end{bmatrix}}^{\text{Integral}} \int_0^{t_f} \mathbf{x}_{s1e}(t)\,dt \tag{5.81}$$

where the constants in Equation (5.81) are the proportional and integral controller gains.

The block diagram of the PI-P cascaded controller is shown in Figure 5.47. The block diagram is quite similar to the one using the PID except the PI and P controller are used instead. As observed in the cascaded control system, there is an inner-loop and outer-loop control. For the inner loop, the velocity states are used as inputs to the P controller. As for the outer loop, the position states are used for the PI controller. The PI controller gains for outer loop are given as: $K_{px} = 2$; $K_{py} = 2$; $K_{pz} = 3$; $K_{p\psi} = 2$ and $K_{ix} = K_{iy} = K_{iz} = K_{i\psi} = 0.5$. The P controller gains for the inner loop are as follows: $K_{pu} = 2$; $K_{pv} = 3$; $K_{pw} = 3$; $K_{pr} = 2$. The tuning of the PI/P gains can be performed using the numerical optimization tool such as the Simulink Design Optimization Toolbox as mentioned earlier.

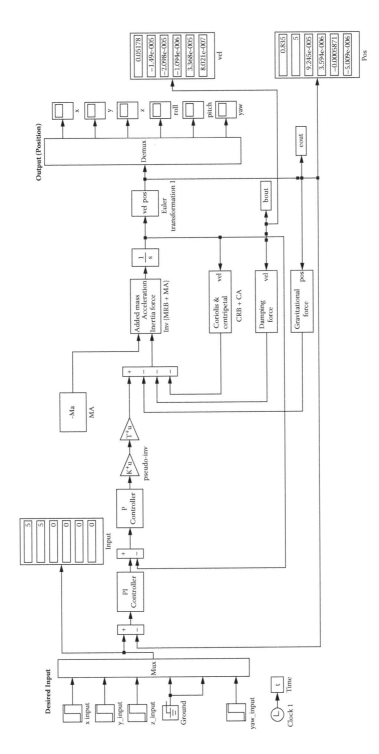

FIGURE 5.47 Cascaded controller using Simulink.

The position and velocity time responses of the ROV can be seen in Figures 5.48 and 5.49. The ROV is able to position itself to the desired inputs. However, it can be observed that the position and velocity are quite oscillatory during the transient stage and there are some overshoots in the surge and sway position responses. The velocities do not exceed 0.7 m/s during the maneuvering. The sway velocity of the ROV

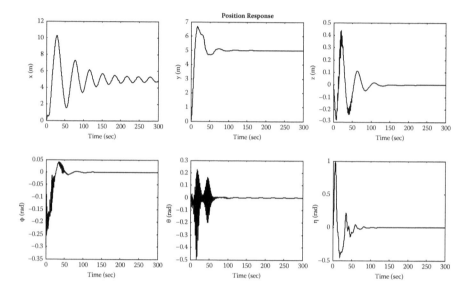

FIGURE 5.48   Position time responses of cascaded controller $(x_d = y_d = 5m)$.

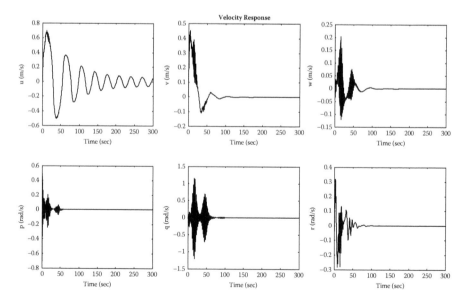

FIGURE 5.49   Velocity time responses of cascaded controller.

decreases to zero after it has reached the target positions at around 100 s while the surge velocity remains quite oscillatory until 300 s. As seen in the figures, the roll and pitch DOFs are stable without any control action.

5.4   LINEAR ROV CONTROL SYSTEMS DESIGN

During the station keeping, the controller has to keep the vehicle at about the equilibrium position or orientation. Under this situation, the vehicle dynamics can be linearized about this equilibrium position as shown in Section 5.2 As the results of the linearization about the equilibrium point: $u_o = 0.5$m/s, $w_o = 0.5$m/s, $\psi_o = 0.75$rad, the linear time invariant model (of twelve orders) for both velocity $\mathbf{x}_1$ and position $\mathbf{x}_2$ can be written in the state-space model as shown in (5.45).

For velocity control, the sixth-orders state-space model is used instead. By controlling the velocity of the ROV, the position of the ROV can be controlled indirectly. The state-space equation for the position states can be obtained by performing Euler's transformation on the velocity states. Equation (5.45) can be reduced as follows.

$$\dot{\mathbf{x}} = \mathbf{A}_{11}\mathbf{x} + \mathbf{B}_1\tau_A \tag{5.82}$$

where $\mathbf{A}$ and $\mathbf{B}$ are constant matrices defined by $\mathbf{A}_{11} = -\mathbf{M}^{-1}(\mathbf{C} + \mathbf{D})$ and

$$\mathbf{B}_1 = \begin{bmatrix} \mathbf{M}^{-1} \\ \mathbf{0} \end{bmatrix},$$

respectively. Here,

$$\mathbf{A}_{11} = \begin{bmatrix} -0.2042 & 0.0000 & 0.0875 & 0.0000 & -0.1183 & -0.0000 \\ -0.0000 & -0.5934 & 0.0000 & 0.2065 & 0.0000 & -0.7597 \\ 0.0883 & -0.0001 & -3.5971 & -0.0000 & 2.0790 & 0.0000 \\ 0.0002 & -2.8467 & -0.0007 & -0.8451 & -0.0000 & 0.5951 \\ 3.4252 & -0.0028 & -14.6789 & -0.0000 & -0.0692 & 0.0001 \\ -0.0006 & 14.4847 & 0.0027 & -0.0802 & 0.0000 & -0.2384 \end{bmatrix}$$

$$\mathbf{B}_1 = \begin{bmatrix} -0.0107 & -0.0107 & -0.0000 & -0.0000 \\ 0.0009 & -0.0009 & -0.0156 & 0.0156 \\ 0.0001 & 0.0001 & -0.0314 & -0.0314 \\ -0.0089 & 0.0089 & 0.1313 & -0.1313 \\ 0.0065 & 0.0065 & -0.0017 & -0.0017 \\ -0.0972 & 0.0971 & 0.0068 & -0.0068 \end{bmatrix}$$

As seen in the nonlinear control of the ROV, we did not consider the thruster dynamics due to the following reasons. In Chapter 4, the thruster's time constant is about 0.5 second and is relatively shorter than the ROV response time (due to large inertia). The thruster dynamic can be ignored as it is not as dominant as compared to the ROV dynamic. Besides, in order not to lose the insight of the ROV's nonlinear behavior, the thrusters' dynamics are not considered. Unfortunately, this will not be true in a real system. The thrust response is affected by the dynamics of the actuators and the drive system. There will also be a loss of thrust efficiency due to disturbances in the water inflow to the thruster blades, caused by thrust-to-thrust and thrust-to-ROV surface interactions, and ROV velocities. In this section, the dynamics model of the actuator will be considered in the linear controller design.

All thruster model parameters were measured experimentally using the test rig as shown in Figure 4.52. By using a simple first-order thruster model dynamic of the thrust over the voltage input, the response can be approximated as:

$$\mathbf{F_T} = \frac{0.97}{0.02s + 1} \; \mathbf{I}_4 \Delta f_T \tag{5.83}$$

where $\mathbf{I}_4$ is a 4×4 identity matrix and

$$\Delta f_T = \begin{cases} 1.00 & \text{for input} = 40\text{volt} \\ 0.82 & \text{for input} = 30\text{volt} \\ 0.60 & \text{for input} = 20\text{volt} \\ 0.33 & \text{for input} = 10\text{volt} \end{cases}$$

The value of Δg_T between each voltage interval is assumed to vary linearly, that is, $\Delta g_T = 0.91$ for 35 volts. Two different models for forward and reverse thrust are used. As shown in Figure 4.54, only the forward thrust in response to 20, 30, and 40 volts are shown. The model steady-state responses are almost in good agreement with those observed in the experiments. The approximated transient response, however, does not match very well during the transient stage. However, it is observed that most oscillations diminish within 0.5 s that is, relatively shorter than the ROV dynamics response time and hence the steady-state value is relatively important than the transient responses. The open-loop system for the ROV can be seen in Figure 5.50. As observed, there are four inputs and the dynamics of the ROV are defined by the state-space matrices $\{\mathbf{A}_{11},\mathbf{B}_1,\mathbf{C}_1,\mathbf{D}_1\}$ where $\mathbf{C}_1$ is an identity matrix and $\mathbf{D}_1$ is a null matrix. Note that the position responses are obtained through the linear Euler's transformation matrix in (5.44).

5.4.1 INHERENT PROPERTIES OF LINEAR ROV SYSTEM

Before designing a controller for the RRC ROV, inherent properties such as open-loop stability, right-half plane (RHP) zero, and system interactions have to be examined. Since the RRC ROV is unlikely to travel more than $u_o = 0.5$m/s, $w_o = 0.5$m/s, $\psi_o = 1.57$rad($=90°$), the following conditions are used. The open-loop RRC ROV system is evaluated at the equilibrium condition: $u_o = 0.5$m/s, $w_o = 0.5$m/s, $\psi_o = 0.75$rad.

Open-Loop Stability Test. First, by examining eigenvalues of the linear system model, it is found that the system is completely stable, since there are no eigenvalues with positive real parts (for all velocities range: $u_o \leq 0.5$m/s, $w_o \leq 0.5$m/s, $\psi_o \leq 0.75$rad). Besides, wide disparity in the eigenvalues indicates that the ROV system consists of a very fast (i.e., thruster dynamic) and slow modes (i.e., ROV

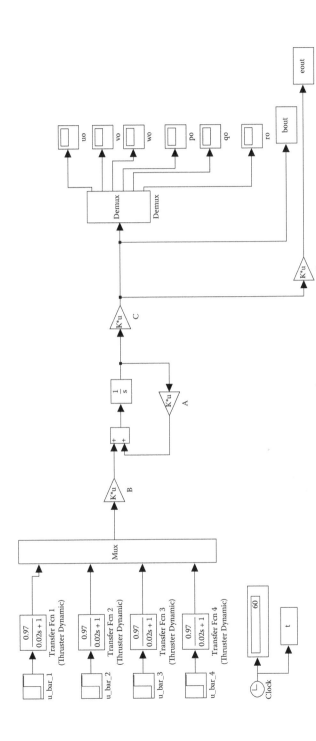

FIGURE 5.50 Linear ROV system in open loop.

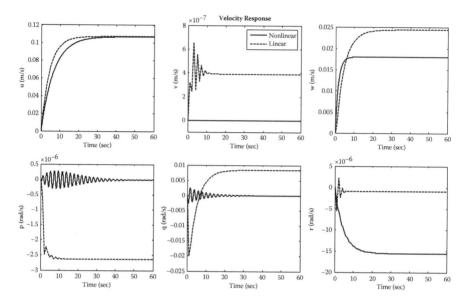

FIGURE 5.51   Linear ROV system open-loop response.

dynamic): $-1.8233 \pm 5.2698i, -0.5005 \pm 3.3995i, -0.2239, -0.6759, -50, -50, -50, -50$. This phenomenon can also be observed in the open-loop time response in Figure 5.51 where all outputs settled to their steady-state values as time increased. The responses match the nonlinear system quite well except for the heave velocity. The error is around 0.01 m/s or 2% of the maximum velocity of the vehicle. The sway, roll, pitch, and yaw rate are quite small in magnitude and the errors are negligible. The MATLAB commands used are as follows:

```
        % State-space matrices
AA=[-inv(M)*(C+D)]        % 6x6
BB=[inv(M)*T]        % 6x4
CC=eye(6)           % 6x6
DD=zeros(6,4)       % 6x4

        % System matrix for ROV
sys_rov=ss(AA,BB,CC,DD);
[Ass,Bss,Css,Dss]=ssdata(sys_rov);

        % Transfer function matrix of 1st order Thruster model
GT=tf([0.97],[0.02 1])
G44=[GT 0 0 0; 0 GT 0 0; 0 0 GT 0; 0 0 0 GT];
invT=pinv(T);
G44_thruster=G44*invT;
sys_thruster=canon(G44_thruster,'modal')
[AT2,BT2,CT2,DT2]=ssdata(sys_thruster);

        % Eigenvalues
eig_AA_rov=eig(Ass)
```

LHP/RHP Zeros Test. Since RHP zeros, in particular, are undesirable due to their effect on the system behavior such as nonminimum phase phenomena and high gain instability in the system time responses, a knowledge of their presence is essential. By examining the open-loop RRC ROV system model, no zeros were detected.

```
%transmission zero
trans_zeros_rov=tzero(sys_rov)
```

Roll and Pitch Stabilizable. Vehicle roll and pitch motions are self-regulating as they correspond to the two directions for which the vehicle has some degree of self-restoring capabilities provided by the buoyancy and gravitational force vectors forming a restoring couple in roll and pitch directions. As long as the roll and pitch motions are not excessive and when the thrusts from all the thrusters reduce to zero, the gravitational and buoyancy moments will eventually restore the ROV to rest state in these two directions. On the other hand, the surges, sway, heave, and yaw motions are not self-regulating and hence left to be control (see Section 5.4.2). For the reader's information, the roll and pitch response are plotted as shown in Figure 5.51.

Interaction Test. Due to the highly coupled nature of the RRC ROV, dynamics system interaction analysis on the linear model was performed. An interaction test using Rosenbrock's row diagonal dominance [81] with Gershgorin discs superimposed on the diagonal elements of the system frequency response was performed. These plots shown in Figure 5.52 indicates that the system is highly interactive over all frequencies (1–100 rad/s), as observed from the increasing size of the discs in diagonal elements in all rows. The increasing size of the disc implies that the off-diagonal terms are more dominant than its diagonal terms. Note that in order to achieve diagonal dominance, either row or column dominance is sufficient.

Minimization of Interaction. To minimize the control effort required for a highly coupled system, preconditioning (i.e., to obtain a more diagonal dominance) on the ROV system is favorable to subsequently facilitate controller design. By using the Direct Nyquist Array with the Gershgorin discs superimposed, the higher the interaction, the larger is the diameter of the Gershgorin disc at each frequency.

Edmunds Scaling. Various design scaling methods such as Edmunds scaling [82], one-norm [37] and Perron–Frobenius (PF) scaling [30] are compared at a scaling frequency, w = 0.1 rad/s. Similarly, the row dominance ratio of each scaling is compared and found that the Edmunds scaling is smaller compared to the two scaling methods. Therefore, the Edmunds scaling is chosen. This is reflected in less concentration of Gershgorin discs at the origin of the plot, as shown in Figure 5.53. This Edmunds scaling algorithm gives the same scaling independently of the ordering of the inputs and outputs with pre- (**Pre$_e$**) and post- (**Post$_e$**) compensators that provide both scaling and I/O pairings as indicated by the resulting nonunity permutation matrices in (5.84) and (5.85).

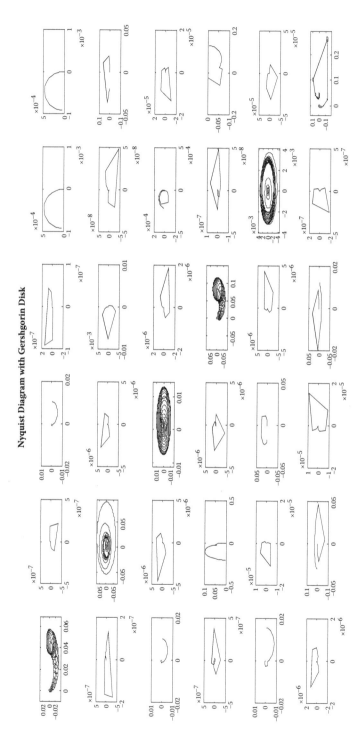

Nyquist Diagram with Gershgorin Disk

FIGURE 5.52   Direct Nyquist Array with Gershgorin discs before Edmunds scaling.

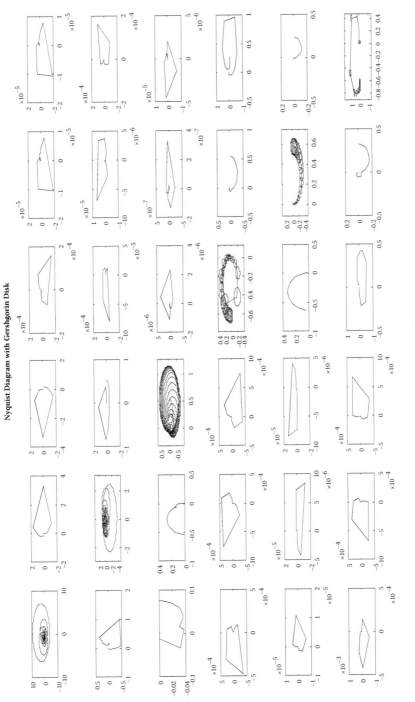

FIGURE 5.53 Direct Nyquist Array with Gershgorin discs after Edmunds scaling.

```
      % Evaluate system matrix at w1
w1=0.1;
Gw1=evalfr(sys_tot,w1);
         % One-norm scaling
[pre1,post1,Gs1]=scale(Gw1,2);
         % Edmund scaling (gs=post*g*pre)
[pre1e,post1e,gs1e,blocks1e,block1e]=normalise(Gw1);
         % Perron-Frobenius scaling
[pfval1p,pre1p,post1p,gs1p]=speron(Gw1);
         % Check whether open loop system is decoupled?
sys_tot_1=post1*sys_tot*pre1;
sys_tot_e=post1e*sys_tot*pre1e;
sys_tot_p=post1p*sys_tot*pre1p;
         % Gershogrin disk
figure 1)
G_tot=post1e*G_tot*pre1e;
run gerdisk_66
         % Step response
figure 2)
step(sys_tot_e)
         % Block diagonal dominance (a/f Edmund scaling)
run block_diag
figure 3);
semilogx(s,ratio_sv,'r');
title('Block Diagonal Dominance Measure of the System');
xlabel('Frequency (rad/s)');
ylabel('Gain');
drawnow
```

$$
\textbf{Post}_e =
\begin{bmatrix}
0 & 30.7505 & 0 & 0 & 0 & 0 \\
0 & 0 & 0 & 0 & 0 & 3.0016 \\
0 & 0 & 0 & 0.4472 & 0 & 0 \\
0 & 0 & 27.7808 & 0 & 0 & 0 \\
6.4536 & 0 & 0 & 0 & 0 & 0 \\
0 & 0 & 0 & 0 & 10.6608 & 0
\end{bmatrix}
\tag{5.84}
$$

$$
\textbf{Pre}_e =
\begin{bmatrix}
0 & 0 & 0 & 0 & 1.9543 & 0 \\
0 & 5.9556 & 0 & 0 & 0 & 0 \\
0 & 0 & 0 & 0 & 0 & 3.4537 \\
0 & 0 & 14.3728 & 0 & 0 & 0 \\
0 & 0 & 0 & 106.5182 & 0 & 0 \\
1.282 & 0 & 0 & 0 & 0 & 0
\end{bmatrix}
\tag{5.85}
$$

As shown in the step responses in Figure 5.54, the input-output reordering using the Edmunds scaling resulted in two diagonal blocks of larger amplitude, composed

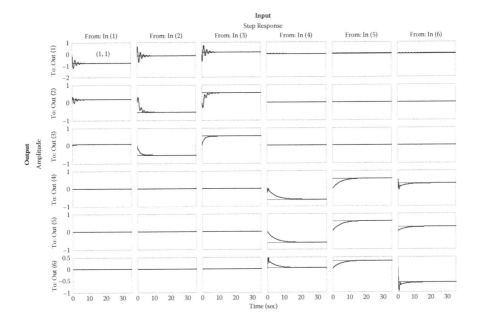

FIGURE 5.54   Step response after Edmunds scaling.

of the linear and angular velocities (reordered as: sway, surge, heave, yaw rate, roll rate, and pitch rate) without interference with each other.

Note that the row refers to the output and the column refers to the commanded optimal distribution of control input. For example, the first step response diagram in position (1,1) of Figure 5.54 indicates the result of the first input on the first output response in surge.

Block Diagonal Dominance. In Figure 5.54, it can be seen that the ROV system can somehow be grouped into two blocks of 3×3 in size. To obtain a good row diagonal dominance may be difficult and therefore, a block diagonal dominance concept can be adopted.

Instead of examining each individual element in each row (column) for row (column) diagonal dominance, a group of elements of different sizes known as a *block* can be tested for diagonal dominance. This is known as the *block diagonal dominance measure* of the system. Suppose that the TFM of the RRC ROV is partitioned into several interconnected subsystems $G_{ij}^{k \times k}(s)$:

$$
\begin{bmatrix}
G_{ii}^{3 \times 3} & G_{ij}^{3 \times 3} \\
G_{ji}^{3 \times 3} & G_{jj}^{3 \times 3}
\end{bmatrix}
\tag{5.86}
$$

where superscript 3×3 and subscript i, j refer to the size of the partitioned block and elements in the block, respectively. The matrix $G(s)$ is considered to be block diagonal dominant if the gain of the diagonal blocks dominates the gain [48] of

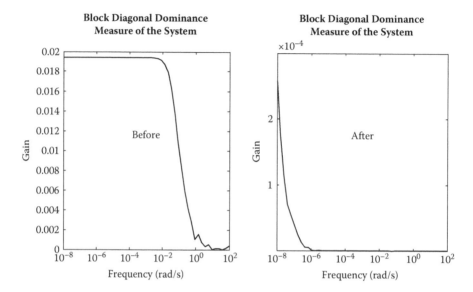

FIGURE 5.55 Block diagonal dominance measure of the system before (left) and after (right) Edmunds scaling using $\underline{\sigma}(\mathbf{G})$.

the off-diagonal blocks. To test this block diagonal dominance of $\mathbf{G(s)}$, the gains of the subblocks of $\mathbf{G(s)}$ are determined as follows. For diagonal block $\mathbf{G_{ii}}$:-

$$\underline{\sigma}(\mathbf{G_{ii}}) = \min_{x \neq 0} \frac{\|\mathbf{G_{ii}x}\|}{\|\mathbf{x}\|}$$

$$= \text{smallest singular value of } \mathbf{G_{ii}}$$

(5.87)

To verify that the RRC ROV system is now block-diagonal dominant, Figure 5.55 shows the block dominance measure of the system before and after the Edmunds scaling using $\underline{\sigma}(\mathbf{G})$. Figure 5.55 indicates that the system becomes more block diagonal dominant, as the value decreases from 1.9×10^{-2} to 2.5×10^{-5}. Hence, the first 3×3 blocks containing the linear velocities are decoupled from the remaining blocks of angular velocities.

5.4.2 LQG/LTR Controller Design

The LQG design, which uses an optimal state estimator, destroys the robustness margin that is achieved in the LQR case. However, with the use of LTR [35] technique, the robustness properties can be recovered. When the plant is a minimum phase system and the cost function weights are chosen so that $\mathbf{R} = \mathbf{I}$ and $\mathbf{Q} = q^2 \mathbf{B_1 B_1^T}$, it has been shown that for the open-loop transfer function for the LQG problem approaches, that the LQR problem as $q \rightarrow \infty$. This suggests a method in control system design, which is known as the *LQG with LTR*. The following steps are adopted for the LQG/LTR design.

Solve the algebraic Riccati equation (ARE) for $\mathbf{P_f}$ using:

$$\mathbf{A_{11}^T P_f} + \mathbf{P_f A_{11}} + \mathbf{Q} - \mathbf{P_f B_1 R^{-1} B_1^T P_f} = 0 \tag{5.88}$$

Determine the optimal state-feedback gain, F (using lqr.m) using:

$$\mathbf{F} = \mathbf{R^{-1} B_1^T P_f} \tag{5.89}$$

Solve the filter algebraic Riccati equation (FARE) for $\mathbf{P_e}$ using:

$$\mathbf{A_{11}^T P_e} + \mathbf{P_e A_{11}} + \mathbf{\Gamma^T} - \mathbf{P_e C_1^T V_1^{-1} C_1 P_e} = 0 \tag{5.90}$$

where the $\mathbf{V_1} = \mathbf{E}[\mathbf{v(t)v^T(t)}]$ represents the measurement noise matrix and the E is a statistical expectation operator.

Determine the optimal state estimator gain, $\mathbf{F_e}$

$$\mathbf{F_e} = \mathbf{P_e C_1^T V_1^{-1}} \tag{5.91}$$

Weight Selection: The Q matrix is typically selected as $\mathbf{C^T C}$, which reflects weighting of the states on the outputs. The matrix $\mathbf{Q_e}$ was selected to be a diagonal scalar matrix of different weights on each state. For the R and $\mathbf{R_e}$ matrices, these were selected to be diagonal matrices of different weights on each control input. The weighting matrices used for the diagonal terms can be seen in the MATLAB function below.

```
      % LQR
[Aall,Ball,Call,Dall]=ssdata(post1e*sys_tot*pre1e);
[len,width]=size(Ball);
[len2,width2]=size(Call');
Q=Call'*Call*4;    Q(7:10,7:10)=diag([1,1,1,1 ]*1)
r=diag([0.0002, 0.02,0.01, 0.3,0.1,0.8])*0.001
f=lqr(Aall,Ball,Q,r)

[Acont,Bcont,Ccont,Dcont]=reg(Aall,Ball,Call,Dall,f,zeros(len,
width));
sys_lqr=ss(Acont,Bcont,Ccont,Dcont);

       % LQG
Qe=1*eye(len)
re=1*eye(width2)
fe=lqr(Aall',Call',Qe,re)
fe=fe'
[Acont,Bcont,Ccont,Dcont]=reg(Aall,Ball,Call,Dall,f,fe);
sys_lqg=ss(Acont,Bcont,Ccont,Dcont);

     % LQG/LTR
Qe1=diag([0.2,0.2,0.3, 0.4, 0.5,      0.1 0.2 0.3 0.3 0.5])*10;
Qe=Qe1+(0.01*Ball*Ball');
re=diag([0.0002, 0.002,0.001,     0.3,0.1,0.8])*1;

fe=lqr(Aall',Call',Qe,re);
fe=fe'

      % Form KLQG/LTR
[Acont,Bcont,Ccont,Dcont]=reg(Aall,Ball,Call,Dall,f,fe);
sys_lqgltr=ss(Acont,Bcont,Ccont,Dcont);
```

In the last part of the MATLAB commands, the reg.m helps to determine the LQG/LTR controller matrix such that it can be used in series with the ROV state-space equation. The Simulink block diagram for RRC ROV can be shown in Figure 5.56.

As mentioned, the pre- and postcompensators are used to reduce the interaction of the ROV system. A step response has been simulated for reference input, $\begin{bmatrix} 0.5 & 0 & 0.5 & 0 & 0.75 \end{bmatrix}^T$ as shown in Figures 5.57 and 5.58. It can be observed that the responses are able to regulate about the desired set points as shown. Note that the velocities are rearranged due to the I/O pairing resulting from

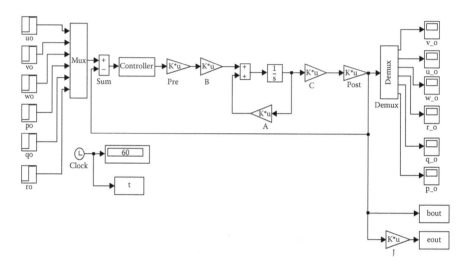

FIGURE 5.56   LQG/LTR Simulink block diagram for linear ROV control.

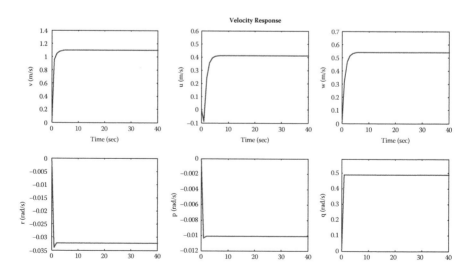

FIGURE 5.57   Velocity time response of LQG/LTR controller.

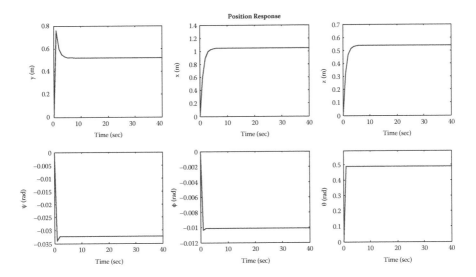

FIGURE 5.58 Position time response of LQG/LTR controller.

the Edmunds scaling algorithm. The position of the ROV was plotted using Euler's transformation on the vehicle's velocity data. As observed in the velocity and position plots, to move the ROV in both the surge and heave simultaneously results in some pitching.

5.4.3 H-Infinity Controller Design

After the dynamic modeling and decoupling are performed on the ROV, the next task is to select a control structure to be used for the H-Infinity (H_∞) [35,43,45] control. Typically, the control structure used is single loop in nature. However, it is not efficient if the DOF is of the twelfth order in the case of ROV RRC. Furthermore, due to the ROV operation requirement, both the velocity and position must be controlled separately during the station keeping (or localized inspection). This inevitably leads to a cascade control structure consisting of an inner loop used for velocity control and outer loop for position control. While the ROV is performing the station keeping, the outer loop is trying to maintain its current position.

As shown in Figure 5.59, in the control structure, the pre- and postcompensators from the Edmunds scaling and reordering routine are included. This results in a decentralized control with only $\begin{bmatrix} u & v & w & r \end{bmatrix}^T$, respectively. Note that the pitch and roll velocity are not included in the feedback due to the self-stabilizing in the roll and pitch angles provided by the buoyancy and gravitational force vectors that form a restoring couple in these directions. Hence, the outputs selected are the position coordinates and the yaw angle, which are not self-stabilizable. These are variables that have no natural equilibrium and hence cannot be left unattended for a long period of time without control.

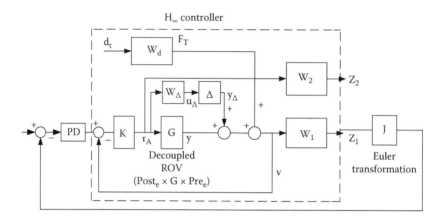

FIGURE 5.59 H_∞ block diagram with PD controller.

The H_∞ controller design used in the inner-loop control (as shown in Figure 5.59) of the vehicle's velocity is shown only. Due to the simplicity in the PD controller for outer-loop control, it is not shown in this section.

The H_∞ controller must be able to perform in the presence of disturbance and parametric uncertainty. In signal terms, the objective is to minimize the maximum value of exogenous output (labeled as) z due to exogenous input signal w. It is desired to find a controller that minimizes the maximum norm or singular value ($\|\cdot\|_\infty$) of closed-loop transfer function, H subjected to a parametric uncertainty bound of $\|\Delta\|_\infty \leq 1$. The Δ is a diagonal additive uncertainty that is used to represent the error due to the linearization and ROV modeling:

$$\Delta = \mathrm{diag}\{\Delta_i\} = \begin{bmatrix} \Delta_1 & & & \\ & \Delta_2 & & \\ & & \ddots & \\ & & & \Delta_6 \end{bmatrix} \tag{5.92}$$

The procedure for designing a H_∞ controller for the RRC ROV is shown as follows.

Consider a feedback system with diagonal additive uncertainty. The $\mathbf{W}_\Delta (= \mathbf{W}_3)$ is a normalization weight for the uncertainty, $\mathbf{W}_d (= \mathbf{W}_4)$ is a weight for the disturbance and $\mathbf{W}_1$ to $\mathbf{W}_2$ are performance weight. The generalized plant has inputs and outputs derived as:

$$\mathbf{w} = \begin{bmatrix} \mathbf{u}_\Delta \\ \mathbf{d}_t \\ \tau_A \end{bmatrix}; \quad \mathbf{z} = \begin{bmatrix} \mathbf{y}_\Delta \\ \mathbf{z}_2 \\ \mathbf{z}_1 \\ \mathbf{v} \end{bmatrix} \tag{5.93}$$

By writing down the equations or simply inspecting Figure 5.59:

$$\mathbf{y}_\Delta = \mathbf{W}_3 \tau_A$$

$$\mathbf{z}_2 = \mathbf{W}_2 \tau_A$$

$$\mathbf{z}_1 = \mathbf{W}_1 \mathbf{W}_4 \mathbf{d}_t + \mathbf{W}_1 \mathbf{u}_\Delta + \mathbf{W}_1 \mathbf{G} \tau_A \tag{5.94}$$

$$\mathbf{v} = -\mathbf{W}_4 \mathbf{d}_t - \mathbf{u}_\Delta - \mathbf{G} \tau_A$$

The generalized plant P from w to z have the form:

$$\mathbf{P} = \begin{bmatrix} \mathbf{0} & \mathbf{0} & \mathbf{W}_3 \\ \mathbf{0} & \mathbf{0} & \mathbf{W}_2 \\ \mathbf{W}_1 & \mathbf{W}_1\mathbf{W}_4 & \mathbf{W}_1\mathbf{G} \\ -\mathbf{I} & -\mathbf{W}_4 & -\mathbf{G} \end{bmatrix} \tag{5.95}$$

Note that the transfer function from $\mathbf{u}_\Delta$ to $\mathbf{y}_\Delta$ (upper left element in P) is zero because $\mathbf{u}_\Delta$ has no direct effect on $\mathbf{y}_\Delta$ (except through K). With that, to derive closed-loop transfer function, H known as Linear Fractional Transformation (LFT), first partition the P to be compatible with K, that is:

$$\mathbf{P}_{11} = \begin{bmatrix} \mathbf{0} & \mathbf{0} \\ \mathbf{0} & \mathbf{0} \\ \mathbf{W}_1 & \mathbf{W}_1\mathbf{W}_4 \end{bmatrix}; \quad \mathbf{P}_{12} = \begin{bmatrix} \mathbf{W}_3 \\ \mathbf{W}_2 \\ \mathbf{W}_1\mathbf{G} \end{bmatrix};$$

$$\mathbf{P}_{21} = \begin{bmatrix} -\mathbf{I} & -\mathbf{W}_4 \end{bmatrix}; \quad \mathbf{P}_{22} = -\mathbf{G} \tag{5.96}$$

and then find $\mathbf{H} = \mathbf{F}_1(\mathbf{P},\mathbf{K})$ using $\mathbf{P}_{11} + \mathbf{P}_{12}\mathbf{K}(\mathbf{I}-\mathbf{P}_{22}\mathbf{K})^{-1}\mathbf{P}_{21}$:

$$\mathbf{H} = \mathbf{P}_{11} + \mathbf{P}_{12}\mathbf{K}(\mathbf{I} - \mathbf{P}_{22}\mathbf{K})^{-1}\mathbf{P}_{21}$$

$$= \begin{bmatrix} -\mathbf{W}_3\mathbf{KS} & -\mathbf{W}_3\mathbf{W}_4\mathbf{KS} \\ -\mathbf{W}_2\mathbf{KS} & -\mathbf{W}_2\mathbf{W}_4\mathbf{KS} \\ \mathbf{W}_1\mathbf{S} & \mathbf{W}_1\mathbf{W}_4\mathbf{S} \end{bmatrix} \tag{5.97}$$

where the sensitivity $\mathbf{S} = (\mathbf{I} + \mathbf{GK})^{-1}$ and complementary sensitivity $\mathbf{T} = \mathbf{GK}(\mathbf{I} + \mathbf{GK})^{-1}$.

The upper left block, $\mathbf{H}_{11}$ shown in Figure 5.60 is the transfer function from $\mathbf{u}_\Delta$ to $\mathbf{y}_\Delta$. This is the transfer function, for evaluating robust stability due to parametric uncertainty:

$$\mathbf{H}_{11} = \mathbf{W}_\Delta \mathbf{K}(\mathbf{I}+\mathbf{GK})^{-1} \tag{5.98}$$

The block diagram for the H_∞ controller can be seen in Figure 5.61. The respective inverse-weighting functions or the desired upper bound on the magnitude of the sensitivity functions can be seen in Figure 5.62. The H_∞ controller is designed to

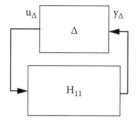

FIGURE 5.60 H_{11} block for evaluating robust stability due to parametric uncertainty.

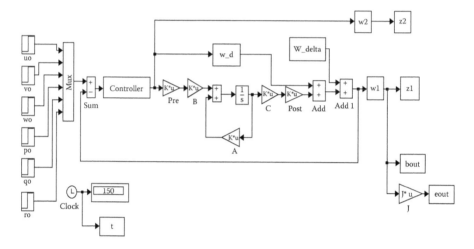

FIGURE 5.61 Simulink block diagram for H_∞ controller.

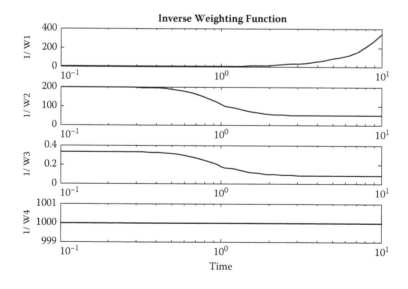

FIGURE 5.62 Inverse-weighting function for H_∞ controller.

minimize the error due to the presence of disturbance and parametric uncertainty. The MATLAB commands for building the weighting function (only W_4 is shown) and H_∞ controller design can be seen below.

```
% W4
n41=0.001*1;d41=1;
n42=0.000090*1;d42=1;
n43=0.0010*1;d43=1;
n44=0.000040*1;d44=1;
n45=0.000010*1;d45=1;
n46=0.00020*1;d46=1;
[aw41,bw41,cw41,dw41]=tf2ss(n41,d41);
[aw42,bw42,cw42,dw42]=tf2ss(n42,d42);
[aw43,bw43,cw43,dw43]=tf2ss(n43,d43);
[aw44,bw44,cw44,dw44]=tf2ss(n44,d44);
[aw45,bw45,cw45,dw45]=tf2ss(n45,d45);
[aw46,bw46,cw46,dw46]=tf2ss(n46,d46);
aw4=daug(aw41,aw42,aw43,aw44,aw45,aw46);
bw4=daug(bw41,bw42,bw43,bw44,bw45,bw46);
cw4=daug(cw41,cw42,cw43,cw44,cw45,cw46);
dw4=daug(dw41,dw42,dw43,dw44,dw45,dw46);
WW1=ss(aw1,bw1,cw1,dw1);WW2=ss(aw2,bw2,cw2,dw2);
WW3=ss(aw3,bw3,cw3,dw3);WW4=ss(aw4,bw4,cw4,dw4);
        % plot inverse weighting function
ww=logspace(-3,3,100);


figure (1);
Wt1=tf(n11,d11)
Wt2=tf(n21,d21)
Wt3=tf(n31,d31)
Wt4=tf(n41,d41)
[mw_1,phw_1,ww]=bode(inv(Wt1));
[mw_2,phw_2,ww]=bode(inv(Wt2));
[mw_3,phw_3,ww]=bode(inv(Wt3));
[mw_4,phw_4,ww]=bode(inv(Wt4));
for i =1 : length(ww)
  mw1(i)=mw_1(:,:,i); phw1(i)=phw_1(:,:,i);
  mw2(i)=mw_2(:,:,i); phw2(i)=phw_2(:,:,i);
  mw3(i)=mw_3(:,:,i); phw3(i)=phw_3(:,:,i);
  mw4(i)=mw_4(:,:,i); phw4(i)=phw_4(:,:,i);
end
subplot(4,1,1); semilogx(ww,mw1(1:length(ww))); ylabel('1 /
W1'); title(' Inverse weighting function')
subplot(4,1,2); semilogx(ww,mw2(1:length(ww))); ylabel('1 /
W2');
subplot(4,1,3); semilogx(ww,mw3(1:length(ww))); ylabel('1 /
W3');
subplot(4,1,4); semilogx(ww,mw4(1:length(ww))); ylabel('1 /
W4'); xlabel('Time')
```

```
        %form Generalized Plant-P
GGG=[zeros(6)                    zeros(6)      WW3;
    zeros(6)                     zeros(6)      WW2;
    WW1                          WW1*WW4       WW1*G_tot2;
    -eye(6)                      -WW4          -G_tot2];

        %Separate the input w and reference input r from
control input u
[AA,BB,CC,DD]=ssdata(GGG);
BB1=BB(:,1:6);
BB2=BB(:,7:12);
CC1=CC(1:18,:);
CC2=CC(19:24,:);
DD11=DD(1:18,1:6);
DD12=DD(1:18,7:12);
DD21=DD(19:24,1:6);
DD22=DD(19:24,7:12);
        %form state space realization of Generalized Plant
pp=pck(AA,[BB1 BB2],[CC1;CC2],[DD11 DD12;DD21 DD22]);
PH22 = rct2lti(mksys(AA,BB1,BB2,CC1,CC2,DD11,DD12,DD21,DD22,
'tss'));
        % Hinf-control
[ss_hinf,ss_cl]=hinf(PH22);
[Acont,Bcont,Ccont,Dcont]=ssdata(ss_hinf);
```

The weighting function is used to shape the sensitivity of the closed-loop system, H to the desired level. A typical type of performance function used is the low-pass filter, high-pass filter and a constant weight. The tuning of the weighting function, irrespective of the type used, is performed iteratively. The sequence is to tune the first entry of the weighting function so that the nominal performance is still enforced. Retain this value for the first entry and repeat the procedure for the second entry, and so on.

$\mathbf{W}_1$. From the frequency response of the disturbance, the disturbance is dominant up to 1 rad/s. To reduce the disturbance up to this frequency, $\mathbf{W}_1$ is selected to be a high-pass filter at the frequency range. Different gains for each entry were selected, sequentially.

$$\mathbf{W}_1 = \frac{10s+100}{s+10}\,\mathrm{diag}\{0.001 \quad 0.0001 \quad 0.01 \quad 0.0001 \quad 3 \quad 0.01\} \tag{5.99}$$

$\mathbf{W}_2$. A low-pass filter was selected to shape the input sensitivity of the closed-loop system. A similar first-order, low-pass filter was used in each channel with a corner frequency of around 0.3 rad/s, in order to limit the input magnitudes at high frequencies and thereby limit the closed-loop bandwidth.

$$\mathbf{W}_2 = \frac{s+0.05}{50s+10}\,\mathrm{diag}\{10 \quad 10 \quad 10 \quad 1 \quad 1 \quad 100\} \times 10^{-10} \tag{5.100}$$

$\mathbf{W_3}\Delta$. Again a low-pass filter was designed for shaping the additive uncertainty to the ROV and hence shaping the input sensitivity of the closed-loop system. It limits the size of the input to the ROV. Here, Δ is an identity matrix.

$$\mathbf{W_3} = \frac{s + 0.05}{50s + 10} \text{diag}\{600 \quad 100 \quad 100 \quad 100 \quad 100 \quad 100\} \tag{5.101}$$

$\mathbf{W_4}\Delta$. The weight here was chosen to be sufficiently small in order to prevent the appearance of some badly damped modes in the closed-loop system due to disturbance.

One of the difficulties in advanced control techniques, based on optimization, is the high order of the optimal controller. It is usually necessary to reduce the order of the controller so that it can be easily implemented; for example, using a Hankel singular value or a Schur balanced truncation to measure the state controllability and observability at different controller's order.

$$\mathbf{W_4} = \text{diag}\{0.001 \quad 0.00009 \quad 0.001 \quad 0.00004 \quad 0.00001 \quad 0.0002\} \tag{5.102}$$

In the $\mathbf{S} = (\mathbf{1} + \mathbf{GK})^{-1}$ plots in Figure 5.63 (left), the weighting function used has shaped the sensitivity of the system below the 0-dB margin. Since the singular value of the sensitivity function is low, it shows that the closed-loop system with a disturbance signal has less influence on the system output. Conversely, a large value of $\mathbf{S}$ means the system has a poor stability margin.

The MATLAB commands used to plot the sensitivity curves can be shown below.

```
% W1W4S
figure (1)
I_GK=inv(eye(6)+G_tot*ss_hinf)*WW1*WW4;
[sv_I_GK]=sigma(I_GK,ww);
loglog(ww,sv_I_GK,'b-');
title('W1W4S ');
xlabel('Frequency (rad/s)');
ylabel('Gain');


% W2W4KS
figure (2)
KI_GK=ss_hinf*inv(eye(6)+G_tot*ss_hinf)*WW2*WW4;
[sv_I_GK]=sigma(KI_GK,ww);
loglog(ww,sv_I_GK,'b-');
title('W2W4KS ');
xlabel('Frequency (rad/s)');
ylabel('Gain');
```

From the input sensitivity, $\mathbf{KS} = \mathbf{K}(\mathbf{1} + \mathbf{GK})^{-1}$ plot in Figure 5.63 (right), it can be seen the effect of the input disturbances is quite negligible on the ROV output. The $\mathbf{KS}$ after applying the shaping function is well below the margin of 0 dB. The robust

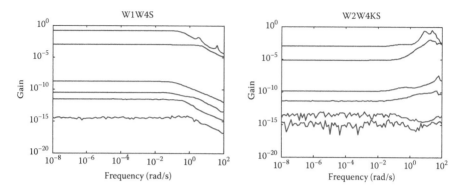

FIGURE 5.63 Sensitivity plots for H_∞ controller.

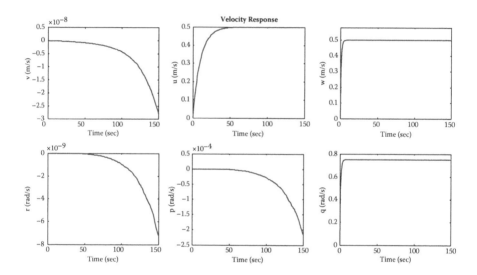

FIGURE 5.64 Velocity time response of H_∞ controller.

stability due to parametric uncertainty as indicated is below the margin of 0 dB and hence is robust against parametric uncertainty.

As observed in Figures 5.64 and 5.65, the response is regulated about the desired position set points and its velocity components are small. This represents the station-keeping condition whereby the velocity is kept at zero while the position is allowed to vary. Note that the velocities and its corresponding positions are rearranged as the result of the Edmund's scaling and reordering routine.

Lastly, comparisons of the controllers (for both the linear and nonlinear ROV model) can be performed. They can be compared against criteria such as: the sum of squared error on the position and velocity outputs and the time-domain performance such as settling time and overshoot in %. Due to some constraint, this will be done in the next edition of the book.

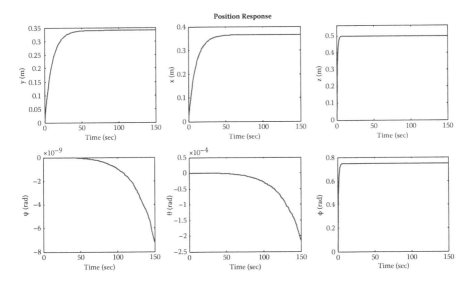

FIGURE 5.65 Position time response of H$_\infty$ controller.

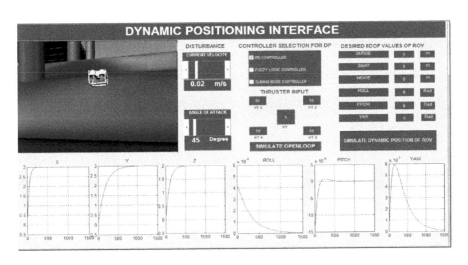

FIGURE 5.66 Velocity time response of the ROV's controllers.

Besides, the graphical user interface (GUI) of the ROV control system tool can be designed to keep all the input ports as well as the final output graphs within one interactive interface screen to make it user friendly. The user can visualize the ROV dynamics in the virtual reality environment, to provide a better understanding of the dynamic positioning performance of each controller. The illustrations (Figure 5.66) show the final GUI of the 3D virtual reality visualization of the ROV's positioning.

References

1. A. Hurwitz. "Über die Bedingungen, unter welchen eine Gleichung nur Wurzeln mit negativen reellen Teilen besitzt." *Math. Ann.* vol. 46, 273–280, in German, 1895.
2. E. J. Routh. *Dynamics of a System of Rigid Bodies.* London: Macmillan, 1905.
3. A. M. Lyapunov. "Probleme General de la Stabilite du Movement." *Ann. Fac. Sci.* Toulouse vol. 9:203–474, 1907.
4. H. Nyquist. "Regeneration Theory." *Bell System Technical Journal* vol. 11:126–147, 1932.
5. H. W. Bode. *Network Analysis and Feedback Amplifier Design.* Princeton, NJ: Van Nostrand, 1945.
6. W. R. Evans. "Graphical Analysis of Control Systems." *Transactions of the AIEE* vol. 61:547–551, 1948.
7. J. C. Maxwell. "On Governors." *Proceedings of the Royal Society of London* vol. 16, 1868.
8. N. Wiener. *The Extrapolation, Interpolation and Smoothing of Stationary Time Series.* John Wiley: New York, 1949.
9. R. Bellman. *Dynamic Programming.* Princeton, NJ: Princeton University Press, 1957.
10. L. S. Pontryagin, V. G. Boltyanskii, R. V. Gamkrelidze and E. F. Mishchenko. *The Mathematical Theory of Optimal Processes.* New York: John Wiley & Sons, 1962.
11. R. E. Kalman. "On the General Theory of Control Systems." *Proceedings of the First International Congress IFAC, Moscow 1960: Automatic and Remote Control.* London: Butterworth & Co., pp. 481–492, 1961.
12. R. E. Kalman and R. S. Bucy. "New Results in Linear Filtering and Prediction Theory." *J. of Basic Eng., Trans. of the Am. Soc. of Mech. Eng.*, 95–108, 1961.
13. M. Athans. The Role and Use of the Stochastic Linear-Quadratic-Gaussian Problem in Control System Design, *IEEE Trans. on Automatic Control* vol. 16, no. 6:529–551, 1971.
14. M. G. Safanov. *Stability and Robustness of Multivariable Feedback Systems.* Cambridge: M.I.T. Press, 1980.
15. R. Y. Chiang. "Modern Robust Control Theory." PhD Thesis, Los Angeles: University of Southern California, 1988.
16. G. Rezevski. *Designing Intelligent Machines*, vol. 1, *Perception, Cognition and Execution.* Oxford, UK: Butterworth-Heinemann, 1995.
17. D. O. Hebb. *The Organization of Behavior.* Wiley: New York, 1949.
18. F. Rosenblatt. *Principles of Neurodynamics: Perceptions and the Theory of Brain Mechanisms.* Washington, DC: Spartan Press, 1961.
19. T. Kohonen. *Self-Organization and Associative Memory.* Berlin: Springer-Verlag, 1988.
20. L. A. Zadeh. "Fuzzy Sets." *Information and Control* vol. 8:338–353, 1965.
21. E. H. Mamdani. "Advances in Linguistic Synthesis of Fuzzy Controllers." *Int. J. Man and Mach. Studies* vol. 8, no. 6:669–678, 1976.
22. M. Sugeno, *Industrial Applications of Fuzzy Control.* North-Holland: Elsevier Science Publishers BV, 1985.
23. R. Sutton, and I. M. Jess. "A Design Study of a Self-Organising Fuzzy Autopilot for Ship Control." *IMechE, Proc. Instn. Mech. Engr, Part I, Journal of Systems and Control Engineering* vol. 205:35–47, 1991.
24. R. M. Tong. "Synthesis of Fuzzy Models for Industrial Processes." *Int. J. General Systems* vol. 4:143–162, 1978.
25. J. G. Ziegler and N. B. Nichols. "Optimum Settings for Automatic Controllers." *Trans. ASME* vol. 64:759–768, 1942.

26. G. P. Liu, R. Dixon, and S. Daley. "Multi-Objective Optimal-Tuning PI Controller Design for the ALSTOM Gasifier Problem." *Proc. Instn. Mech. Engrs, Part I, Journal of Systems and Control Engineering* vol. 214, no. 16:395–403, 2000.

27. M. J. Rice, J. A. Rossiter and J. Schuurmans J. "An Advanced Predictive Control Approach to the ALSTOM Gasifier Problem." *Proc. Instn. Mech. Engrs, Part I, Journal of Systems and Control Engineering*, 214(16):405–413, 2000.

28. E. Prempain, I. Postlethwaite and X. D. Sun. "Robust Control of the Gasifier Using a Mixed-Sensitivity H∞ Approach." *Proc. Instn. Mech. Engrs, Part I, Journal of Systems and Control Engineering* vol. 214, no. 16:415–426, 2000.

29. I. A. Griffin, P. Schroder, A. J. Chipperfield and P. J. Fleming. "Multi-Objective Optimization Approach to the ALSTOM Gasifier Problem." *Proc. Instn. Mech. Engrs, Part I, Journal of Systems and Control Engineering* vol. 214, no. 16:453–467, 2000.

30. R. V. Patel and N. Munro. *Multivariable System Theory and Design.* Oxford: Pergamon Press, 1982.

31. E. E. Osborne. "On Preconditioning of Matrices." *J. of Assoc. of Computing Machinery* vol. 7:338–345, 1960.

32. N. Munro and R. S. Mcleod. "Minimal Realisation of Transfer Function Matrices Using the System Matrix." *Proc. IEE* vol. 118, no. 9, 1971.

33. M. B. Jager. "A Review of Methods for Input/Output Selection." *Automatica* vol. 37:487–510, 2000.

34. E. H. Bristol. "On a New Measure of Interactions for Multivariable Process Control." *IEEE Trans. Automatic Control* vol. 11:133–134, 1966.

35. S. Skogestad and I. Postlethwaite. *Multivariable Feedback Control Analysis and Design.* New York: John Wiley, 1995.

36. C. S. Chin. "The Analysis and Model Order Reduction and Control of a Gasifier." MSc Thesis, Manchester: UMIST, 2001.

37. A. I. Mees. "Achieving Diagonal Dominance." *System and Control Letter* vol. 1, no. 3:155–158, 1981.

38. E. J. Davison. "A Computational Method for Finding the Zeros of a Multivariable Linear Time-Invariant System." *Automatica* vol. 6:481–484, 1970.

39. H. Kwakernaak and R. Sivan. *Linear Optimal Control Systems.* New York: Wiley and Sons, 1972.

40. B. D. O. Anderson and J. B. Moore, *Optimal Control-Linear Quadratic Methods.* Englewood Cliffs, NJ: Prentice Hall, 1990.

41. N. A. Lehtomak, N. R. Sandell and M. Athans. "Robustness Results in LQG Based Multivariable Control Designs." *IEEE Trans. Automat. Control* vol. 26, no. 1:75–93, 1981.

42. R. Martin, L. Valavani and M. Athans. "Multivariable Control of a Submarine Using the LQG/LTR Design Methodlogy." Proc. American Control Conference, 18–20 June 1986, Seattle, WA, 1986.

43. H. Kwakernaak. "Robust Control and H∞-Optimization—Tutorial Paper." *Automatica* vol. 29, no. 2:255–273, 1993.

44. B. A. Francis. *A Course in H∞ Control Theory, Lecture Notes in Control and Information Sciences* vol. 88. Berlin: Springer-Verlag, 1987.

45. K. Zhou, J. C. Doyle and K. Glover. *Robust Optimal Control.* Englewood Cliffs, NJ: Prentice Hall, 1996.

46. M. G. Safonov, D. J. N. Limebeer and R. Y. Chiang. "Simplifying the H∞ Theory via Loop Shifting Matrix Pencil and Descriptor Concepts." *Int. J. Control* vol. 50, no. 6:2467–2488, 1989.

47. H. Kwakernaak. "H2-Optimization-Theory and Application to Robust Control Design." IFAC Symposium on Robust Control Design, Czech Republic, pp. 21–23, 2000.

48. L. Postlethwaite, J. M. Edmunds and A. G. J. MacFarlane. "Principal Gains and Principal Phases in the Analysis of Linear Multivariable Feedback Systems." *IEEE Trans. Automatic Control* vol. 26, no. 1:32–46, 1981.

49. J. C. Doyle and G. Stein. "Multivariable Feedback Design: Concepts for a Classical/ Modern synthesis." *IEEE Trans Automatic Control vol.* 26, no. 1:4–16, 1981.

50. E. Lefeber, K. Y. Pettersen and H. Nijmeijer. "Tracking Control of an Underactuated Ship." *IEEE Transactions on Control Systems Technology* vol. 11, no. 1:52–61, 2003.

51. A. Behal, D. M. Dawson, W. E. Dixon and F. Yang. "Tracking and Regulation Control of an Underactuated Surface Vessel with Non-Integrable Dynamics." *IEEE Transactions on Automatic Control* vol. 47, no. 3:495–500, 2002.

52. C. C. Chiok, K. Hariharan and C. L. Teo. "A Comparison of Controller Performance for an Autonomous Underwater Vehicle." Proceedings of the 2nd International Conference on Recent Advances in Mechatronics, Istanbul, Turkey: 402–407, 1999.

53. B. Jalving. "The NDRE-AUV Flight Controls System." *IEEE Journal of Oceanic Engineering* vol. 19, no. 4:497–501, 1994.

54. C. S. Chin, M. W. S. Lau, E. Low and G. G. Seet. "Software for Modelling and Simulation of a Remotely Operated Vehicle." *International Journal of Simulation Modelling* vol. 5, no. 3:114–125, 2006.

55. D. A. Smallwood and L. L. Whitcomb. "Preliminary Experiments in the Adaptive Identification of Dynamically Positioned Underwater Robotic Vehicles." In Proceedings of the 2001 IEEE/RSJ International Conference on Intelligent Robots and Systems, pp. 1803–1810, 2001.

56. D. A. Smallwood and L. L. Whitcomb. "Toward Model Based Dynamic Positioning of Underwater Robotic Vehicles." Proceedings of IEEE/MTS Oceans 2001: 1106–1114, 2001.

57. C. S. Chin, M. W. S. Lau, E. Low and G. G. Seet. "Robust and Decoupled Cascaded Control System of URV for Stabilization and Pipeline Tracking." *Proc. of IMech. Part I: Journal of System & Control Engineering* vol. 222, no. 4: 261–278, 2008.

58. K. D. Do, Z. P. Jiang and J. Pan. "A Global Output Feedback Controller for Simultaneous Tracking and Stabilization of Unicycle-Type Mobile Robot." IEEE *Transactions on Robotic and Automation* vol. 20, no. 3:589–594, 2004.

59. K. D. Do, Z. P. Jiang, J. Pan and H. Nijmeijer. "A Global Output Feedback Controller for Stabilization and Tracking of Underactuated ODIN: A Spherical Underwater Vehicle." *Automatica* vol. 40, no. 1:117–124, 2004.

60. T. I. Fossen. *Guidance and Control of Ocean Vehicles.* New York: John Wiley & Sons Ltd, 1994.

61. Y. H. Eng. "Identification of Hydrodynamic Terms for Underwater Robotic Vehicle." Master First Year Report, Singapore: Robotic Research Center, Mechanical and Aerospace Engineering, NTU, 2007.

62. Y. H. Eng, M. W. S. Lau, E. Low, G. G. L. Seet and C. S. Chin. "A Novel Method to Determine the Hydrodynamic Coefficients of an Eyeball ROV." AIP Conference Proceedings vol. 1089, pp. 11–22, 2009.

63. G. Conte, S. M. Zanoli, D. Scaradozzi and A. Conti. "Evaluation of Hydrodynamics Parameters of a UUV. A Preliminary Study." International Symposium on Control, Communications and Signal Processing, ISCCSP, Hammamet, 2004.

64. J. N. Newman. *Marine Hydrodynamics.* Cambridge: MIT Press, 1977.

65. K. Wendel. "Hydrodynamic Masses and Hydrodynamic Moment of Inertia." Technical Report, TMB Translation 260, 1956. http://hdl.handle.net/1721.3/51294

66. M. S. Triantafyllou and F. S. Hover. *Maneuvering and Control of Marine Vehicles.* Cambridge, Massachusetts: Department of Ocean Engineering, MIT, 2003.

67. W. S. Atkins Consultants. "Best Practices Guidelines for Marine Applications of CFD." MARNET-CFD Report, 2002.

68. B. E. Launder and D. B. Spalding. "The Numerical Computation of Turbulent Flows." *Comp. Methods Appl. Mech. Eng* 269–289, 1974.

69. A. Goodman. "Experimental Techniques and Methods of Analysis Used in Submerged Body Research." Third Symposium on Naval Hydrodynamics, Scheveningen, 1960.

70. C. D. Williams, M. Mackay, C. Perron and C. Muselet. "The NRC-IMD Marine Dynamic Test Facility: A Six-Degree-of-Freedom Forced-Motion Test Apparatus for Underwater Vehicle Testing." International UUV Symposium, Newport RI, 2000.

71. E. Y. K. Ng and C. K. Tan. "Viscous Flow Simulation Around a Moving Projectile and URV." *International Journal of Computer Applications in Technology* vol. 11:350–362, 1998.

72. S. F. Hoerner. *Fluid-Dynamic Drag: Practical Information on Aerodynamic Drag and Hydrodynamic Resistance*. Washington: Hoerner Fluid Dynamics, 1965.

73. T. Prestero. "Verification of a Six-Degree of Freedom Simulation Model for the REMUS Autonomous Underwater Vehicle." Master's thesis, Massachusetts: Massachusetts Institute of Technology and the Woods Hole Oceanographic Institution, 2001.

74. T. Sarkar, P. G. Sayer and S. M. Fraser. "A Study of Autonomous Underwater Vehicle Hull Forms Using Computational Fluid Dynamics." *International Journal for Numerical Methods in Fluids* vol. 25:1301–1313, 1997.

75. C. S. Chin and M. W. S. Lau. "Measurement and Control of a Thruster for Unmanned Robotic Vehicle (URV)." 5th National Undergraduate Research Opportunities Programme (NUROP) Congress, Singapore: NTU, 1999.

76. L. L. Whitcomb and D. R. Yoerger. "Development, Comparison, and Preliminary Experimental Validation of Nonlinear Dynamic Thruster Models." *IEEE Journal of Oceanic Engineering* vol. 24:481–494, 1999.

77. K. J. Åström and T. Hägglund. *PID Controllers: Theory, Design, and Tuning*. Research Triangle Park, NC: Instrum. Soc. Amer., 1995.

78. V. I. Utkin. "Variable Structure System with Sliding Modes." *IEEE Transactions on Automatic Control* vol. 22, no. 2:212–222, 1977.

79. E. Freund. "Decoupling and Pole Assignment in Nonlinear System." *Electronics Letters*, vol. 9, no. 16:373–374, 1973.

80. K. Liu and F. L. Lewis. "Some Issues about Fuzzy Logic Control." In Proceedings of 32nd Conference on Decision and Control, pp. 1743–1748, 1993. San Antorio, Texas.

81. H. H. Rosenbrock. *Computer-Aided Control System Design*. USA: Academic Press, 1974.

82. J. M. Edmunds. "Input and Output Scaling and Reordering for Diagonal Dominance and Block Diagonal Dominance." *Proc. IEE*, Part D vol. 145:523–530, 1998.

83. D. Kim and H. Choi. "Laminar Flow Past a Sphere Rotating in the Stream-Wise Direction." *Journal of Fluid Mechanics,* vol. 461, no. 1:365–386, 2002.

84. T. A. Johnson and V. C. Patel. "Flow Past a Sphere Up to a Reynolds Number of 300." *Journal of Fluid Mechanics*, vol. 378:19–70, 1999.

85. E. Achenbach. "Experiments on the Flow Past Spheres at Very High Reynolds Number." *Journal of Fluid Mechanics*, vol. 54, part 3:565–575, 1972.

86. G. S. Constantinescu, R. Pacheco and K. D. Squires. "Detached-Eddy Simulation of Flow Over a Sphere." *AIAA* paper 2002-0425, Aerospace Sciences Meeting, 2002.

Appendix A1: State-Space Matrices for ALSTOM Gasifier System (Linear)

1.1 STATE-SPACE MATRICES AT 0% OPERATING CONDITION

Matrix A

	Col 1	Col 2	Col 3	Col 4	Col 5	Col 6	Col 7	Col 8	Col 9
Row 1	-5.49E+01	5.35E-03	-1.98E+03	0	0	0	0	0	0
Row 2	2.21E+03	-2.18E-01	8.04E+04	2.78E+00	5.61E+00	5.61E+00	5.61E+00	5.61E+00	1.44E+00
Row 3	-1.61E-05	-8.70E-15	-1.11E-04	7.66E-06	1.55E-05	1.55E-05	1.55E-05	1.55E-05	3.97E-06
Row 4	0	0	0	-7.41E-05	0	0	0	0	0
Row 5	-9.37E-07	-5.08E-16	1.12E-07	5.80E-05	-1.33E-04	1.71E-05	1.71E-05	1.71E-05	1.71E-05
Row 6	-9.77E-07	-5.29E-16	1.17E-07	8.83E-06	1.18E-04	-1.32E-04	1.78E-05	1.78E-05	1.78E-05
Row 7	-9.64E-07	-5.22E-16	1.15E-07	8.71E-06	1.76E-05	1.18E-04	-1.32E-04	1.76E-05	1.76E-05
Row 8	-6.83E-07	-3.70E-16	5.64E-08	6.10E-06	1.23E-05	1.23E-05	1.12E-04	-1.37E-04	1.23E-05
Row 9	-7.25E-06	-3.93E-15	-1.81E-05	3.23E-06	6.53E-06	6.53E-06	6.53E-06	1.07E-04	-1.67E-04
Row 10	-1.53E-05	-8.28E-15	-3.86E-05	2.38E-06	4.81E-06	4.81E-06	4.81E-06	4.81E-06	1.05E-04
Row 11	-2.15E-06	-1.17E-15	-3.89E-06	2.62E-06	5.30E-06	5.30E-06	5.30E-06	5.30E-06	5.30E-06
Row 12	0	0	0	5.00E-01	1.00E+00	1.00E+00	1.00E+00	1.00E+00	1.00E+00
Row 13	-7.05E-07	-3.82E-16	-2.34E-06	0	0	0	0	0	0
Row 14	-9.21E-07	-4.99E-16	-2.03E-06	0	0	0	0	0	0
Row 15	-6.86E-07	-3.71E-16	-8.90E-07	0	0	0	0	0	0
Row 16	-9.44E-08	-5.11E-17	-1.71E-07	0	0	0	0	0	0
Row 17	0	0	0	0	0	0	0	0	0
Row 18	0	0	0	0	0	0	0	0	0
Row 19	0	0	0	0	0	0	0	0	0
Row 20	-2.75E-06	-1.49E-15	-4.98E-06	0	0	0	0	0	0
Row 21	-3.27E-08	-1.77E-17	-5.91E-08	0	0	0	0	0	0
Row 22	0	0	0	0	0	0	0	0	0
Row 23	0	0	0	0	0	0	0	0	0
Row 24	-6.98E-07	-3.78E-16	-8.99E-07	0	0	0	0	0	0
Row 25	0	0	0	0	0	0	0	0	0

	Col 10	Col 11	Col 12	Col 13	Col 14	Col 15	Col 16	Col 17	Col 18
Row 1	0	0	-4.40E+03	-1.50E+06	5.38E+06	-3.22E+06	5.91E+06	9.17E+06	1.18E+07
Row 2	-2.73E+02	-1.11E+03	1.79E+05	5.90E+07	-2.18E+08	1.27E+08	-2.40E+08	-3.71E+08	-4.76E+08
Row 3	-7.54E-04	-3.05E-03	-1.65E-05	-1.78E+00	1.62E-01	-2.71E+00	2.01E-01	1.16E+00	1.93E+00
Row 4	0	0	0	0	0	0	0	0	0
Row 5	1.71E-05	8.94E-06	-1.11E-05	1.58E-02	1.18E-01	-1.06E-01	1.02E-01	1.58E-01	2.03E-01
Row 6	1.78E-05	9.32E-06	-7.76E-06	1.65E-02	1.23E-01	-1.10E-01	1.06E-01	1.65E-01	2.11E-01
Row 7	1.76E-05	9.20E-06	-4.91E-06	1.63E-02	1.22E-01	-1.09E-01	1.05E-01	1.62E-01	2.08E-01
Row 8	1.23E-05	6.43E-06	2.17E-07	9.89E-03	8.46E-02	-7.79E-02	7.32E-02	1.14E-01	1.46E-01
Row 9	6.53E-06	3.41E-06	-2.72E-05	-1.09E+00	-1.94E-01	-1.35E+00	-1.30E-01	3.04E-01	6.52E-01
Row 10	-1.72E-03	2.51E-06	-5.78E-05	-2.39E+00	-4.84E-01	-2.89E+00	-3.38E-01	5.75E-01	1.31E+00
Row 11	1.05E-04	-6.30E-03	-8.45E-06	-2.97E-01	-2.95E-02	-3.93E-01	-1.73E-02	1.10E-01	2.11E-01
Row 12	1.00E+00	5.27E-01	-1.00E+00	0	0	0	0	0	0
Row 13	0	0	-7.16E-06	-2.37E-01	-1.09E-01	-1.57E-01	-9.22E-02	-4.88E-02	-1.38E-02
Row 14	0	0	-6.21E-06	-1.77E-01	-9.36E-02	-2.08E-01	-4.82E-02	7.77E-03	5.30E-02
Row 15	0	0	-2.72E-06	-7.41E-02	-1.51E-03	-1.58E-01	7.47E-03	4.88E-02	8.21E-02
Row 16	0	0	-5.22E-07	-1.51E-02	-3.39E-03	-1.79E-02	-2.58E-02	3.35E-03	7.96E-03
Row 17	0	0	0	0	0	0	0	-2.34E-02	0
Row 18	0	0	0	0	0	0	0	0	-2.34E-02
Row 19	0	0	0	0	0	0	0	0	0
Row 20	0	0	-1.52E-05	-4.38E-01	-9.89E-02	-5.20E-01	-6.90E-02	9.76E-02	2.32E-01
Row 21	0	0	-1.81E-07	-5.21E-03	-1.17E-03	-6.18E-03	-8.19E-04	1.16E-03	2.76E-03
Row 22	0	0	0	0	0	0	0	0	0
Row 23	0	0	0	0	0	0	0	0	0
Row 24	0	0	-2.74E-06	-8.12E-02	1.19E-02	-9.87E-02	7.61E-03	4.97E-02	8.36E-02
Row 25	0	0	0	0	0	0	0	0	0

	Col 19	Col 20	Col 21	Col 22	Col 23	Col 24	Col 25
Row 1	1.49E+07	-1.55E+06	-6.80E+06	-8.31E+06	-1.83E+06	2.45E+06	-2.41E+06
Row 2	-6.02E+08	6.05E+07	2.72E+08	3.33E+08	7.17E+07	-1.02E+08	9.49E+07
Row 3	2.85E+00	-1.98E+00	-3.52E+00	-3.95E+00	-2.06E+00	-1.35E+00	-2.23E+00
Row 4	0	0	0	0	0	0	0
Row 5	2.56E-01	-2.54E-02	-1.15E-01	-1.41E-01	-3.02E-02	-7.14E-02	-4.00E-02
Row 6	2.67E-01	-2.65E-02	-1.20E-01	-1.47E-01	-3.15E-02	-7.45E-02	-4.17E-02
Row 7	2.63E-01	-2.61E-02	-1.18E-01	-1.45E-01	-3.10E-02	-7.35E-02	-4.11E-02
Row 8	1.85E-01	-1.98E-02	-8.51E-02	-1.04E-01	-2.32E-02	-5.22E-02	-3.04E-02
Row 9	1.07E+00	-1.11E+00	-1.81E+00	-2.00E+00	-1.15E+00	-6.27E-01	-1.23E+00
Row 10	2.18E+00	-2.41E+00	-3.87E+00	-4.29E+00	-2.49E+00	-1.33E+00	-2.65E+00
Row 11	3.33E-01	-3.11E-01	-5.18E-01	-5.79E-01	-3.22E-01	-1.88E-01	-3.45E-01
Row 12	0	0	0	0	0	0	0
Row 13	2.78E-02	-1.87E-01	-2.53E-01	-2.70E-01	-1.90E-01	-9.84E-02	-1.97E-01
Row 14	1.07E-01	-1.72E-01	-2.60E-01	-2.83E-01	-1.77E-01	-5.24E-02	-1.86E-01
Row 15	1.22E-01	-8.54E-02	-1.51E-01	-1.68E-01	-8.88E-02	-2.76E-02	-9.59E-02
Row 16	1.34E-02	-1.51E-02	-2.41E-02	-2.65E-02	-1.56E-02	-8.27E-03	-1.66E-02
Row 17	0	0	0	0	0	0	0
Row 18	0	0	0	0	0	0	0
Row 19	-2.34E-02	0	0	0	0	0	0
Row 20	3.92E-01	-4.64E-01	-7.01E-01	-7.71E-01	-4.54E-01	-2.41E-01	-4.82E-01
Row 21	4.65E-03	-5.23E-03	-3.17E-02	-9.16E-03	-5.39E-03	-2.86E-03	-5.73E-03
Row 22	0	0	0	-2.34E-02	-2.34E-02	0	0
Row 23	0	0	0	0	-2.34E-02	0	0
Row 24	1.24E-01	-8.69E-02	-1.53E-01	-1.72E-01	-9.04E-02	-1.13E-01	-9.76E-02
Row 25	0	0	0	0	0	0	-2.34E-02

Matrix B

	Col 1	Col 2	Col 3	Col 4	Col 5	Col 6
Row 1	7.13E+03	9.57E+03	1.26E+03	-8.26E+02	-1.59E+03	3.80E-01
Row 2	-4.64E+05	-1.85E+04	3.92E+03	0	-6.85E+04	6.29E+00
Row 3	-4.85E-01	-8.78E-02	6.34E-01	0	-2.06E-03	1.73E-05
Row 4	0	0	0	0	0	0
Row 5	-1.14E-01	-1.84E-02	1.63E-01	0	1.13E-01	0
Row 6	-1.95E-01	-1.92E-02	1.70E-01	0	1.17E-01	1.25E-17
Row 7	-2.48E-01	-1.90E-02	1.68E-01	0	1.16E-01	6.77E-13
Row 8	-2.48E-01	-1.33E-02	1.17E-01	0	8.09E-02	1.41E-08
Row 9	-1.80E-01	-7.03E-03	6.22E-02	0	4.29E-02	1.03E-05
Row 10	-1.13E-02	-5.18E-03	4.59E-02	0	3.16E-02	2.25E-05
Row 11	-2.69E-04	-5.71E-03	5.05E-02	0	3.48E-02	2.80E-06
Row 12	0	0	0	0	0	0
Row 13	9.17E-05	2.57E-04	2.78E-05	0	3.42E-04	1.91E-06
Row 14	1.20E-04	-6.28E-12	1.03E-02	0	0	1.68E-06
Row 15	8.92E-05	7.07E-03	7.64E-04	0	9.42E-03	7.59E-07
Row 16	1.23E-05	-6.49E-13	2.09E-03	0	0	1.43E-07
Row 17	0	0	0	0	0	0
Row 18	0	0	0	0	0	0
Row 19	0	0	0	0	0	0
Row 20	3.58E-04	2.68E-02	3.00E-04	0	0	4.16E-06
Row 21	4.25E-06	3.19E-04	2.33E-06	0	0	4.94E-08
Row 22	0	0	0	0	0	0
Row 23	0	0	0	0	0	0
Row 24	9.08E-05	3.47E-04	6.95E-03	5.55E-02	2.22E-03	7.73E-07
Row 25	0	0	0	0	0	0

Matrix C

	Col 1	Col 2	Col 3	Col 4	Col 5	Col 6	Col 7	Col 8	Col 9
Row 1	0	0	0	0	0	0	0	0	0
Row 2	0	0	0	0	0	0	0	0	0
Row 3	6.61E-01	3.58E-10	1.18E+00	0	0	0	0	0	0
Row 4	6.55E-04	3.55E-13	-1.71E-13	0	0	0	0	0	0

	Col 10	Col 11	Col 12	Col 13	Col 14	Col 15	Col 16	Col 17	Col 18
Row 1	0	0	0	5.03E+05	7.74E+05	-6.91E+05	2.42E+06	4.28E+06	6.11E+06
Row 2	0	0	1.00E+00	0	0	0	0	0	0
Row 3	0	0	3.60E+00	1.07E+05	2.39E+04	1.27E+05	1.74E+04	-2.19E+04	-5.35E+04
Row 4	0	0	0	1.76E+01	-6.46E+01	3.79E+01	-7.10E+01	-1.10E+02	-1.41E+02

	Col 19	Col 20	Col 21	Col 22	Col 23	Col 24	Col 25
Row 1	7.94E+06	-4.40E+05	-6.27E+05	-1.01E+06	-4.71E+05	-2.83E+05	-5.02E+05
Row 2	0	0	0	0	0	0	0
Row 3	-9.11E+04	1.07E+05	1.71E+05	1.89E+05	1.11E+05	5.88E+04	1.18E+05
Row 4	-1.78E+02	1.81E+01	8.07E+01	9.88E+01	2.14E+01	-2.99E+01	2.83E+01

Matrix D

	Col 1	Col 2	Col 3	Col 4	Col 5	Col 6
Row 1	0	0	0	0	0	0
Row 2	0	0	0	0	0	0
Row 3	-8.59E+01	4.54E-06	0	0	0	0
Row 4	-8.51E-02	4.44E-09	0	0	0	0

1.2   STATE-SPACE MATRICES AT 50% OPERATING CONDITION

Matrix A

	Col 1	Col 2	Col 3	Col 4	Col 5	Col 6	Col 7	Col 8	Col 9
Row 1	-4.17E+01	3.99E-03	-1.63E+03	0	0	0	0	0	0
Row 2	2.26E+03	-2.19E-01	8.95E+04	1.24E+01	2.51E+01	2.51E+01	2.50E+01	1.44E+01	-3.35E+02
Row 3	-1.95E-05	-8.06E-15	-2.78E-04	3.15E-05	6.37E-05	6.37E-05	6.35E-05	3.65E-05	-8.49E-04
Row 4	0	0	0	-1.68E-04	0	0	0	0	0
Row 5	-9.81E-07	-4.05E-16	1.31E-07	1.46E-04	-2.94E-04	4.49E-05	4.49E-05	4.49E-05	4.49E-05
Row 6	-1.02E-06	-4.23E-16	1.33E-07	2.32E-05	2.97E-04	-2.92E-04	4.68E-05	4.68E-05	4.68E-05
Row 7	-1.11E-06	-4.59E-16	-3.10E-07	2.28E-05	4.61E-05	2.96E-04	-2.93E-04	4.61E-05	4.61E-05
Row 8	-7.50E-06	-3.10E-15	-3.01E-05	1.60E-05	3.23E-05	3.23E-05	2.82E-04	-3.60E-04	3.23E-05
Row 9	-1.87E-05	-7.70E-15	-8.21E-05	8.46E-06	1.71E-05	1.71E-05	1.71E-05	2.67E-04	-2.11E-03
Row 10	-3.17E-06	-1.31E-15	-1.16E-05	6.24E-06	1.26E-05	1.26E-05	1.26E-05	1.26E-05	2.63E-04
Row 11	-1.28E-06	-5.30E-16	-2.49E-06	6.87E-06	1.39E-05	1.39E-05	1.39E-05	1.39E-05	1.39E-05
Row 12	0	0	0	5.00E-01	1.00E+00	1.00E+00	1.00E+00	1.00E+00	1.00E+00
Row 13	-2.40E-07	-9.92E-17	-1.71E-06	0	0	0	0	0	0
Row 14	-5.25E-07	-2.17E-16	-1.45E-06	0	0	0	0	0	0
Row 15	-5.42E-07	-2.24E-16	-6.59E-07	0	0	0	0	0	0
Row 16	-6.53E-08	-2.69E-17	-1.29E-07	0	0	0	0	0	0
Row 17	0	0	0	0	0	0	0	0	0
Row 18	0	0	0	0	0	0	0	0	0
Row 19	0	0	0	0	0	0	0	0	0
Row 20	-1.90E-06	-7.85E-16	-3.76E-06	0	0	0	0	0	0
Row 21	-2.26E-08	-9.32E-18	-4.47E-08	0	0	0	0	0	0
Row 22	0	0	0	0	0	0	0	0	0
Row 23	0	0	0	0	0	0	0	0	0
Row 24	-5.95E-07	-2.46E-16	-7.70E-07	0	0	0	0	0	0
Row 25	0	0	0	0	0	0	0	0	0

	Col 10	Col 11	Col 12	Col 13	Col 14	Col 15	Col 16	Col 17	Col 18
Row 1	0	0	-3.62E-03	-9.03E+05	3.32E+06	-1.91E+06	3.79E+06	5.99E+06	7.72E+06
Row 2	-1.52E+03	-3.64E+03	1.98E+05	4.79E+07	-1.82E+08	1.01E+08	-2.07E+08	-3.27E+08	-4.21E+08
Row 3	-3.84E-03	-9.24E-03	4.73E-05	-1.84E+00	7.85E-02	-2.79E+00	2.11E-01	1.24E+00	2.06E+00
Row 4	0	0	0	0	0	0	0	0	0
Row 5	2.60E-05	2.31E-05	-3.27E-05	1.02E-02	9.91E-02	-1.15E-01	9.05E-02	1.42E-01	1.83E-01
Row 6	2.71E-05	2.41E-05	-2.54E-05	1.05E-02	1.03E-01	-1.20E-01	9.44E-02	1.48E-01	1.91E-01
Row 7	2.68E-05	2.38E-05	-1.85E-05	-2.92E-03	9.90E-02	-1.34E-01	9.14E-02	1.50E-01	1.96E-01
Row 8	1.87E-05	1.66E-05	-2.15E-05	-8.64E-01	-1.15E-01	-1.11E+00	-4.13E-02	3.56E-01	6.69E-01
Row 9	9.91E-06	8.80E-06	-1.68E-05	-2.31E+00	-4.39E-01	-2.78E+00	-2.26E-01	7.64E-01	1.55E+00
Row 10	-7.78E-03	6.49E-06	-3.08E-06	-3.67E-01	-4.77E-02	-4.71E-01	-1.76E-02	1.49E-01	2.80E-01
Row 11	1.53E-04	-1.82E-02	-4.92E-06	-1.28E-01	2.58E-03	-1.93E-01	8.97E-03	7.44E-02	1.26E-01
Row 12	5.85E-01	5.20E-01	-1.00E+00	0	0	0	0	0	0
Row 13	0		-5.22E-06	-1.55E-01	-7.84E-02	-3.03E-02	-6.62E-02	-5.10E-02	-3.85E-02
Row 14	0		-4.42E-06	-8.96E-02	-8.94E-02	-1.02E-01	-2.81E-02	1.73E-03	2.57E-02
Row 15	0		-2.01E-06	-3.92E-02	5.12E-03	-1.38E-01	1.32E-02	4.28E-02	6.63E-02
Row 16	0		-3.94E-07	-8.14E-03	-1.89E-03	-9.46E-03	-4.53E-02	2.68E-03	5.59E-03
Row 17	0		0	0	0	0	0	-4.44E-02	0
Row 18	0		0	0	0	0	0	0	-4.44E-02
Row 19	0		0	0	0	0	0	0	0
Row 20	0		-1.15E-05	-2.37E-01	-5.52E-02	-2.76E-01	-2.77E-02	7.82E-02	1.63E-01
Row 21	0		-1.36E-07	-2.82E-03	-6.56E-04	-3.27E-03	-3.29E-04	9.29E-04	1.93E-03
Row 22	0		0	0	0	0	0	0	0
Row 23	0		0	0	0	0	0	0	0
Row 24	0		-2.35E-06	-5.00E-02	1.25E-02	-6.00E-02	1.18E-02	4.43E-02	7.03E-02
Row 25	0		0	0	0	0	0	0	0

	Col 19	Col 20	Col 21	Col 22	Col 23	Col 24	Col 25
Row 1	9.75E+06	-9.49E+05	-4.21E+06	-5.08E+06	-1.13E+06	1.51E+06	-1.44E+06
Row 2	-5.31E+08	5.01E+07	2.27E+08	2.75E+08	6.00E+07	-8.45E+07	7.66E+07
Row 3	3.01E+00	-2.01E+00	-3.53E+00	-3.93E+00	-2.09E+00	-1.47E+00	-2.23E+00
Row 4	0	0	0	0	0	0	0
Row 5	2.31E-01	-2.10E-02	-9.78E-02	-1.18E-01	-2.53E-02	-9.21E-02	-3.25E-02
Row 6	2.41E-01	-2.20E-02	-1.02E-01	-1.24E-01	-2.65E-02	-9.61E-02	-3.40E-02
Row 7	2.50E-01	-3.51E-02	-1.22E-01	-1.45E-01	-4.00E-02	-1.02E-01	-4.81E-02
Row 8	1.04E+00	-8.93E-01	-1.48E+00	-1.63E+00	-9.26E-01	-5.42E-01	-9.80E-01
Row 9	2.46E+00	-2.34E+00	-3.80E+00	-4.17E+00	-2.42E+00	-1.29E+00	-2.56E+00
Row 10	4.34E-01	-3.79E-01	-6.27E-01	-6.94E-01	-3.93E-01	-2.26E-01	-4.16E-01
Row 11	1.86E-01	-1.39E-01	-2.41E-01	-2.72E-01	-1.45E-01	-1.01E-01	-1.55E-01
Row 12	0	0	0	0	0	0	0
Row 13	-2.43E-02	-9.14E-02	-1.08E-01	-1.09E-01	-9.21E-02	-3.34E-02	-9.35E-02
Row 14	5.35E-02	-8.59E-02	-1.25E-01	-1.32E-01	-8.79E-02	7.31E-04	-9.14E-02
Row 15	9.38E-02	-4.75E-02	-8.90E-02	-9.85E-02	-4.97E-02	-1.03E-02	-5.36E-02
Row 16	8.98E-03	-8.21E-03	-1.31E-02	-1.42E-02	-8.46E-03	-4.47E-03	-8.91E-03
Row 17	0	0	0	0	0	0	0
Row 18	0	0	0	0	0	0	0
Row 19	-4.44E-02	0	0	0	0	0	0
Row 20	2.61E-01	-2.83E-01	-3.83E-01	-4.13E-01	-2.47E-01	-1.30E-01	-2.60E-01
Row 21	3.11E-03	-2.84E-03	-4.89E-02	-4.90E-03	-2.93E-03	-1.55E-03	-3.08E-03
Row 22	0	0	0	-4.44E-02	0	0	0
Row 23	0	0	0	0	-4.44E-02	0	0
Row 24	1.01E-01	-5.49E-02	-1.00E-01	-1.11E-01	-5.73E-02	-1.22E-01	-6.15E-02
Row 25	0	0	0	0	0	0	-4.44E-02

Matrix B

	Col 1	Col 2	Col 3	Col 4	Col 5	Col 6
Row 1	-3.60E+03	7.14E+03	8.90E+02	-9.37E+02	-1.30E-03	1.60E-01
Row 2	-6.17E+03	-2.12E+04	4.10E+03	0	-7.71E+04	6.61E+00
Row 3	-5.13E-01	-8.94E-02	6.33E-01	0	-4.10E-03	1.68E-05
Row 4	0	0	0	0	0	0
Row 5	-1.32E-01	-1.88E-02	1.63E-01	0	1.12E-01	7.40E-13
Row 6	-2.36E-01	-1.96E-02	1.70E-01	0	1.17E-01	8.34E-10
Row 7	-3.10E-01	-1.93E-02	1.68E-01	0	1.15E-01	1.19E-07
Row 8	-2.80E-01	-1.35E-02	1.17E-01	0	8.06E-02	7.74E-06
Row 9	-4.26E-02	-7.15E-03	6.22E-02	0	4.27E-02	2.06E-05
Row 10	-1.97E-03	-5.27E-03	4.59E-02	0	3.15E-02	3.25E-06
Row 11	-5.16E-04	-5.81E-03	5.05E-02	0	3.47E-02	1.13E-06
Row 12	0	0	0	0	0	0
Row 13	-2.07E-05	5.12E-04	5.53E-05	0	6.82E-04	9.59E-07
Row 14	-4.53E-05	-6.50E-13	1.03E-02	0	0	8.31E-07
Row 15	-4.68E-05	6.94E-03	7.50E-04	0	9.25E-03	3.99E-07
Row 16	-5.63E-06	-8.12E-14	2.09E-03	0	0	7.57E-08
Row 17	0	0	0	0	0	0
Row 18	0	0	0	0	0	0
Row 19	0	0	0	0	0	0
Row 20	-1.64E-04	2.68E-02	3.00E-04	0	0	2.21E-06
Row 21	-1.95E-06	3.19E-04	2.33E-06	0	0	2.62E-08
Row 22	0	0	0	0	0	0
Row 23	0	0	0	0	0	0
Row 24	-5.14E-05	3.47E-04	6.95E-03	5.55E-02	2.22E-03	4.68E-07
Row 25	0	0	0	0	0	0

Matrix C

	Col 1	Col 2	Col 3	Col 4	Col 5	Col 6	Col 7	Col 8	Col 9
Row 1	0	0	0	0	0	0	0	0	0
Row 2	0	0	0	0	0	0	0	0	0
Row 3	8.62E-01	3.56E-10	1.63E+00	0	0	0	0	0	0
Row 4	6.56E-04	2.71E-13	5.68E-14	0	0	0	0	0	0

	Col 10	Col 11	Col 12	Col 13	Col 14	Col 15	Col 16	Col 17
Row 1	0	0	0	3.99E+05	5.92E+05	-5.03E+05	1.86E+06	3.29E+06
Row 2	0	0	1.00E+00	0	0	0	0	0
Row 3	0	0	4.97E+00	1.13E-05	2.54E+04	1.33E+05	1.56E+04	-2.98E+04
Row 4	0	0	0	1.40E-01	-5.24E+01	2.96E+01	-5.99E+01	-9.44E+01

	Col 18	Col 19	Col 20	Col 21	Col 22	Col 23	Col 24	Col 25
Row 1	4.69E+06	6.09E+06	-3.20E+05	-4.56E+05	-7.32E+05	-3.43E+05	-2.06E+05	-3.66E+05
Row 2	0	0	0	0	0	0	0	0
Row 3	-6.56E+04	-1.08E+05	1.14E+05	1.81E+05	1.99E+05	1.17E+05	6.24E+04	1.24E+05
Row 4	-1.22E+02	-1.54E+02	1.47E+01	6.60E+01	7.97E+01	1.76E+01	-2.43E+01	2.24E+01

Matrix D

	Col 1	Col 2	Col 3	Col 4	Col 5	Col 6
Row 1	0	0	0	0	0	0
Row 2	0	0	0	0	0	0
Row 3	7.44E+01	1.07E-06	0	0	0	0
Row 4	5.66E-02	7.31E-10	0	0	0	0

1.3   STATE-SPACE MATRICES AT 100% OPERATING CONDITION

Matrix A

	Col 1	Col 2	Col 3	Col 4	Col 5	Col 6	Col 7	Col 8	Col 9
Row 1	-3.29E+01	3.10E-03	-1.34E+03	0	0	0	0	0	0
Row 2	2.26E+03	-2.16E-01	9.35E-04	2.81E+01	5.68E+01	5.66E+01	5.22E+01	-6.76E+01	-1.40E+03
Row 3	-2.55E-05	-7.62E-15	-4.69E-04	6.76E-05	1.37E-04	1.36E-04	1.26E-04	-1.63E-04	-3.38E-03
Row 4	0	0	0	-2.43E-04	0	0	0	0	0
Row 5	-1.05E-06	-3.14E-16	1.50E-07	2.35E-04	-4.16E-04	7.37E-05	7.37E-05	7.37E-05	7.37E-05
Row 6	-1.17E-06	-3.51E-16	-3.02E-07	3.80E-05	4.77E-04	-4.14E-04	7.68E-05	7.68E-05	7.68E-05
Row 7	-3.56E-06	-1.06E-15	-1.44E-05	3.75E-05	7.58E-05	4.76E-04	-4.35E-04	7.58E-05	7.58E-05
Row 8	-2.31E-05	-6.89E-15	-1.33E-04	2.62E-05	5.30E-05	5.30E-05	4.53E-04	-1.01E-03	5.30E-05
Row 9	-1.20E-05	-3.59E-15	-6.80E-05	1.39E-05	2.81E-05	2.81E-05	2.81E-05	4.28E-04	-7.17E-03
Row 10	-2.06E-06	-6.15E-16	-8.75E-06	1.02E-05	2.07E-05	2.07E-05	2.07E-05	2.07E-05	4.21E-04
Row 11	-1.25E-06	-3.73E-16	-2.64E-06	1.13E-05	2.28E-05	2.28E-05	2.28E-05	2.28E-05	2.28E-05
Row 12	0	0	0	5.00E-01	1.00E+00	1.00E+00	1.00E+00	1.00E+00	1.00E+00
Row 13	-1.36E-07	-4.06E-17	-1.73E-06	0	0	0	0	0	0
Row 14	-4.60E-07	-1.37E-16	-1.45E-06	0	0	0	0	0	0
Row 15	-5.46E-07	-1.63E-16	-6.80E-07	0	0	0	0	0	0
Row 16	-6.27E-08	-1.87E-17	-1.34E-07	0	0	0	0	0	0
Row 17	0	0	0	0	0	0	0	0	0
Row 18	0	0	0	0	0	0	0	0	0
Row 19	0	0	0	0	0	0	0	0	0
Row 20	-1.83E-06	-5.45E-16	-3.90E-06	0	0	0	0	0	0
Row 21	-2.17E-08	-6.48E-18	-4.63E-08	0	0	0	0	0	0
Row 22	0	0	0	0	0	0	0	0	0
Row 23	0	0	0	0	0	0	0	0	0
Row 24	-6.16E-07	-1.84E-16	-8.52E-07	0	0	0	0	0	0
Row 25	0	0	0	0	0	0	0	0	0

	Col 10	Col 11	Col 12	Col 13	Col 14	Col 15	Col 16	Col 17	Col 18
Row 1	0	0	-2.98E-03	-6.00E-05	2.18E-06	-1.25E-06	2.57E-06	4.08E+06	5.26E+06
Row 2	-3.28E+03	-6.52E+03	2.07E-05	4.00E-07	-1.51E+08	8.34E+07	-1.78E+08	-2.82E+08	-3.63E+08
Row 3	-7.90E-03	-1.57E-02	1.07E-04	-2.17E+00	-5.59E-02	-3.17E+00	1.70E-01	1.34E+00	2.26E+00
Row 4	0	0	0	0	0	0	0	0	0
Row 5	3.99E-05	3.79E-05	-5.97E-05	7.77E-03	8.85E-02	-1.18E-01	8.37E-02	1.32E-01	1.70E-01
Row 6	4.16E-05	3.96E-05	-5.13E-05	-6.29E-04	9.02E-02	-1.33E-01	8.62E-02	1.40E-01	1.82E-01
Row 7	4.10E-05	3.90E-05	-4.44E-05	-2.64E-01	2.91E-02	-4.41E-01	5.40E-02	2.18E-01	3.45E-01
Row 8	2.87E-05	2.73E-05	-1.26E-05	-2.40E+00	-4.63E-01	-2.92E+00	-1.94E-01	8.71E-01	1.70E+00
Row 9	1.52E-05	1.45E-05	2.65E-05	-1.23E+00	-2.24E-01	-1.50E+00	-8.34E-02	4.71E-01	9.04E-01
Row 10	-1.55E-02	1.07E-05	-2.11E-06	-1.89E-01	-1.46E-02	-2.60E-01	3.97E-03	9.74E-02	1.70E-01
Row 11	2.29E-04	-3.01E-02	-4.95E-06	-1.05E-01	3.96E-03	-1.65E-01	1.07E-02	6.53E-02	1.07E-01
Row 12	5.46E-01	5.20E-01	-1.00E+00	0	0	0	0	0	0
Row 13	0	0	-5.27E-06	-1.41E-01	-6.86E-02	2.71E-03	-5.75E-02	-4.82E-02	-4.02E-02
Row 14	0	0	-4.42E-06	-6.82E-02	-9.47E-02	-7.64E-02	-2.22E-02	1.71E-03	2.09E-02
Row 15	0	0	-2.08E-06	-3.12E-02	5.55E-03	-1.41E-01	1.41E-02	4.06E-02	6.15E-02
Row 16	0	0	-4.09E-07	-6.43E-03	-1.60E-03	-7.34E-03	-5.74E-02	2.56E-03	5.06E-03
Row 17	0	0	0	0	0	0	0	-5.68E-02	0
Row 18	0	0	0	0	0	0	0	0	-5.68E-02
Row 19	0	0	0	0	0	0	0	0	0
Row 20	0	0	-1.19E-05	-1.87E-01	-4.65E-02	-2.14E-01	-1.70E-02	7.44E-02	1.47E-01
Row 21	0	0	-1.42E-07	-2.22E-03	-5.53E-04	-2.54E-03	-2.02E-04	8.84E-04	1.75E-03
Row 22	0	0	0	0	0	0	0	0	0
Row 23	0	0	0	0	0	0	0	0	0
Row 24	0	0	-2.60E-06	-4.21E-02	1.05E-02	-4.96E-02	1.21E-02	4.21E-02	6.59E-02
Row 25	0	0	0	0	0	0	0	0	0

	Col 19	Col 20	Col 21	Col 22	Col 23	Col 24	Col 25
Row 1	6.64E+06	-6.23E+05	-2.76E+06	-3.33E-06	-7.38E+05	1.01E+06	-9.64E+05
Row 2	-4.58E+08	4.13E+07	1.88E+08	2.27E+08	4.93E+07	-7.23E+07	6.48E+07
Row 3	3.33E+00	-2.31E+00	-3.97E+00	-4.41E+00	-2.40E+00	-1.67E+00	-2.58E+00
Row 4	0	0	0	0	0	0	0
Row 5	2.14E-01	-1.84E-02	-8.67E-02	-1.05E-01	-2.21E-02	-1.01E-01	-2.93E-02
Row 6	2.32E-01	-2.79E-02	-1.04E-01	-1.25E-01	-3.21E-02	-1.10E-01	-4.01E-02
Row 7	4.95E-01	-2.92E-01	-5.24E-01	-5.86E-01	-3.05E-01	-2.53E-01	-3.29E-01
Row 8	2.67E+00	-2.43E+00	-3.93E+00	-4.32E+00	-2.51E+00	-1.37E+00	-2.67E+00
Row 9	1.41E+00	-1.25E+00	-2.03E+00	-2.23E+00	-1.29E+00	-6.99E-01	-1.37E+00
Row 10	2.55E-01	-1.98E-01	-3.33E-01	-3.71E-01	-2.05E-01	-1.31E-01	-2.19E-01
Row 11	1.57E-01	-1.13E-01	-1.97E-01	-2.23E-01	-1.18E-01	-9.03E-02	-1.27E-01
Row 12	0	0	0	0	0	0	0
Row 13	-3.13E-02	-6.81E-02	-7.43E-02	-7.13E-02	-6.81E-02	-1.45E-02	-6.86E-02
Row 14	4.29E-02	-6.46E-02	-9.23E-02	-9.56E-02	-6.59E-02	1.51E-02	-6.86E-02
Row 15	8.58E-02	-3.78E-02	-7.22E-02	-7.93E-02	-3.95E-02	-6.59E-03	-4.31E-02
Row 16	7.95E-03	-6.46E-03	-1.03E-02	-1.10E-02	-6.65E-03	-3.49E-03	-7.04E-03
Row 17	0	0	0	0	0	0	0
Row 18	0	0	0	0	0	0	0
Row 19	-5.68E-02	0	0	0	0	0	0
Row 20	2.31E-01	-2.45E-01	-3.01E-01	-3.19E-01	-1.94E-01	-1.02E-01	-2.05E-01
Row 21	2.75E-03	-2.24E-03	-6.03E-02	-3.79E-03	-2.30E-03	-1.21E-03	-2.44E-03
Row 22	0	0	0	-5.68E-02	0	0	0
Row 23	0	0	0	0	-5.68E-02	0	0
Row 24	9.34E-02	-4.63E-02	-8.50E-02	-9.26E-02	-4.82E-02	-1.32E-01	-5.22E-02
Row 25	0	0	0	0	0	0	-5.68E-02

Matrix B

	Col 1	Col 2	Col 3	Col 4	Col 5	Col 6
Row 1	2.02E+03	5.66E+03	6.75E+02	-9.32E+02	-1.13E+03	1.03E-01
Row 2	-3.57E+05	-2.33E+04	4.19E+03	0	-8.30E+04	7.83E+00
Row 3	-5.22E-01	-9.12E-02	6.33E-01	0	-6.56E-03	1.89E-05
Row 4	0	0	0	0	0	0
Row 5	-1.50E-01	-1.92E-02	1.63E-01	0	1.12E-01	4.97E-10
Row 6	-2.79E-01	-2.00E-02	1.70E-01	0	1.16E-01	7.50E-08
Row 7	-3.67E-01	-1.97E-02	1.68E-01	0	1.15E-01	2.32E-06
Row 8	-1.87E-01	-1.38E-02	1.17E-01	0	8.02E-02	2.06E-05
Row 9	-1.36E-02	-7.30E-03	6.22E-02	0	4.25E-02	1.05E-05
Row 10	-7.97E-04	-5.38E-03	4.59E-02	0	3.13E-02	1.60E-06
Row 11	-3.19E-04	-5.93E-03	5.05E-02	0	3.45E-02	8.64E-07
Row 12	0	0	0	0	0	0
Row 13	8.37E-06	8.19E-04	8.86E-05	0	1.09E-03	7.27E-07
Row 14	2.83E-05	-1.51E-13	1.03E-02	0	0	6.27E-07
Row 15	3.36E-05	6.79E-03	7.34E-04	0	9.05E-03	3.14E-07
Row 16	3.86E-06	-1.79E-14	2.09E-03	0	0	5.94E-08
Row 17	0	0	0	0		0
Row 18	0	0	0	0		0
Row 19	0	0	0	0		0
Row 20	1.12E-04	2.68E-02	3.00E-04	0	0	1.73E-06
Row 21	1.33E-06	3.19E-04	2.33E-06	0	0	2.05E-08
Row 22	0	0	0	0	0	0
Row 23	0	0	0	0	0	0
Row 24	3.79E-05	3.47E-04	6.95E-03	5.55E-02	2.22E-03	3.93E-07
Row 25	0	0	0	0	0	0

Matrix C

	Col 1	Col 2	Col 3	Col 4	Col 5	Col 6	Col 7	Col 8	Col 9
Row 1	0	0	0	0	0	0	0	0	0
Row 2	0	0	0	0	0	0	0	0	0
Row 3	1.06E+00	3.15E−10	2.10E+00	0	0	0	0	0	0
Row 4	6.45E−04	1.93E−13	1.14E−13	0	0	0	0	0	0

	Col 10	Col 11	Col 12	Col 13	Col 14	Col 15	Col 16	Col 17	Col 18
Row 1	0	0	0	3.28E−05	4.75E+05	−3.92E+05	1.49E+06	2.64E+06	3.77E+06
Row 2	0	0	1.00E+00	0	0	0	0	0	0
Row 3	0	0	6.41E+00	1.17E+05	2.75E+04	1.37E+05	1.48E+04	−3.38E+04	−7.17E+04
Row 4	0	0	0	1.17E+01	−4.29E+01	2.41E+01	−5.07E+01	−8.04E+01	−1.04E+02

	Col 19	Col 20	Col 21	Col 22	Col 23	Col 24	Col 25
Row 1	4.90E+06	−2.49E+05	−3.56E+05	−5.70E+05	−2.67E+05	−1.60E+05	−2.85E+05
Row 2	0	0	0	0	0	0	0
Row 3	−1.16E+05	1.17E+05	1.86E+05	2.04E+05	1.21E+05	6.43E+04	1.28E+05
Row 4	−1.31E+02	1.20E+01	5.41E+01	6.51E+01	1.43E+01	−2.04E+01	1.87E+01

Matrix D

	Col 1	Col 2	Col 3	Col 4	Col 5	Col 6
Row 1	0	0	0	0	0	0
Row 2	0	0	0	0	0	0
Row 3	−6.49E+01	3.34E−07	0	0	0	0
Row 4	−3.97E−02	1.96E−10	0	0	0	0

Appendix A2: LQR Simulation Model and Results

2.1 LQR DESIGN

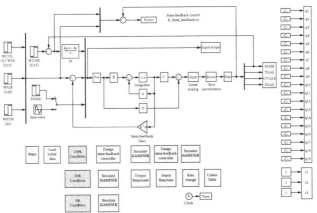

2.1.1 100% LOAD CONDITION (STEP PRESSURE DISTURBANCE)

Outputs

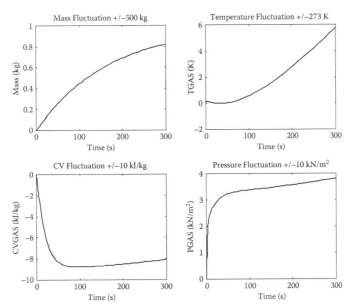

Inputs

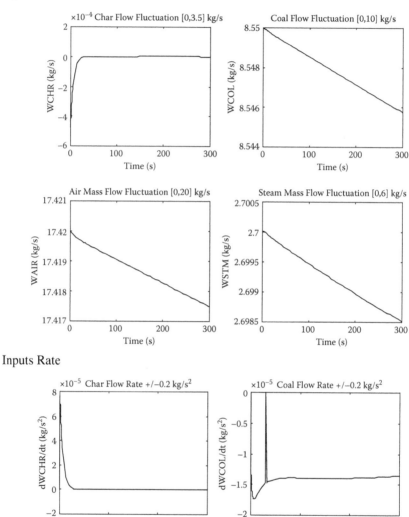

Inputs Rate

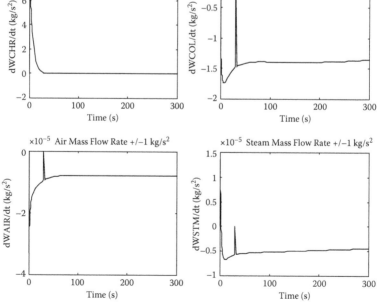

2.1.2 100% LOAD CONDITION (SINUSOIDAL PRESSURE DISTURBANCE)

Outputs

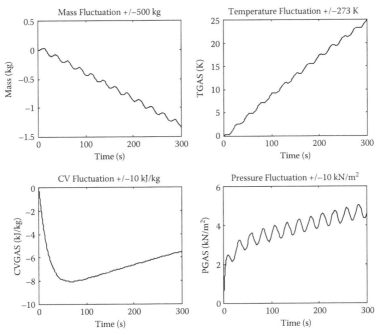

Inputs

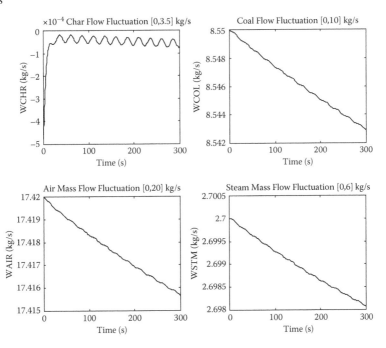

Inputs Rate

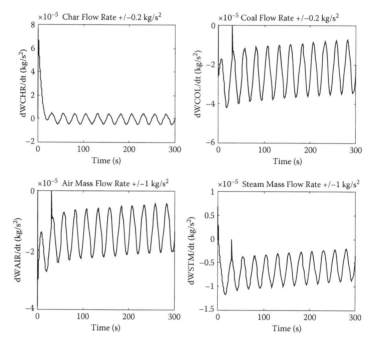

2.1.3   50% Load Condition (Step Pressure Disturbance)

Outputs

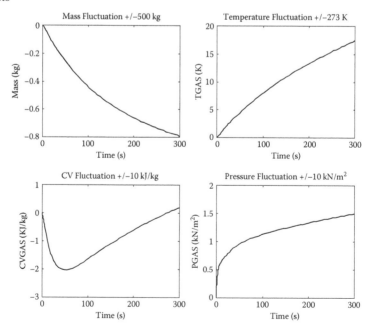

Inputs

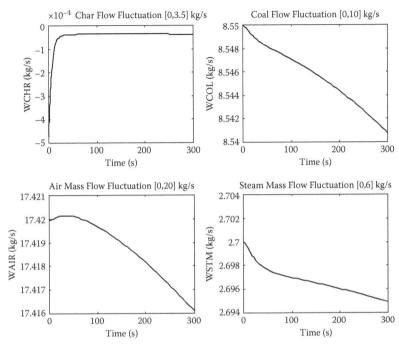

Inputs Rate

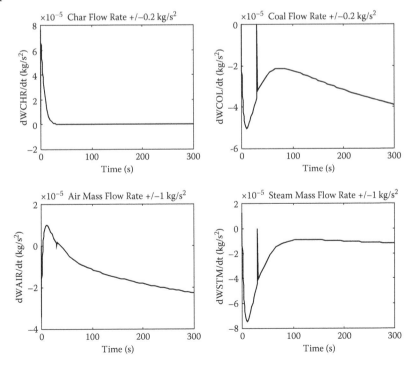

2.1.4 50% Load Condition (Sinusoidal Pressure Disturbance)

Outputs

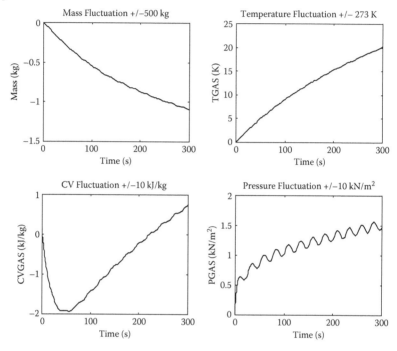

Inputs

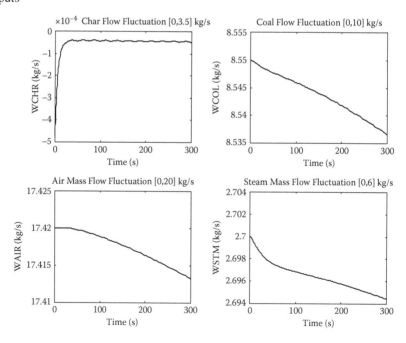

Inputs Rate

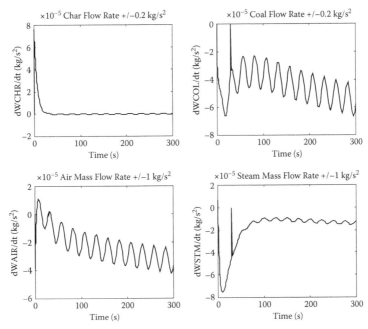

2.1.5 0% Load Condition (Step Pressure Disturbance)

Outputs

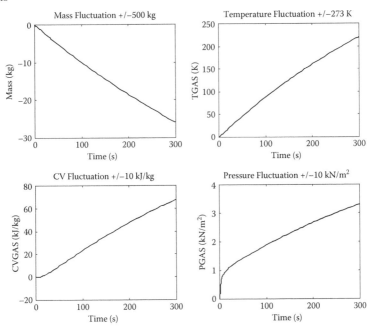

Inputs

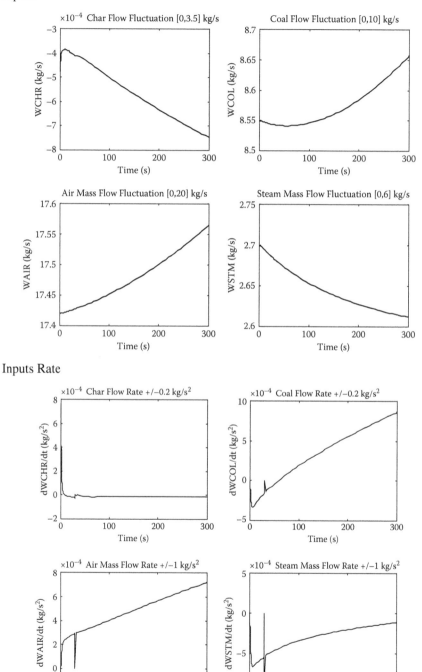

Inputs Rate

2.1.6   0% LOAD CONDITION (SINUSOIDAL PRESSURE DISTURBANCE)

Outputs

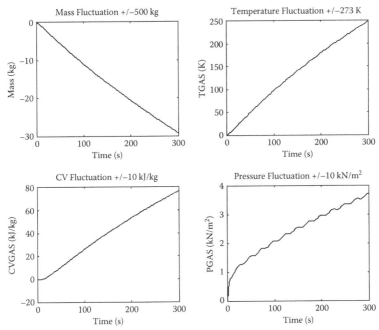

Inputs

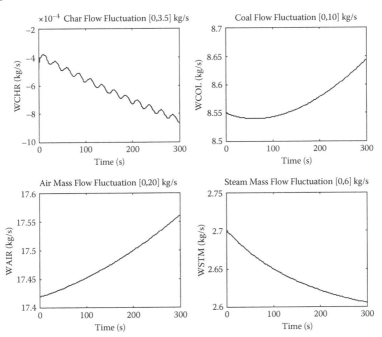

Inputs rate

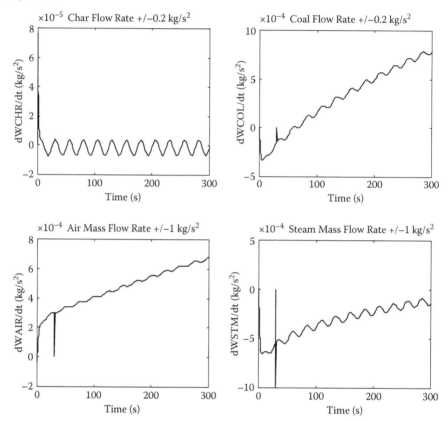

2.2 LQR

2.2.1 100% LOAD CONDITION (STEP PRESSURE DISTURBANCE)

	Min Value	Max Value	Peak Rate	IAE
WCHR	**−4.7700 × 10⁻⁴**	1.5119 × 10⁻⁶	7.6498 × 10⁻⁵	-
WCOL	8.5458	8.55	0	-
WAIR	17.4175	17.42	0	-
WSTM	2.6985	2.7	1.3656 × 10⁻⁵	-
MASS	10000	10000.8212	-	-
TGAS	1223.1898	1228.9918	-	-
CVGAS	4351.2286	4360	-	472.3469
PGAS	2000	2003.8156	-	224.8128

2.2.2 100% Load Condition (Sinusoidal Pressure Disturbance)

	Min Value	Max Value	Peak Rate	IAE
WCHR	-4.7700×10^{-4}	-1.7396×10^{-5}	7.6497×10^{-5}	-
WCOL	8.5429	8.55	0	-
WAIR	17.4157	17.42	0	-
WSTM	2.6981	2.70	6.9101×10^{-6}	-
MASS	9998.6541	10000.0313	-	-
TGAS	1223.2	1248.1685	-	-
CVGAS	4351.9592	4360	-	1363.3051
PGAS	2000	2005.0432	-	766.5956

2.2.3 50% Load Condition (Step Pressure Disturbance)

	Min Value	Max Value	Peak Rate	IAE
WCHR	-4.7700×10^{-4}	-3.63×10^{-5}	7.65×10^{-5}	-
WCOL	8.5408	8.55	0	-
WAIR	17.4161	17.4201	9.9864×10^{-6}	-
WSTM	2.6949	2.7	1.2282×10^{-5}	-
MASS	9999.2083	10000	-	-
TGAS	1181.1	1198.5348	-	-
CVGAS	4487.9694	4490.2063	-	70.5674
PGAS	1550	1551.501	-	75.0549

2.2.4 50% Load Condition (Sinusoidal Pressure Disturbance)

	Min Value	Max Value	Peak Rate	IAE
WCHR	-4.7700×10^{-4}	-3.9317×10^{-5}	7.6497×10^{-5}	-
WCOL	8.5366	8.55	0	-
WAIR	17.4133	17.42	1.1154×10^{-5}	-
WSTM	2.6944	2.7	6.8866×10^{-6}	-
MASS	9998.9062	10000	-	-
TGAS	1181.1	1201.3161	-	-
CVGAS	4488.0709	4490.7516	-	165.6168
PGAS	1550	1551.5657	-	166.1125

2.2.5   0% Load Condition (Step Pressure Disturbance)

	Min Value	Max Value	Peak Rate	IAE
WCHR	-7.4766×10^{-4}	-3.8566×10^{-4}	7.6493×10^{-5}	-
WCOL	8.5412	8.6575	0.000869	-
WAIR	17.42	17.5655	0.0007213	-
WSTM	2.6124	2.7	1.0909×10^{-5}	-
MASS	9974.1313	10000	-	-
TGAS	1115.1	1335.7561	-	-
CVGAS	4709.9085	4777.688	-	1835.8117
PGAS	1120	1123.3136	-	136.6727

2.2.6   0% Load Condition (Sinusoidal Pressure Disturbance)

	Min Value	Max Value	Peak Rate	IAE
WCHR	-8.6478×10^{-4}	-3.7995×10^{-4}	7.6492×10^{-5}	-
WCOL	8.54	8.6443	0.0007906	-
WAIR	17.42	17.5613	0.0006831	-
WSTM	2.6054	2.7	6.8642×10^{-6}	-
MASS	9970.8712	10000	-	-
TGAS	1115.1	1363.7049	-	-
CVGAS	4709.9134	**4786.5249**	-	6092.4963
PGAS	1120	1123.7313	-	423.2837

Appendix A3:
LQG Simulation Model

3.1 LQG

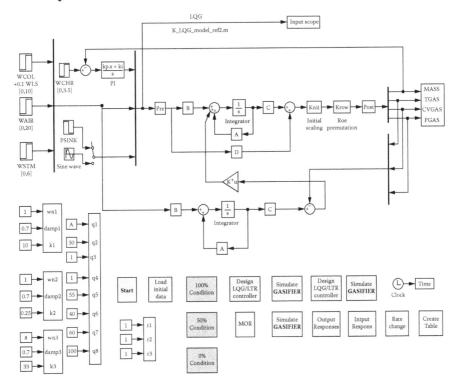

3.1.1 100% LOAD CONDITION (STEP PRESSURE DISTURBANCE)

Outputs

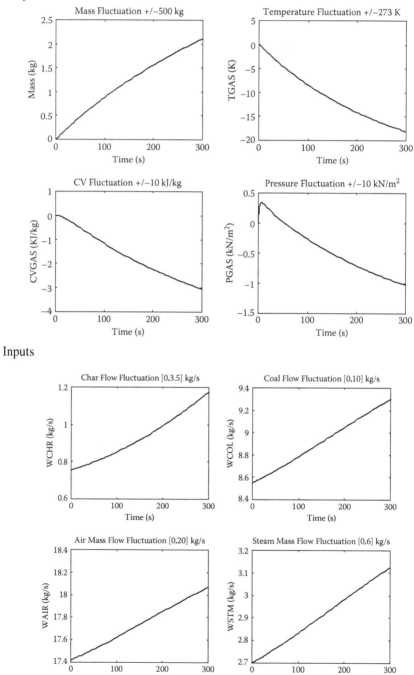

Inputs

Inputs Rate

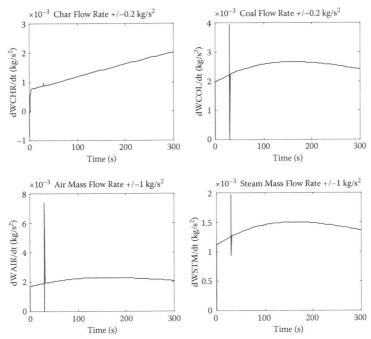

3.1.2   100% LOAD CONDITION (SINUSOIDAL PRESSURE DISTURBANCE)

Outputs

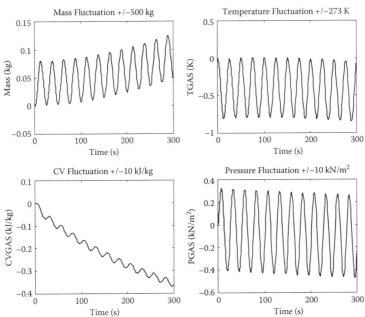

Inputs

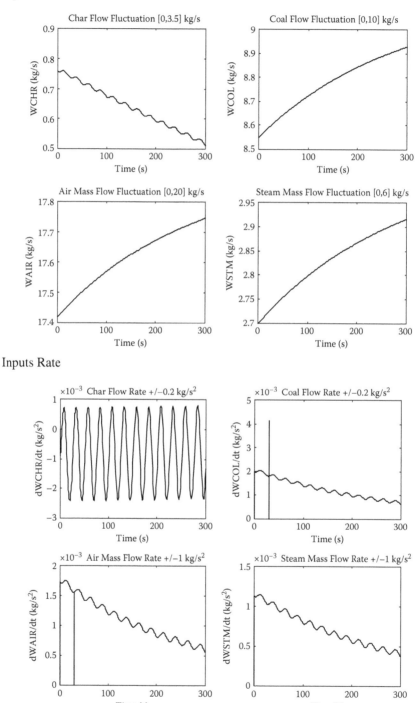

Inputs Rate

3.1.3   50% Load Condition (Step Pressure Disturbance)

Outputs

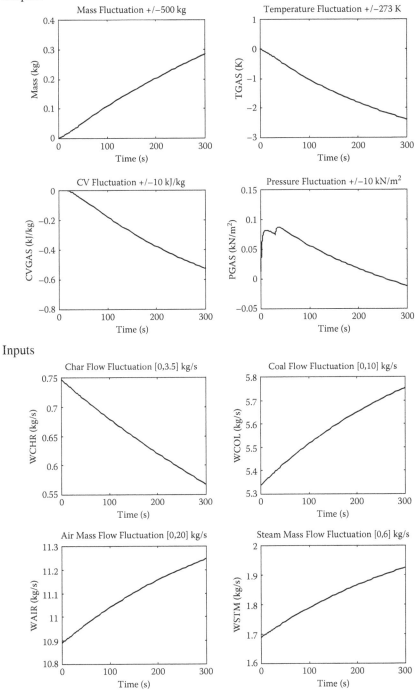

Inputs

Inputs Rate

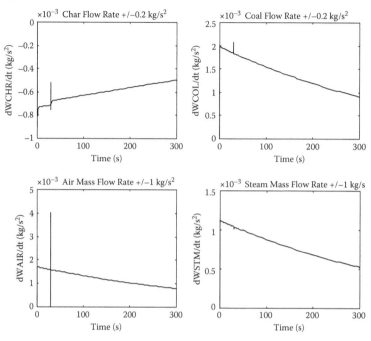

3.1.4   50% Load Condition (Sinusoidal Pressure Disturbance)

Outputs

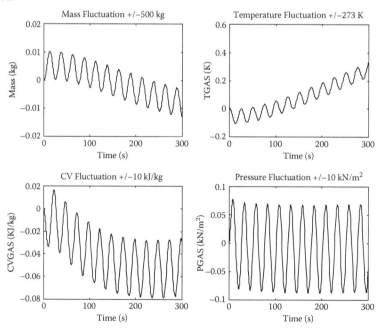

Inputs

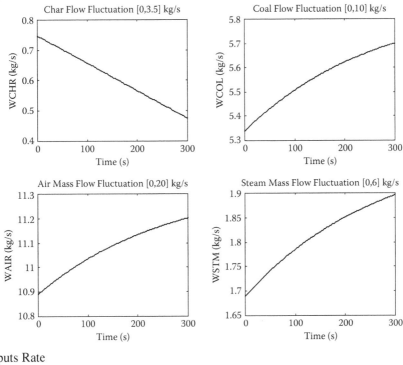

Inputs Rate

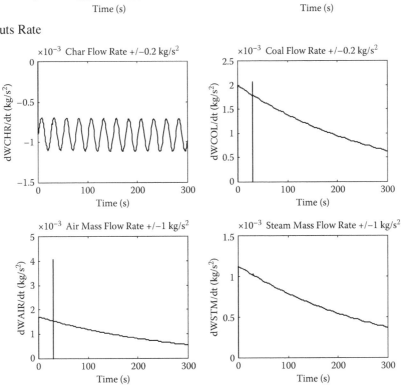

3.1.5   0% LOAD CONDITION (STEP PRESSURE DISTURBANCE)

Outputs

Inputs

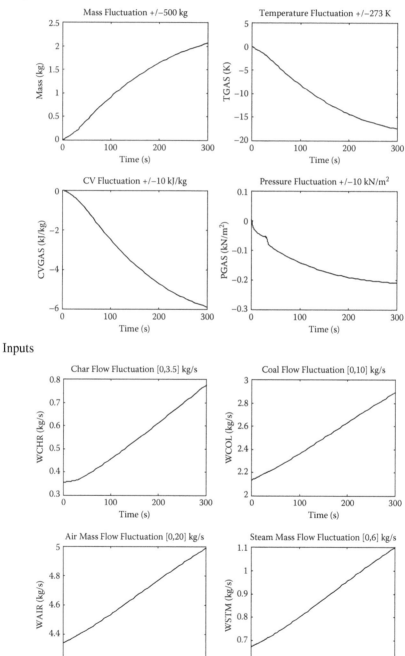

Inputs Rate

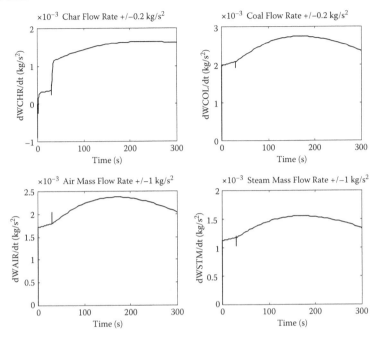

3.1.6   0% Load Condition (Sinusoidal Pressure Disturbance)

Outputs

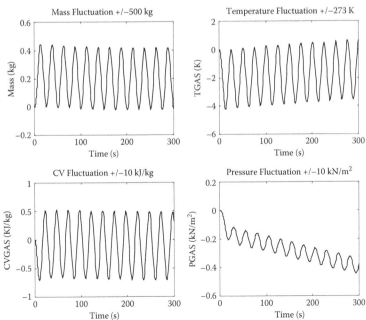

Inputs

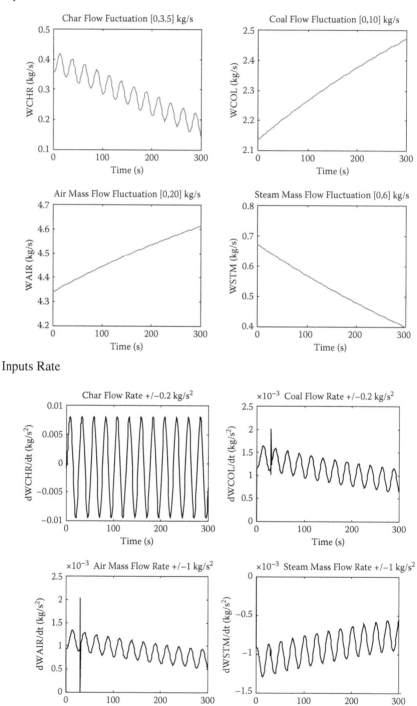

Inputs Rate

3.2   LQG

3.2.1   100% Load Condition (Step Pressure Disturbance)

	Min Value	Max Value	Peak Rate	IAE
WCHR	0.7566	1.176	0.0020161	-
WCOL	8.55	9.3007	0.003937	-
WAIR	17.42	18.0682	0.0074074	-
WSTM	2.7	3.1248	0.0019685	-
MASS	10000	10002.113	-	-
TGAS	1205.122	1223.201	-	-
CVGAS	4356.9623	4360	-	87.1286
PGAS	1998.9762	2000.3477	-	34.0927

3.2.2   100% Load Condition (Sinusoidal Pressure Disturbance)

	Min Value	Max Value	Peak Rate	IAE
WCHR	0.51196	0.7589	0.00079787	-
WCOL	8.55	8.9277	0.0041322	-
WAIR	17.42	17.7464	0.0017517	-
WSTM	2.7	2.916	0.0011483	-
MASS	10000	10000.1244	-	-
TGAS	1222.3527	1223.2092	-	-
CVGAS	4359.6311	4360	-	60.9311
PGAS	1999.5315	2000.3264	-	82.4273

3.2.3   50% Load Condition (Step Pressure Disturbance)

	Min Value	Max Value	Peak Rate	IAE
WCHR	0.56766	0.7469	0	-
WCOL	5.34	5.7535	0.0020833	-
WAIR	10.89	11.2473	0.0040323	-
WSTM	1.69	1.926	0.0011287	-
MASS	10000	10000.2847	-	-
TGAS	1178.7137	1181.101	-	-
CVGAS	4489.4251	4490	-	18.0448
PGAS	1549.9764	1550.0754	-	4.0316

3.2.4   50% LOAD CONDITION (SINUSOIDAL PRESSURE DISTURBANCE)

	Min Value	Max Value	Peak Rate	IAE
WCHR	0.47463	0.7469	0	-
WCOL	5.34	5.7003	0.0020661	-
WAIR	10.89	11.2014	0.004065	-
WSTM	1.69	1.8962	0.0011287	-
MASS	9999.987	10000.0103	-	-
TGAS	1180.9972	1181.4307	-	-
CVGAS	4489.9215	4490.0168	-	12.4874
PGAS	1549.9128	1550.0778	-	17.4643

3.2.5   0% LOAD CONDITION (STEP PRESSURE DISTURBANCE)

	Min Value	Max Value	Peak Rate	IAE
WCHR	0.35641	0.77665	0.0016446	-
WCOL	2.136	2.8904	0.002747	-
WAIR	4.34	4.9915	0.0023718	-
WSTM	0.672	1.0993	0.0015545	-
MASS	10000	10002.0612	-	-
TGAS	1097.6528	1115.101	-	-
CVGAS	4704.1268	4710	-	182.0897
PGAS	1119.7919	1120.0021	-	9.2571

3.2.6   0% LOAD CONDITION (SINUSOIDAL PRESSURE DISTURBANCE)

	Min Value	Max Value	Peak Rate	IAE
WCHR	0.1458	0.41819	0.0080681	-
WCOL	2.136	2.4724	0.0020161	-
WAIR	4.34	4.6128	0.0020325	-
WSTM	0.40033	0.672	0	-
MASS	9999.9764	10000.4428	-	-
TGAS	1110.8944	1115.7418	-	-
CVGAS	4709.286	4710.5219	-	99.7212
PGAS	1119.5581	1120	-	63.9433

Appendix A4: LQG/LTR Simulation Model and Results

4.1 LQG/LTR

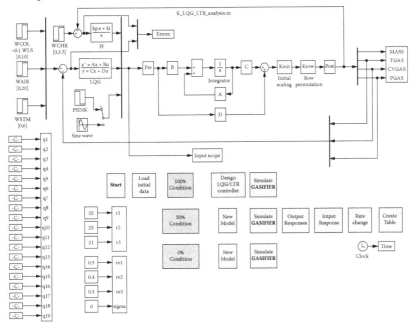

4.1.1   100% LOAD CONDITION (STEP PRESSURE DISTURBANCE)

Outputs

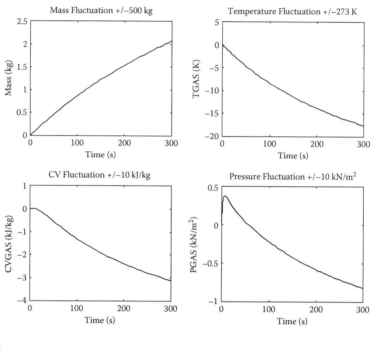

Inputs

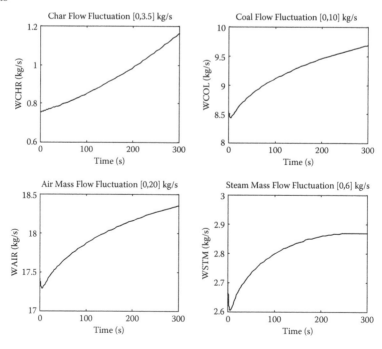

Inputs Rate

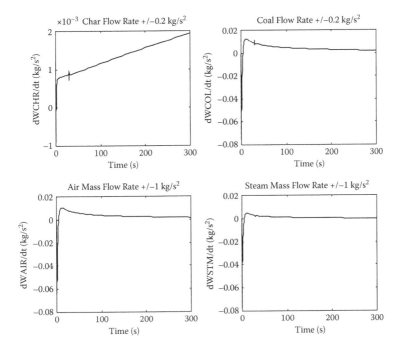

4.1.2   100% LOAD CONDITION (SINUSOIDAL PRESSURE DISTURBANCE)

Outputs

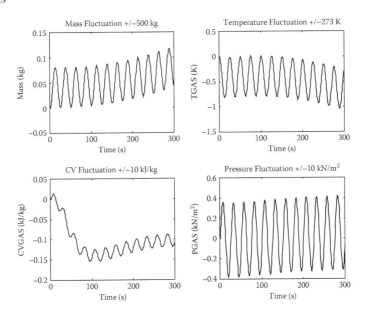

Inputs

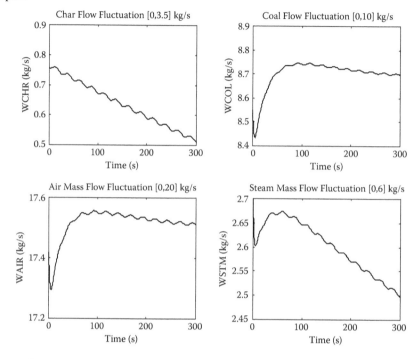

Inputs Rate

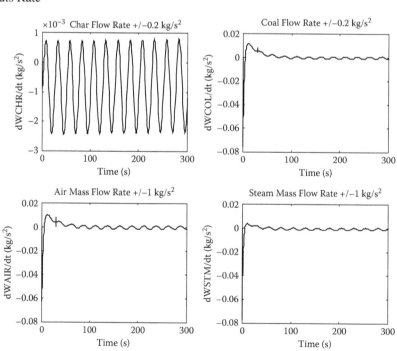

4.1.3 50% Load Condition (Step Pressure Disturbance)

Outputs

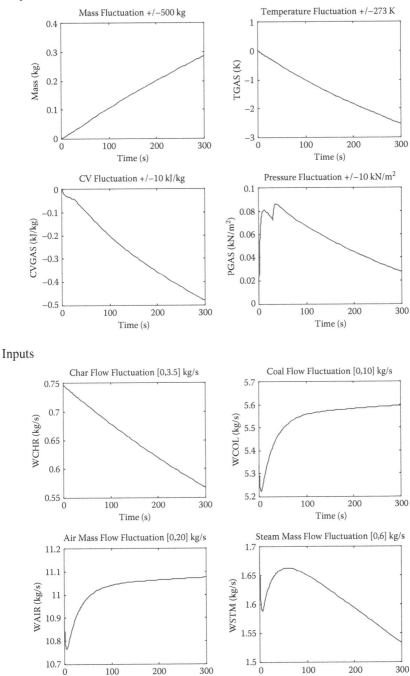

Inputs

Inputs Rate

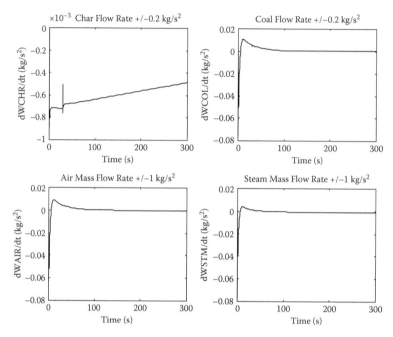

4.1.4   50% Load Condition (Sinusoidal Pressure Disturbance)

Outputs

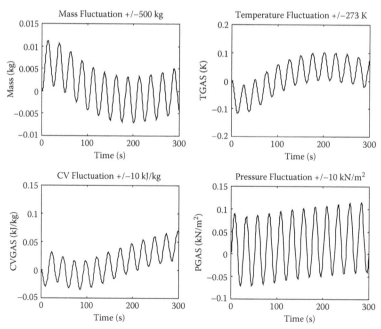

Inputs

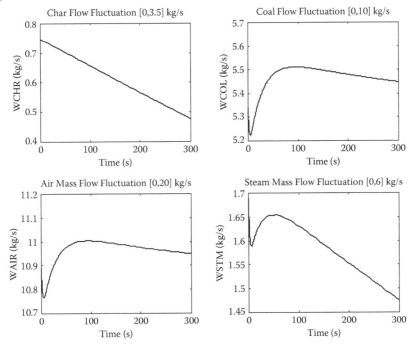

Inputs Rate

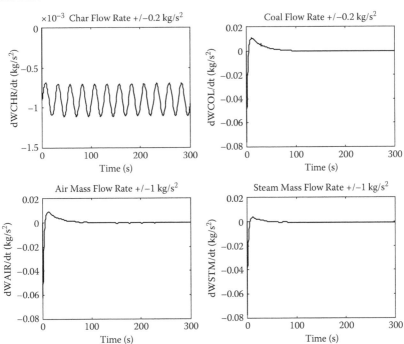

4.1.5   0% LOAD CONDITION (STEP PRESSURE DISTURBANCE)

Outputs

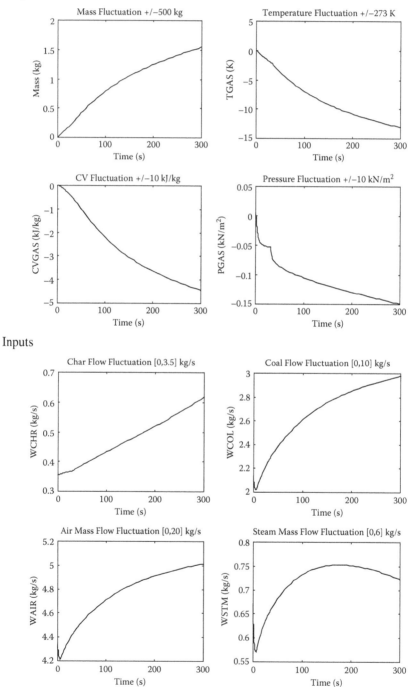

Inputs

Inputs Rate

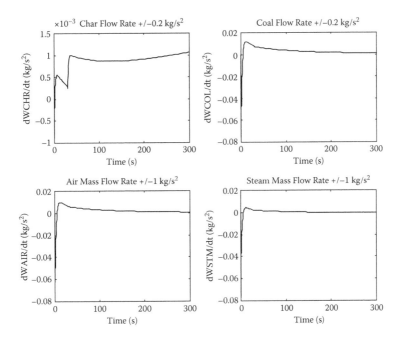

4.1.6 0% Load Condition (Sinusoidal Pressure Disturbance)

Outputs

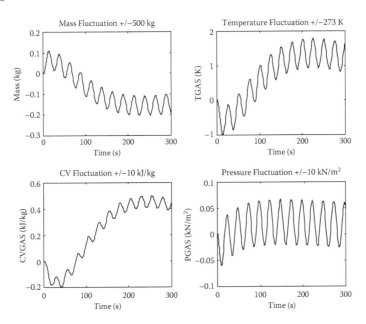

Inputs

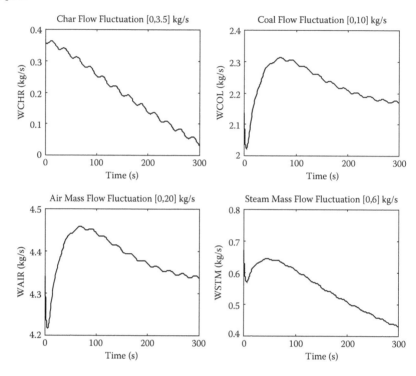

Inputs Rate

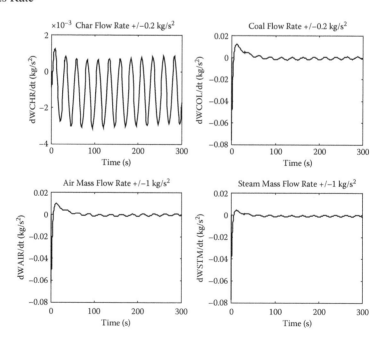

4.2   LQG/LTR

4.2.1   100% LOAD CONDITION (STEP PRESSURE DISTURBANCE)

	Min Value	Max Value	Peak Rate	IAE
WCHR	0.7566	1.1642	0.0019411	-
WCOL	8.4384	9.6856	0.012571	-
WAIR	17.2967	18.3568	0.010571	-
WSTM	2.6046	2.8707	0.004725	-
MASS	10000	10002.0697	-	-
TGAS	1205.666	1223.201	-	-
CVGAS	4356.9089	4360.0069	-	92.1937
PGAS	1999.1871	2000.3769	-	30.3756

4.2.2   100% LOAD CONDITION (SINUSOIDAL PRESSURE DISTURBANCE)

	Min Value	Max Value	Peak Rate	IAE
WCHR	0.50958	0.75895	0.00079027	-
WCOL	8.4364	8.7462	0.011932	-
WAIR	17.2955	17.557	0.010357	-
WSTM	2.4973	2.7	0.0041309	-
MASS	10000	10000.1171	-	-
TGAS	1222.1601	1223.2092	-	-
CVGAS	4359.8462	4360.0129	-	32.6429
PGAS	1999.6219	2000.4166	-	82.4584

4.2.3   50% LOAD CONDITION (STEP PRESSURE DISTURBANCE)

	Min Value	Max Value	Peak Rate	IAE
WCHR	0.56758	0.7469	0	-
WCOL	5.2252	5.5959	0.010624	-
WAIR	10.7654	11.0739	0.0087843	-
WSTM	1.533	1.69	0.0038518	-
MASS	10000	10000.2864	-	-
TGAS	1178.5693	1181.101	-	-
CVGAS	4489.5212	4490	-	14.8512
PGAS	1550.0007	1550.0862	-	5.3209

4.2.4   50% Load Condition (Sinusoidal Pressure Disturbance)

	Min Value	Max Value	Peak Rate	IAE
WCHR	0.47639	0.7469	0	-
WCOL	5.225	5.5121	0.010582	-
WAIR	10.7654	11.0051	0.0088516	-
WSTM	1.4749	1.69	0.003744	-
MASS	9999.993	10000.0111	-	-
TGAS	1180.9829	1181.2024	-	-
CVGAS	4489.9655	4490.0692	-	7.5863
PGAS	1549.9288	1550.1149	-	18.2621

4.2.5   0% Load Condition (Step Pressure Disturbance)

	Min Value	Max Value	Peak Rate	IAE
WCHR	0.35645	0.61826	0.0010679	-
WCOL	2.0221	2.9787	0.011858	-
WAIR	4.2173	5.0158	0.0099027	-
WSTM	0.57046	0.75339	0.0043931	-
MASS	10000	10001.5496	-	-
TGAS	1102.0184	1115.101	-	-
CVGAS	4705.5685	4710	-	147.0668
PGAS	1119.851	1120.0021	-	7.1202

4.2.6   0% Load Condition (Sinusoidal Pressure Disturbance)

	Min Value	Max Value	Peak Rate	IAE
WCHR	0.031332	0.36337	0.001255	-
WCOL	2.0215	2.3134	0.012207	-
WAIR	4.2167	4.4587	0.010136	-
WSTM	0.42924	0.672	0.0045974	-
MASS	9999.7969	10000.1093	-	-
TGAS	1114.1138	1116.8896	-	-
CVGAS	4709.8036	4710.4975	-	65.3319
PGAS	1119.9405	1120.066	-	7.5825

Appendix A5: H$_2$ Simulation Model and Results

5.1   H$_2$ OPTIMIZATION DESIGN

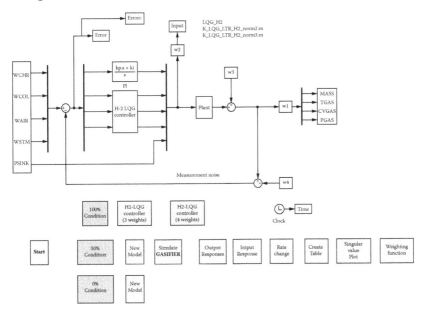

5.1.1 100% LOAD CONDITION (STEP PRESSURE DISTURBANCE)

Outputs

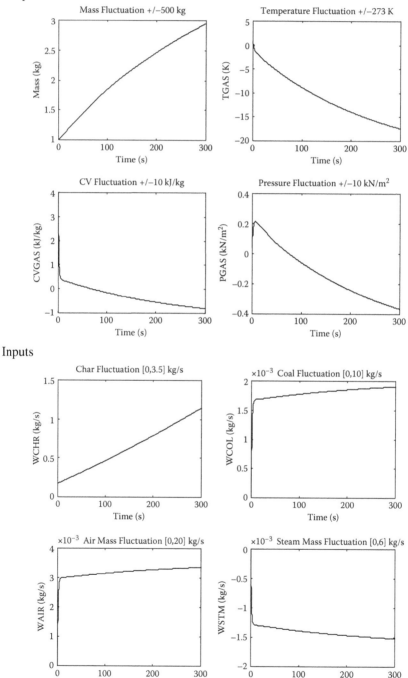

Inputs

Inputs Rate

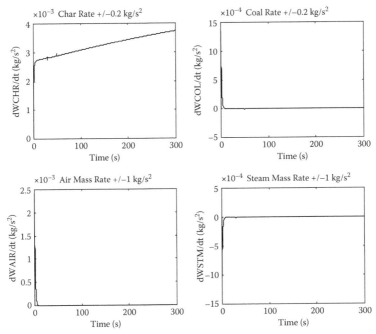

5.1.2 100% LOAD CONDITION (SINUSOIDAL PRESSURE DISTURBANCE)

Outputs

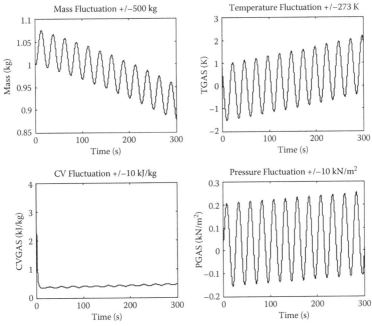

Inputs

Inputs Rate

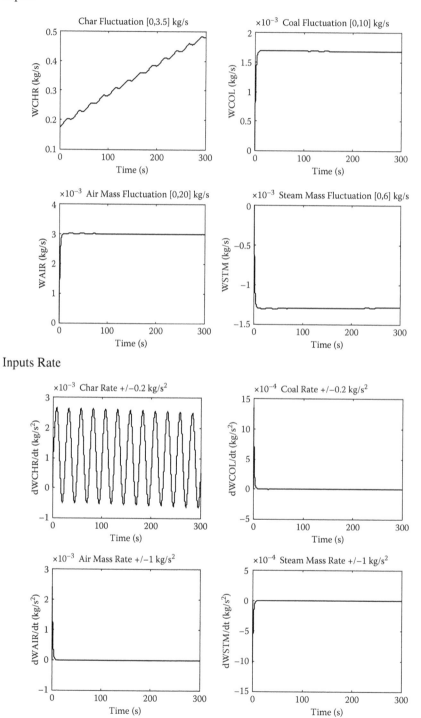

5.1.3 50% Load Condition (Step Pressure Disturbance)

Outputs

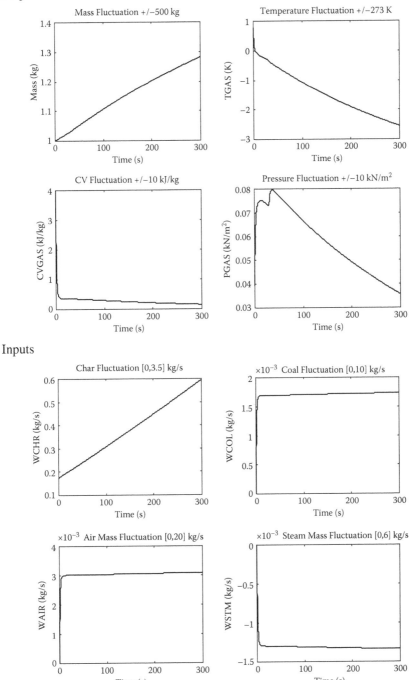

Inputs

Inputs Rate

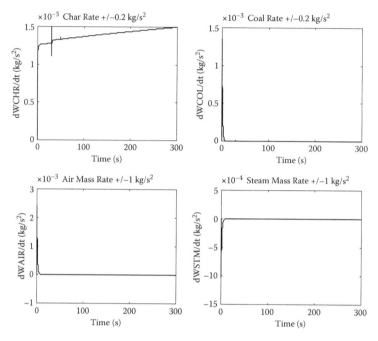

5.1.4   50% LOAD CONDITION (SINUSOIDAL PRESSURE DISTURBANCE)

Outputs

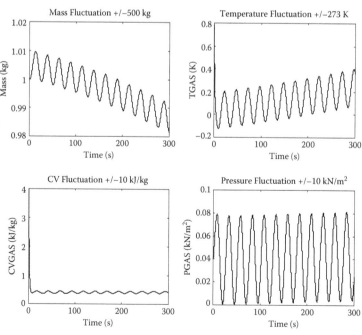

Inputs

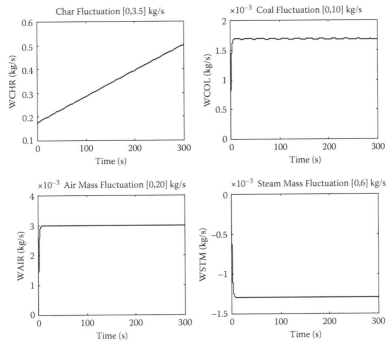

Inputs Rate

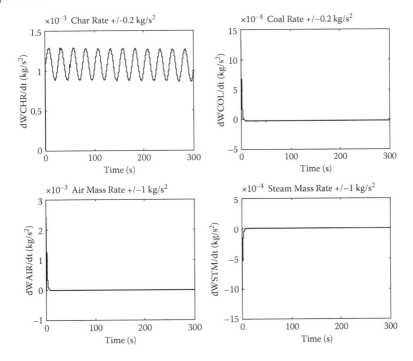

5.1.5   0% Load Condition (Step Pressure Disturbance)

Outputs

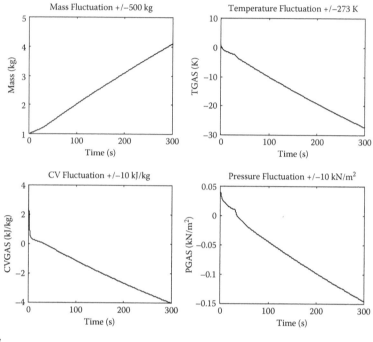

Inputs

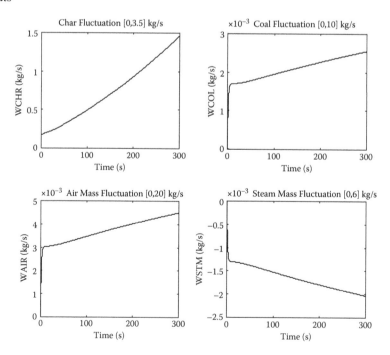

Inputs Rate

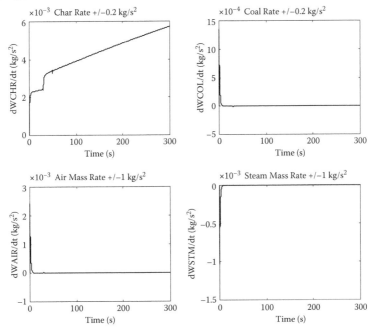

5.1.6 0% LOAD CONDITION (SINUSOIDAL PRESSURE DISTURBANCE)

Outputs

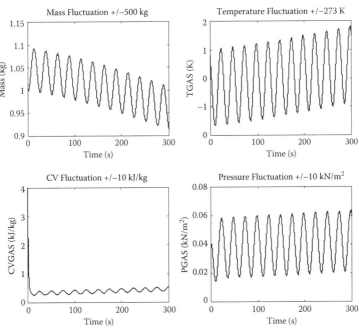

Inputs

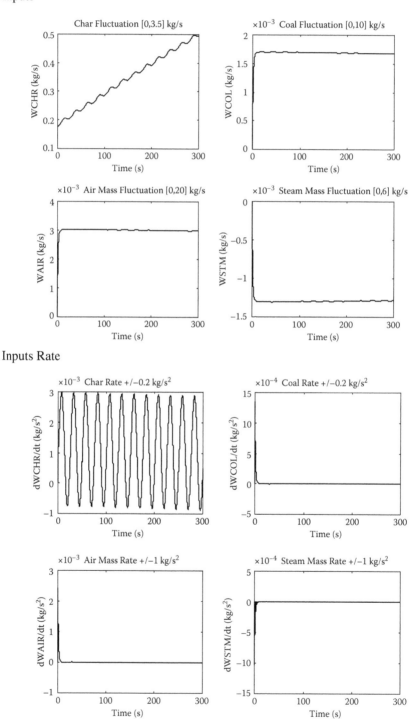

Inputs Rate

5.2   H₂ OPTIMIZATION DESIGN

5.2.1   100% LOAD CONDITION (STEP PRESSURE DISTURBANCE)

	Min Value	Max Value	Peak Rate	IAE
WCHR	0.1749	1.1469	0.003737	-
WCOL	0	0.002355	0.001663	-
WAIR	0	0.004181	0.002971	-
WSTM	**−0.00183**	0	0	-
MASS	10001	10002.9502	-	-
TGAS	1205.8293	1223.988	-	-
CVGAS	4359.205	4364	-	1304.5798
PGAS	1999.6353	2000.2179	-	552.0836

5.2.2   100% LOAD CONDITION (SINUSOIDAL PRESSURE DISTURBANCE)

	Min Value	Max Value	Peak Rate	IAE
WCHR	0.1749	0.4831	0.002657	-
WCOL	0	0.002085	0.001663	-
WAIR	0	0.003725	0.002971	-
WSTM	**−0.00158**	0	4.3788×10^{-7}	-
MASS	10000.876	10001.0757	-	-
TGAS	1221.6882	1225.3818	-	-
CVGAS	4360.3532	4364	-	1409.3729
PGAS	1999.8456	2000.2542	-	362.1591

5.2.3   50% LOAD CONDITION (STEP PRESSURE DISTURBANCE)

	Min Value	Max Value	Peak Rate	IAE
WCHR	0.1749	0.5947	0.0015	-
WCOL	0	0.002137	0.001663	-
WAIR	0	0.003815	0.002971	-
WSTM	**−0.00164**	0	7.3242×10^{-6}	-
MASS	10001	10001.2818	-	-
TGAS	1178.5523	1181.888	-	-
CVGAS	4490.1487	4494	-	885.696
PGAS	1550.0358	1550.0797	-	174.329

5.2.4   50% LOAD CONDITION (SINUSOIDAL PRESSURE DISTURBANCE)

	Min Value	Max Value	Peak Rate	IAE
WCHR	0.1749	0.5010	0.001302	-
WCOL	0	0.002083	0.001663	-
WAIR	0	0.003720	0.002971	-
WSTM	**−0.00159**	0	4.8828×10^{-6}	-
MASS	10000.9815	10001.0097	-	-
TGAS	1180.9782	1181.888	-	-
CVGAS	4490.3579	4494	-	1374.4187
PGAS	1550.0005	1550.0813	-	124.2167

5.2.5   0% LOAD CONDITION (STEP PRESSURE DISTURBANCE)

	Min Value	Max Value	Peak Rate	IAE
WCHR	0.1749	1.4583	0.005729	-
WCOL	0	0.003138	0.001663	-
WAIR	0	0.005567	0.002971	-
WSTM	**−0.0025**	0	0	-
MASS	10001	10004.0946	-	-
TGAS	1087.8245	1115.888	-	-
CVGAS	4706.0004	4714	-	5827.3806
PGAS	1119.8549	1120.0403	-	216.1661

5.2.6   0% LOAD CONDITION (SINUSOIDAL PRESSURE DISTURBANCE)

	Min Value	Max Value	Peak Rate	IAE
WCHR	0.1749	0.4984	0.002992	-
WCOL	0	0.002099	0.001663	-
WAIR	0	0.003749	0.002971	-
WSTM	**−0.0016**	0	8.138×10^{-6}	-
MASS	10000.914	10001.0929	-	-
TGAS	1113.4416	1116.9121	-	-
CVGAS	4710.2581	4714	-	1337.1389
PGAS	1120.0141	1120.0635	-	120.2022

Appendix A6: H$_\infty$ Simulation Model and Results

6.1 H$_\infty$ OPTIMIZATION DESIGN

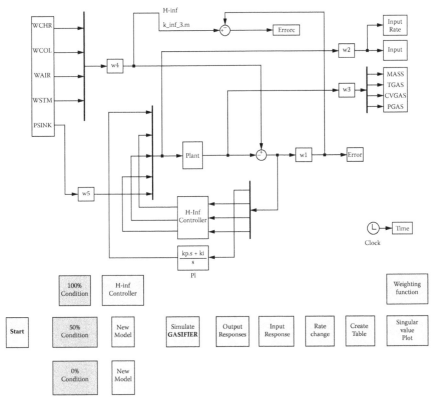

6.1.1 100% LOAD CONDITION (STEP PRESSURE DISTURBANCE)

Outputs

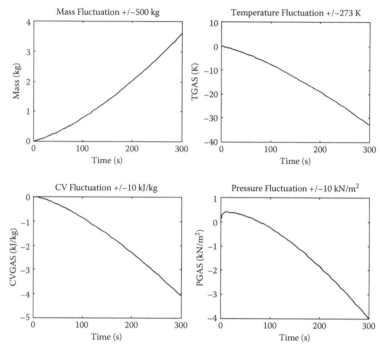

Inputs

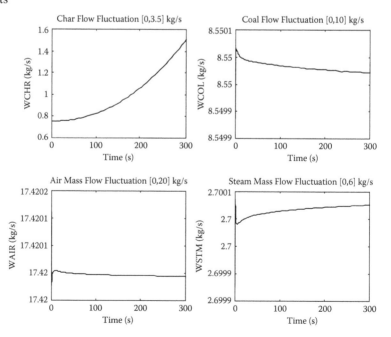

Inputs Rate

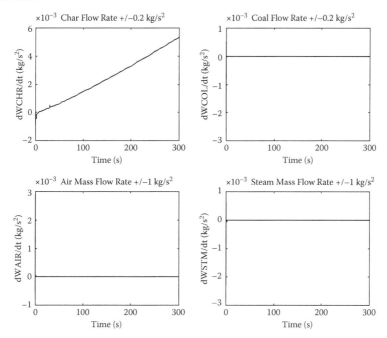

6.1.2 100% LOAD CONDITION (SINUSOIDAL PRESSURE DISTURBANCE)

Outputs

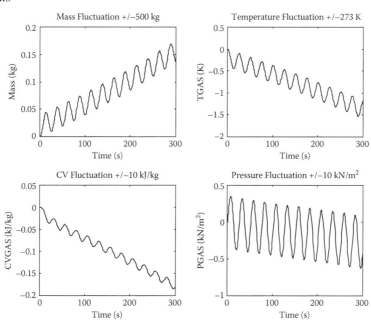

Inputs

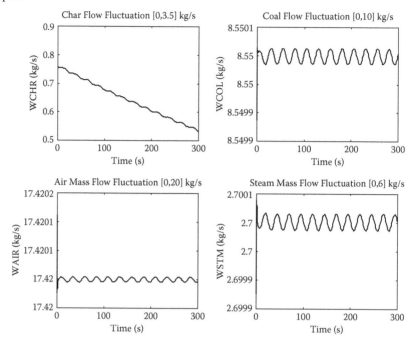

Inputs Rate

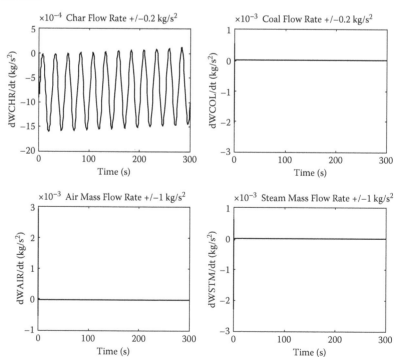

6.1.3   50% LOAD CONDITION (STEP PRESSURE DISTURBANCE)

Outputs

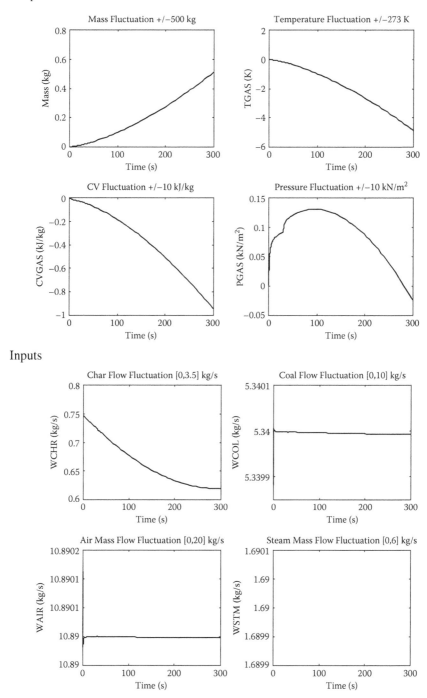

Inputs

Inputs Rate

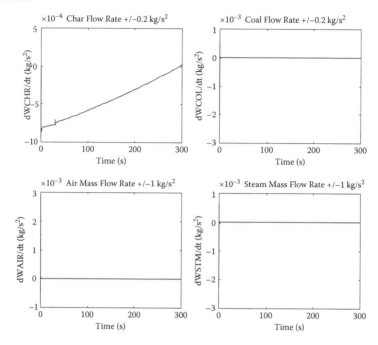

6.1.4   50% LOAD CONDITION (SINUSOIDAL PRESSURE DISTURBANCE)

Outputs

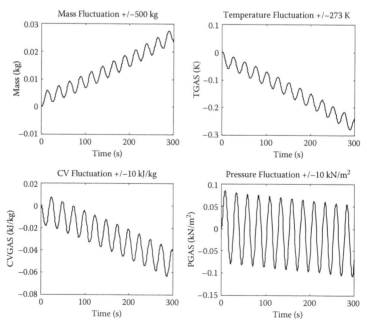

Inputs

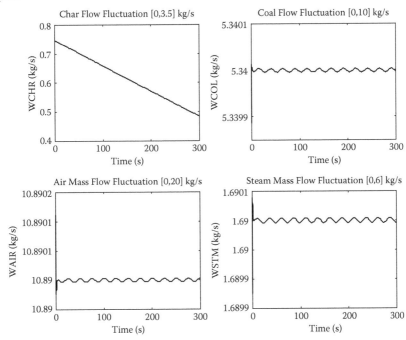

Inputs Rate

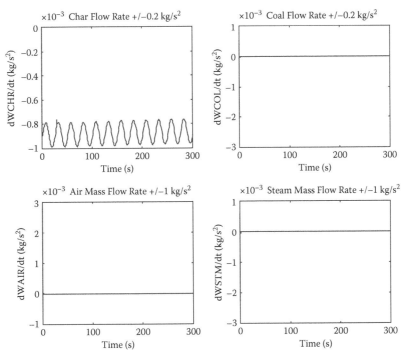

6.1.5 0% LOAD CONDITION (STEP PRESSURE DISTURBANCE)

Outputs

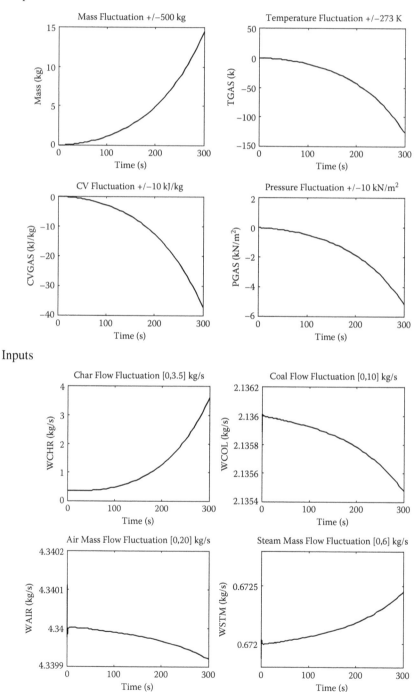

Inputs

Inputs Rate

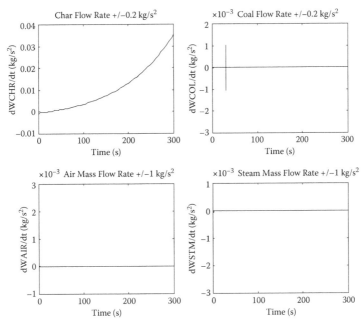

6.1.6 0% LOAD CONDITION (SINUSOIDAL PRESSURE DISTURBANCE)

Outputs

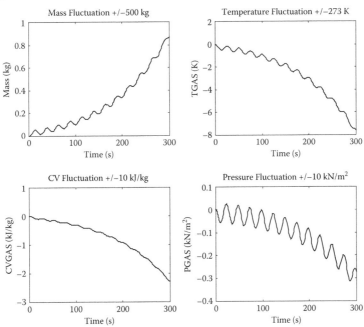

Inputs

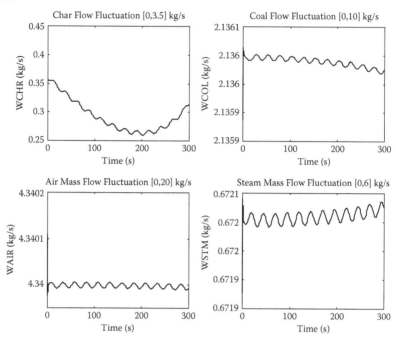

Inputs Rate

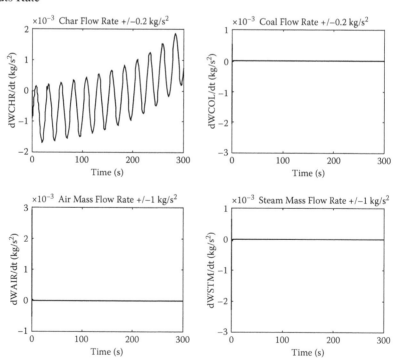

6.2 H$_\infty$ OPTIMIZATION DESIGN

6.2.1 100% LOAD CONDITION (STEP PRESSURE DISTURBANCE)

	Min Value	Max Value	Peak Rate	IAE
WCHR	0.75591	1.5089	0.005398	-
WCOL	8.5499	8.55	0.0005236	-
WAIR	17.42	17.4201	0.002917	-
WSTM	2.6999	2.7	0.0006149	-
MASS	10000	10003.598	-	-
TGAS	1190.4203	1223.201	-	-
CVGAS	4355.9209	4360	-	89.7984
PGAS	1996.0166	2000.3969	-	84.0701

6.2.2 100% LOAD CONDITION (SINUSOIDAL PRESSURE DISTURBANCE)

	Min Value	Max Value	Peak Rate	IAE
WCHR	0.5337	0.7569	0.0001159	-
WCOL	8.5499	8.55	0.0005176	-
WAIR	17.42	17.4201	0.0029304	-
WSTM	2.6999	2.7	0.0006226	-
MASS	10000	10000.1696	-	-
TGAS	1221.6775	1223.201	-	-
CVGAS	4359.8156	4360	-	27.5372
PGAS	1999.3739	2000.3542	-	81.9694

6.2.3 50% LOAD CONDITION (STEP PRESSURE DISTURBANCE)

	Min Value	Max Value	Peak Rate	IAE
WCHR	0.61957	0.7469	2.8021×10^{-5}	-
WCOL	5.3399	5.34	0.0004928	-
WAIR	10.89	10.8901	0.0029008	-
WSTM	1.6899	1.69	0.0006027	-
MASS	10000	10000.5122	-	-
TGAS	1176.2317	1181.101	-	-
CVGAS	4489.0588	4490	-	21.1846
PGAS	1549.9758	1550.1306	-	7.873

6.2.4   50% Load Condition (Sinusoidal Pressure Disturbance)

	Min Value	Max Value	Peak Rate	IAE
WCHR	0.48467	0.7469	0	-
WCOL	5.3399	5.34	0.0005172	-
WAIR	10.89	10.8901	0.002930	-
WSTM	1.6899	1.69	0.0006219	-
MASS	10000	10000.0275	-	-
TGAS	1180.822	1181.101	-	-
CVGAS	4489.9369	4490.0075	-	8.6733
PGAS	1549.8927	1550.0853	-	19.3906

6.2.5   0% Load Condition (Step Pressure Disturbance)

	Min Value	Max Value	Peak Rate	IAE
WCHR	0.3531	**3.6088**	0.03549	-
WCOL	2.1355	2.136	0.001016	-
WAIR	4.3399	4.3401	0.002901	-
WSTM	0.67187	0.67245	0.0006209	-
MASS	10000	10014.4802	-	-
TGAS	988.3508	1115.101	-	-
CVGAS	4672.7681	**4710**	-	559.1913
PGAS	1114.8435	1120.0029	-	83.8559

6.2.6   0% Load Condition (Sinusoidal Pressure Disturbance)

	Min Value	Max Value	Peak Rate	IAE
WCHR	0.25978	0.3569	0.001855	-
WCOL	2.1359	2.136	0.0005154	-
WAIR	4.34	4.3401	0.002931	-
WSTM	0.67188	0.67204	0.0006264	-
MASS	10000	10000.8688	-	-
TGAS	1107.5482	1115.101	-	-
CVGAS	4707.7114	4710	-	175.0044
PGAS	1119.6877	1120.0264	-	25.435

Index

Milton Keynes UK
Ingram Content Group UK Ltd.
UKHW031141141024
449569UK00024B/1162